Advances in Experimental Medicine and Biology

Volume 1230

More information about this series at http://www.springer.com/series/5584

J. Miguel Oliveira • Rui L. Reis
Editors

Biomaterials- and Microfluidics-Based Tissue Engineered 3D Models

Springer

Editors
J. Miguel Oliveira
3B's Research Group
University of Minho
Guimarães, Portugal

Rui L. Reis
3B's Research Group
University of Minho
Guimarães, Portugal

ISSN 0065-2598 ISSN 2214-8019 (electronic)
Advances in Experimental Medicine and Biology
ISBN 978-3-030-36587-5 ISBN 978-3-030-36588-2 (eBook)
https://doi.org/10.1007/978-3-030-36588-2

This Springer imprint is published by the registered company Springer Nature Switzerland AG.
The registered company address is: Gewerbestrasse 11, 6330 Cham, Switzerland

Preface

In recent years, we have assisted a new impetus in tissue engineering and personalized regenerative medicine research due to adoption of emerging technologies for fabrication of functional microtissues, which best emulate human condition and allow minimizing animal experimentation. In this respect, advanced biomaterials and microfluidics are truly game changers. This book on *biomaterials- and microfluidics-based tissue-engineered 3D models* provides different chapters concerning the fundamental aspects, challenges, and progress on the biomaterials and microfluidics applications in the development of 3D tissue-engineered in vitro models. The type and role of biomaterials and microfluidics systems considered in the designing of in vitro models are overviewed. The microfluidics platforms used in processing of biomaterials are succinctly addressed, while current challenges relating tissue-, organ-, and body-on-a-chip models are explored in depth and relevant examples given. Other hot topics discussed relate to the current applications of the 3D models in the central nervous system, angiogenesis, and drug discovery and development. The future trends in the field are concisely provided. This book will serve as a core reference for multidisciplinary students and a wide range of established researchers and professionals working in tissue engineering and precision medicine fields.

Guimarães, Portugal J. M. Oliveira

Contents

Microfluidic Devices and Three Dimensional-Printing Strategies for *in vitro* Models of Bone

1

F. Raquel Maia, Rui L. Reis, Vitor M. Correlo, and Joaquim M. Oliveira

Abstract

Bone is a complex and highly dynamic tissue, which has been worldwide studied, from fundamental biology to tissue engineering fields. Even so, current *in vitro* models do not truly replicate the native bone tissue environment. For so, new and improved *in vitro* tissue models are necessary to obtain more reliable data, not only in a development point of view, but also to fasten the translation of new drugs into the clinics. In this reasoning, tissue-engineering strategies were applied to develop mimetic and three-dimensional (3D) microenvironments, which were associated with microfluidic devices for the development of more complex and realistic systems. Such devices mimic blood vessels that are present in the native tissue, thus enabling the study of complex biological mechanism as such as bone angiogenesis. More recently, 3D printing has been pursued to produce more intricate microfluidic devices and engineered tissues in a single step. The ability to print spatially controlled structures composed of different biomaterials, growth factors and cells caught the attention of scientists for the development of more efficient *in vitro* models. Additionally, it allows obtaining microfluidic devices and/or engineered tissues with the desired architecture within a small amount of time and with reduced costs. Recently, the use of high-resolution scanning boosted the production of patient-specific implants. Despite the difficulties associated with 3D printed structures that still need to be overcome, it has been proven to be a valuable tool to accomplish a new generation of 3D bioprinted bone-on-a-chip platforms.

Keywords

Bone tissue engineering · Microfluidic devices · Three-dimensional models · Bioprinting · *In vitro* models

F. R. Maia (✉) · R. L. Reis · V. M. Correlo
J. M. Oliveira
3B's Research Group, I3Bs – Research Institute on Biomaterials, Biodegradables and Biomimetics, University of Minho, Headquarters of the European Institute of Excellence on Tissue Engineering and Regenerative Medicine, Guimarães, Portugal

ICVS/3B's PT Government Associate Lab, Braga, Guimarães, Portugal

The Discoveries Centre for Regenerative and Precision Medicine, Headquarters at University of Minho, Guimarães, Portugal
e-mail: raquel.maia@i3bs.uminho.pt

J. M. Oliveira, R. L. Reis (eds.), *Biomaterials- and Microfluidics-Based Tissue Engineered 3D Models*, Advances in Experimental Medicine and Biology 1230,
https://doi.org/10.1007/978-3-030-36588-2_1

Highlights

- Microfluidic devices and 3D printing have great potential for the development of highly realistic bone tissue *in vitro* models.
- The most promising approaches to obtain bone tissue *in vitro* models that better mimic the native tissue are reviewed.
- Relevant examples of the approaches reviewed are highlighted.

1.1 Introduction

Bone is a very complex and highly dynamic tissue continually renewing and controlled by several transcription regulatory cascades. Bone tissue is composed mainly of four different types of cells, osteoblasts, osteoclasts, osteocytes and bone lining cells (Florencio-Silva et al. 2015). Osteoblasts and osteoclasts are involved in the bone remolding process, where osteoblasts are responsible for bone tissue neoformation, while osteoclasts are responsible for bone tissue resorption. Osteocytes act as mechanosensors and are involved in the initiation of the bone remodeling process (Florencio-Silva et al. 2015). In the case of bone lining cells, there is evidence that these cells are a major source of osteoblasts in adulthood (Matic et al. 2016). This intricate scenario has been studied worldwide to try to develop new therapeutic approaches to bone diseases.

Currently, it is common knowledge that two-dimensional (2D) culture *in vitro* models do not truly replicate bone tissues (Duval et al. 2017). This led scientist to develop new *in vitro* models using Tissue Engineering approaches. For example, Lima et al. (Lima et al. 2014b) developed a scaffold with micro organized and architected structures that support cell adhesion and proliferation. In fact, tissue-engineering can involve the combination of cells and scaffolds resulting in a three-dimensional (3D) environment that better mimic the reality of bone tissue (Lima et al. 2014a; Bouet et al. 2015). Amongst the different 3D strategies followed to develop new biomaterials to guide bone tissue formation, osteoblasts are usually used, while other cells like osteoclast are ignored. For so, new studies have been pursued to understand the behavior of osteoclast to new biomaterials (Maia et al. 2018). Within this setting, where 3D and several types of cells are considered it is possible to obtain more reliable results. Even so, most 3D *in vitro* models were cultivated under static conditions, which resulted in poor nutrient and waste exchange. To overcome these issues, bioreactors were developed allowing not only to improve nutrients perfusion and waste elimination but also comprise mechanical stimuli (Bouet et al. 2015).

Despite the incredible progress achieved, most models only partly mimic the *in vivo* scenario. In fact, it lacks the presence of vasculature structures, and other features, as such as specific architectures and tissue-tissue interaction. For so, innovative technological platforms and techniques have become available, as the microfluidic devices (Aziz et al. 2017) and 3D printing strategies (Bhattacharjee et al. 2016; Jariwala et al. 2015). Within microfluidic devices it is possible to obtain intricate multi-compartment co-cultures, connected by micro-channels that mimic the vasculature (Carvalho et al. 2018; Jusoh et al. 2015). All this, using low reagents' amounts and cell numbers. In the case of 3D printing, it has allowed pushing forward the development of *in vitro* models of bone, enabling the production of 3D printed microfluidic devices and/or tissues with the desired complexity in a single step.

This chapter aims to overview the recent development on 3D bone tissue engineered *in vitro* models and highlight the most relevant examples of the progress achieved. For so, four different topics will be explored: (i) microfluidic-based *in vitro* models of bone; (ii) techniques used to obtain 3D printed microfluidic devices; (iii) 3D bioprinting-based bone tissue models; and (iv) 3D bioprinted microfluidic *in vitro* models.

1.2 Microfluidic-Based in vitro Models of Bone

Microfluidic devices have been used to mimic the physiological environment of tissues and organs. In fact, the ability to mimic blood vessels has step forward the development of efficient 3D *in vitro* models. Moreover, such devices have the capability to spatial and temporally control fluids and physical features. Additionally, they allow screening several conditions through parallelization, using reduce amounts of reagents/cells. Altogether, these features boost its use in combination with *in vitro* models' strategies for drug discovery as well as in a development point of view. In this reasoning, microfluidic devices have been showing to be powerful tools that allow conducting experiments with high sensitivity and within small amounts of time.

One of the most appealing characteristics of microfluidic devices is the possibility to study complex biological mechanisms *in vitro,* such as bone angiogenesis, as demonstrated by Jusoh and colleagues (Jusoh et al. 2015). They developed a vascular network within a matrix composed of fibrin and hydroxyapatite to obtain a microfluidic-based *in vitro* model of vascularized bone tissue (Fig. 1.1). The designed microfluidic presented channels separated by gaps using columns (Fig. 1.1c), through where, blood vessels sprouted, invading the mineralized matrix (Fig. 1.1d).

As demonstrated, such platform offers new relevant models for different areas, namely, screening of new drugs, new biomaterials and study biological mechanisms of disease (e.g. cancer). For example, new *in vitro* models of bone were developed to study the interaction of bone cells with newly developed biomaterials aiming to speed up the development of novel orthopedic implants (Lee et al. 2012). In a different study, Hao et al. (2018) showed the development of a microfluidic-based *in vitro* model of breast cancer bone metastasis composed of a simultaneous-growth-and-dialysis mechanism. The developed mechanism allowed the exchange of nutrients and waste continuously through the dialysis membrane. Moreover, they were able to obtain a long-term culture of highly mineralized osteoblastic tissue. Additionally, this microfluidic tool allowed the co-culture of metastatic human breast cancer cells and osteoblastic tissue, observing similar results as *in vivo* studies. As example, researchers observed that metastasis-

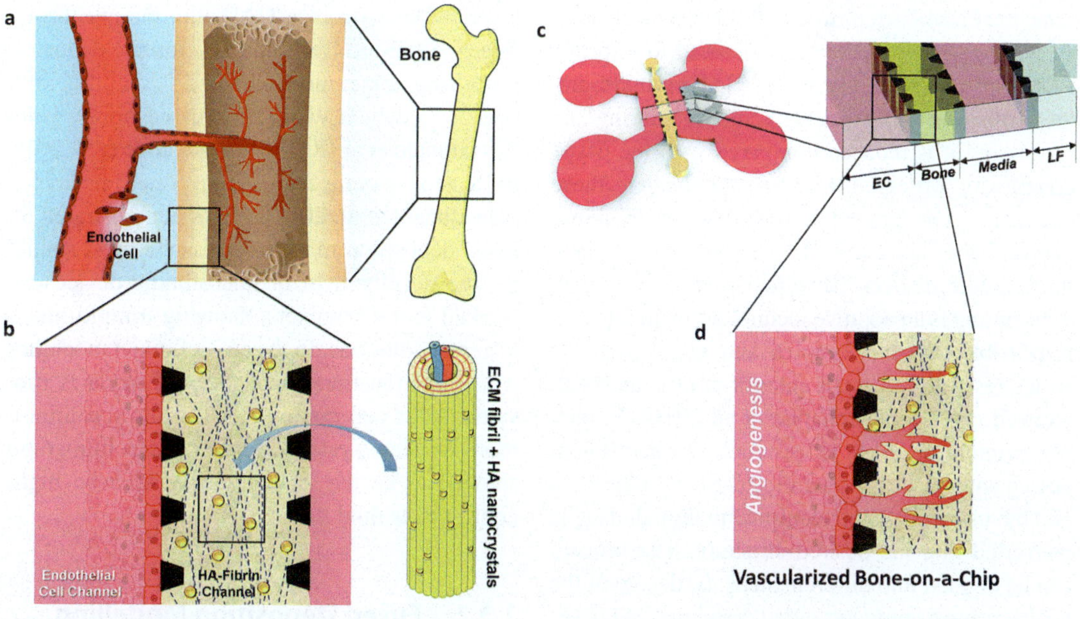

Fig. 1.1 Microfluidic based vascularized bone model. (**a**) Scheme of bone angiogenesis; (**b**) Fibrin matrix with hydroxyapatite (HA) to mimic the entrapment of blood vessel; (**c**) scheme of developed microfluidic device showing channels with endothelial cells (EC) separated by columns from fibrin matrix with HA to mimic bone tissue; (**d**) image showing blood vessels sprouting towards mimic bone tissue. (Adapted with permission from (Jusoh et al. 2015). Copyright © 2015, Royal Society of Chemistry)

suppressed breast cancer cells proliferated at slower rates in opposite to 2D cultured cells, where it was observed higher proliferation rates. Moreover, those cells did not invade the matrix of osteoblastic tissue. In the case of normal metastatic breast cancer cells, after 2 weeks they observed that cells were capable to invade the osteoblastic tissue matrix and degraded it, reaching distant locations. Moreover, Veerla et al. studied a drug delivery system within a microfluidic channel for osteoblastic differentiation (Veerla et al. 2018). For that, researchers developed graphene oxide/alginate beads loaded with risedronate. The channel with the beads showed controlled delivery of risedronate inducing osteoblasts differentiation, without the need for additional supplements.

Yet, although the research associated costs were very much reduced, the production of such device using the traditional soft lithography technique can be very time consuming and expensive. In fact, the development process involves several steps including the device's design using an engineering drawing software, the fabrication of a master mold according to the design and its replication using polydimethylsiloxane (PDMS). After the curing time, the PDMS is removed and cut into the desired shape. Finally, oxygen plasma is used to improve PDMS-glass bound (Ho et al. 2015). Furthermore, the structural complexity achieved during its processing it is reduced compared with the tissue complexity observed *in vivo*. Altogether, these drawbacks encouraged a new line of thinking, namely the use of 3D-printing to produce microfluidic devices (Bhattacharjee et al. 2016). 3D-printing is an additive technology with which it is possible to print complex and highly realistic structures within a small amount of time and with reduced costs (Bhattacharjee et al. 2016). In fact, 3D printing enables to produce a microfluidic device with the desired architecture in a single step, i.e. by using just one equipment digital data is converted into the desired structure. The process can be divided into different steps: (i) design of the device using a computer-aided drawing (CAD); (ii) convert this into standard triangulation language (STL) file; (iii) move the STL file into the printer, where it is divided into 2D sequential cross-sectional slices; and finally, (iv) printed layer-by-layer (Fig. 1.2) (Ho et al. 2015; Waheed et al. 2016). Furthermore, 3D printing allows the standardization of procedures which results in more reproducible devices as compared with soft lithography. Even so, the cost, resolution and materials necessary to print the microfluidic devices need to be evaluated to guarantee optimum results.

1.3 Techniques to 3D - Printing Microfluidic Devices

Nowadays, the fabrication of microfluidic devices is predominantly obtained by casting-based 3D molding approaches (soft lithography) of PDMS. Nevertheless, this technique presents some constraints, such as difficulty to fully automatize the molding process, namely curing, assembly and bonding. In this sense, a process that was supposed to be low cost is, in fact, expensive and time-consuming. In this reasoning, a new technique is being used to overcome the difficulties along the manufacturing process. 3D printing allows obtaining microfluidic devices, printing layer by layer from a CAD file, mainly through three different techniques: fused deposition modelling, inkjet modelling and stereolithography (Fig. 1.3), which will be discussed briefly below (Macdonald et al. 2017; Bhattacharjee et al. 2016). In a study comparing these three techniques (Macdonald et al. 2017), it was demonstrated that fused deposition modelling was the most suitable for the development of microfluidic devices that worked as micromixers, showing a rapid mix of different solutions. In the case of inkjet modelling, it was demonstrated that this technique is more suitable for cell culture since it allowed printing more intricate systems. Finally, stereolithography showed to be the technique that allowed higher control over fluid flow.

1.3.1 Fused Deposition Modelling

Fused deposition modeling is the technique more extensively used. In fact, this technique

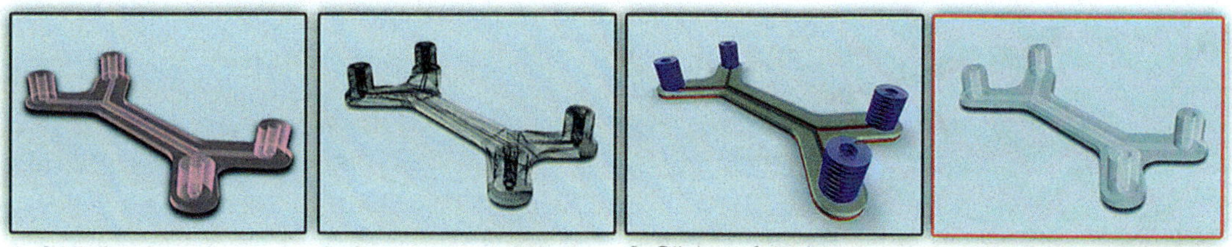

Fig. 1.2 3D printing of a microfluidic device. The different steps of 3D printing: (1) production of computer-aided drawing file; (2) conversion of CAD file into a standard triangulation language (STL) file; (3) conversion of STL file into individual layers; (4) printing of microfluidic device in a layer-by-layer manner. (Reprinted with permission from (Waheed et al. 2016) - Published by The Royal Society of Chemistry)

relies on the extrusion, layer-by-layer, of a high variety of thermoplastic materials through a high temperature nozzle (Fig. 1.3a) (Waheed et al. 2016; Macdonald et al. 2017). During extrusion, the thickness of polymer thread depends on the distance from the nozzle to the substrate, and on the ratio between flow rate and print speed, ranging between 0.1 and 0.3 mm. Most of the materials compatible with this technique are biocompatible, such as polycaprolactone (PCL), polylactic acid (PLA) and polycarbonate (PC) (Salentijn et al. 2017). Additionally, fused deposition modeling support the printing of composite materials, with different physicochemical properties, and extrude different materials at the same time, which makes it very appealing for different tissue engineering applications. Despite the possibility to extrude an extensive variety of bioinks, the more viscous ones guarantee the integrity of the printed structures, preventing its collapse (Yi et al. 2017). Moreover, this technique allows the placement of other materials inside of the printing structure in the middle of the run (Salentijn et al. 2017). Nevertheless, fused deposition-modeling present some restrictions related with the rough surface obtained, and with the accuracy of dimensions. For so, the resulting microfluidic devices present fusion lines and blank spaces between the printed layers, which affects the resolution obtained (Waheed et al. 2016; Macdonald et al. 2017). Finally, the velocity of printing is slower than inkjet modelling and stereolithography.

1.3.2 Inkjet Modelling

Inkjet modelling can produce microfluidic devices through continuous or dropwise deposition (Fig. 1.3b) (Waheed et al. 2016; Macdonald et al. 2017). During the continuous deposition, the printing is performed using lower viscosity inks at higher velocity as compared with drop deposition. Nevertheless, through drop deposition it is possible to achieve higher placement accuracy, resulting in reproducible microfluidic devices. To obtain the printed device, jetting heads deposits the material into trays, while a UV light that is coupled to the jetting head, polymerize the material as it is being printed (Rafał and Krzysztof 2015). To print the next layer, the tray moves and a new process begins until getting the entire structure (Waheed et al. 2016; Macdonald et al. 2017). Additionally, it allows the fast printing of intricate microfluidic devices with several materials, providing physicochemical features accordingly with the final approach. Nevertheless, this technique presents some limitations, including the removal of the support material and the amount of waste produced during the printing process. Also, amongst the low number of materials used to print using this technique (≈ 100), it is still necessary to develop materials with biocompatible features for biomedical devices applications.

1.3.3 Stereolithography

Stereolithography is a technique that allows to print a curable material layer-by-layer using two

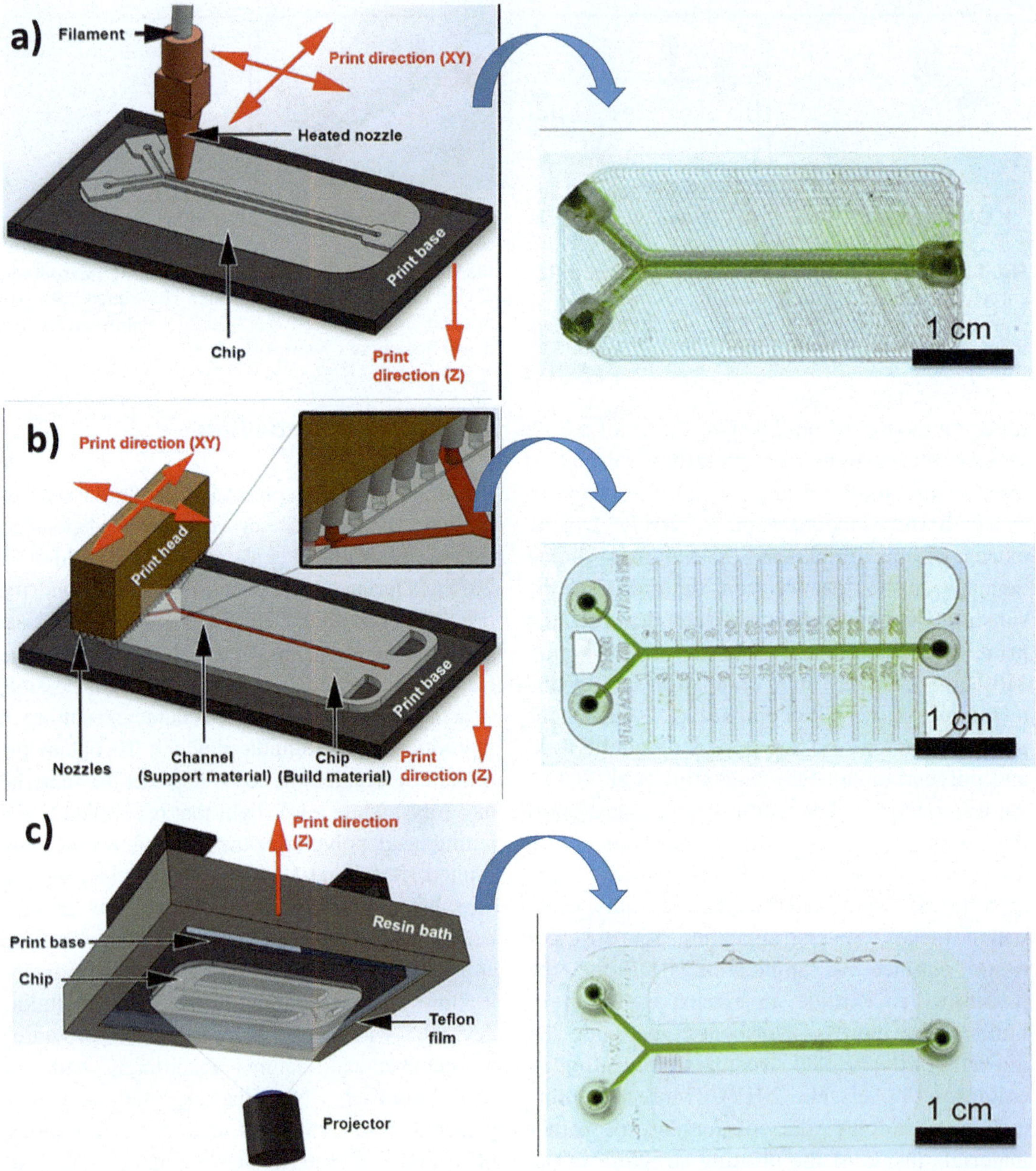

Fig. 1.3 3D-printing strategies and images of microfluidic devices printed. (**a**) Fused deposition modelling, where thermoplastics are extruded through a heated nozzle, in an XY manner until complete the layer; (**b**) Inkjet modelling, where the material is extruded through nozzles in a continuous or dropwise manner. Between each layer, the material is polymerized; (**c**) Stereolithography, where a photopolymerizable promoter is projected onto a platform immersed in a photopolymer. Scale bars = 1 cm. (Adapted with permission from (Macdonald et al. 2017). Copyright © 2017 American Chemical Society)

different approaches, free surface approach and constrained surface approach (Fig. 1.3c) (Waheed et al. 2016; Macdonald et al. 2017). The printing process in both approaches just differs in the configuration of printing, being the first configuration printed on top of a platform, while the second configuration printed hanging from a platform. The materials used are photopolymerized resins, as epoxides, which after the first layer being completed printed needs to be photopolymerized by laser or UV light. Then, until the structure is complete, the platform moves, a new layer is printed and traced by the laser to be polymerized. This technique only needs a small amount of material which allow reducing the associated costs. Since there are two main variables along the process, photopolymerizable promoter and resin features, the resolution achieved also varies with them (Waheed et al. 2016; Macdonald et al. 2017). One of the main challenges resides in the clearance of uncured resin and the fact that it is only possible to print one material at a time.

1.4 3D – Bioprinting Based Bone Tissue Models

3D bioprinting strategies have been pursued in last few years as one of the most promising technology to be used in tissue engineering approach (Jariwala et al. 2015). In fact, it enables the development of highly reproducible and spatially controlled constructs with different materials, growth factors and/or other molecules of interest as reported by Faramarzi et al. (Faramarzi et al. 2018). They studied the development of a new bioink composed of alginate and platelet-rich plasma, which contains several growth factors that can improve tissue regeneration. Also, the developed bioink could be crosslinked upon reaction with native calcium and allowed the controlled release of growth factors. Interestingly, this bioink allowed the use of patient specific platelet-rich plasma which ultimately would minimize the host immune response. In a different study, Cui et al. (2016) printed a vascularized bone construct with immobilized bone morphogenetic protein 2 (BMP2) and vascular endothelial growth factor (VEGF) peptides. Additionally, the printed construct presented a hard structure surrounded by a soft matrix to mimic native bone. The results have shown that, upon perfusion, it was possible to obtain an osteogenic construct with a vascular network.

Additionally, 3D printing allows the easy production of constructs with different sizes, geometries and porosities. With this in mind, Cox et al. (2015) investigated the production of a 3D printed scaffold composed of hydroxyapatite and poly(vinyl)alcohol (PVA) through fused deposition modeling for bone tissue engineering applications. In a first stage, researchers evaluated the precursor flowability and observed that this is a very important feature during printing, since it influences the mechanical strength, microstructure and porosity. The obtained scaffold exhibited a porosity around 55%, with pore channel size of around 1460–1750 μm and micropores size around 10–60 μm and rough topography, which are very promising features for bone tissue engineering applications. In a different approach, He et al. (2015) produced a scaffold through extrusion-based rapid prototyping technique using gels, which is similar to fused deposition modeling. For that, they used 15% PVA/calcined goat spongy bone mixed with a biphasic ceramic powder. In this experiment, the obtained scaffolds presented a uniform pore size and porosity around 68%. Moreover, they have shown that, when rabbit bone marrow stromal cells were cultured on top of developed scaffolds, these cells were able to spread and proliferate. Finally, they injected *in vivo* via the marginal ear vein the leachate of bone scaffold and observed no acute toxicity, pyrogenic reaction or stimulation. Using a different strategy, Boga et al. (2018) develop a scaffold that emulated the native hollow bone through fused deposition modeling, using a mixture of tri-calcium phosphate, to mimic the inorganic phase, and alginic acid, to mimic the organic phase. Finally, they modified the mixture with graphene oxide to enhance the mechanical and biological properties. Then, human osteoblasts were cultured on the designed scaffolds and the construct was evaluated *in vitro*. The results showed calcium deposition and an increase of alkaline phosphatase activity after 21 days of culture, demonstrating its potential to improve bone regeneration. In a different strategy to mimic bone vasculature, Byambaa et al. (2017) printed a bone-like structure with a perfusable vascular lumen. After processing, the structure was colonized with human umbilical vein endothelial cells and bone marrowderived human mesenchymal stem cells with the intent to promote different environments capable to support vasculogenesis and osteogenesis. The culture was maintained during 21 days showing that the printed structure was capable to support cell proliferation and viability.

Yet, a new 3D - printing approach emerged to answer the demand for effective engineered tissues, the bioprinting of 3D cell-laden tissue constructs (Adepu et al. 2017; Wu et al. 2016). This technique allows to print (e.g. through jetting and extrusion techniques), not only materials, but also, of cells embedded within chemically or thermally crosslinkable hydrogels (bioink) (Ho et al. 2015). In fact, it allows controlling cellular distribution within the developed matrices. In the work of Wang et al. (2016), it is reported the development of a 3D hybrid construct composed of alginate, gelatin and adiposederived stem cells. Upon the bioprinting process, researchers have shown that the encapsulated cells remained viable and proliferative. Additionally, osteogenic differentiation assays were also conducted *in vitro*, showing cell differentiation along the osteogenic lineage, and subcutaneously *in vivo*, showing ectopic bone formation. In brief, this study demonstrated the efficacy of using 3D bioprinting constructs in bone tissue engineering.

Even so, despite the excitement about the possibility to print complex structures with different cell types and mechanical properties (Fedorovich et al. 2011; Arrigoni et al. 2017; Wang et al. 2014), there is a clear necessity to optimize some conditions as the development of adequate hydrogel type (bioink) and needle diameter to maintain cell viability (Fedorovich et al. 2008). Bendtsen and Wei (2017) studied the optimization of a bioink composed of alginate-polyvinyl alcohol-hydroxyapatite with collagen to promote cell proliferation for bone tissue engineering applications. For that they printed cells encapsulated within different bioink formulations. The results showed that the amount of collagen and level of crosslinking influenced cell proliferation. In this reasoning, the 3D printed constructs with higher amount of collagen provided more anchorage points, while the calcium used for the crosslinking of the bioink provided the right mechanical properties to enable cell migration. Although natural hydrogels are favored due to its similarities with extracellular matrix, among the most used there are still some drawbacks to overcome (Adepu et al. 2017). In fact, there is a limited choice concerning the number of hydrogels for bioprinting of cells and compatible mechanisms for hydrogels' crosslinking. To tackle these issues Costa et al. (2017) studied silk fibroin bioink crosslinked through an enzymatic reaction for 3D bioprinting. The silk fibroin used in this study is produced by silk-worm *Bombyx mori* and has gained attention due to its features, such as mechanical properties, biocompatibility, degradability, water-based processing, and accessibility of chemical groups for functional modifications. The results showed that human adiposederived stem cells cultured on top of the developed matrices were viable and were able to proliferate. Nevertheless, it is difficult to guarantee a specific cellular distribution the developed matrices.

In the case of bone, it is easy to understand the advantages of using bioprinting technique. In fact, constructs for bone tissue approaches needs to be compliant enough to allow, e.g. cell proliferation, migration and formation of blood vessels, but also strong enough to support the mechanical stress that bone tissue is subjected to. Additionally, as described previously, several types of cells should be taken into consideration. Bioprinting emerged as a technique that allowed including several of the desired features in bone constructs production in a simple and controlled way. As an example, Cunniffe et al. (2017) proposed 3D bioprinted bone tissue for the development of personalized gene activated implant to treat complex bone defects. They developed a bioink composed of alginate with RGD and nano-hydroxyapatite with plasmid DNA to obtain an engineered non-viral gene activated matrix. Bone marrow-derived mesenchymal stem cells were added to this bioink and printed simultaneously with a polycaprolactone supporting mesh. The authors were able to detect a sustained expression of

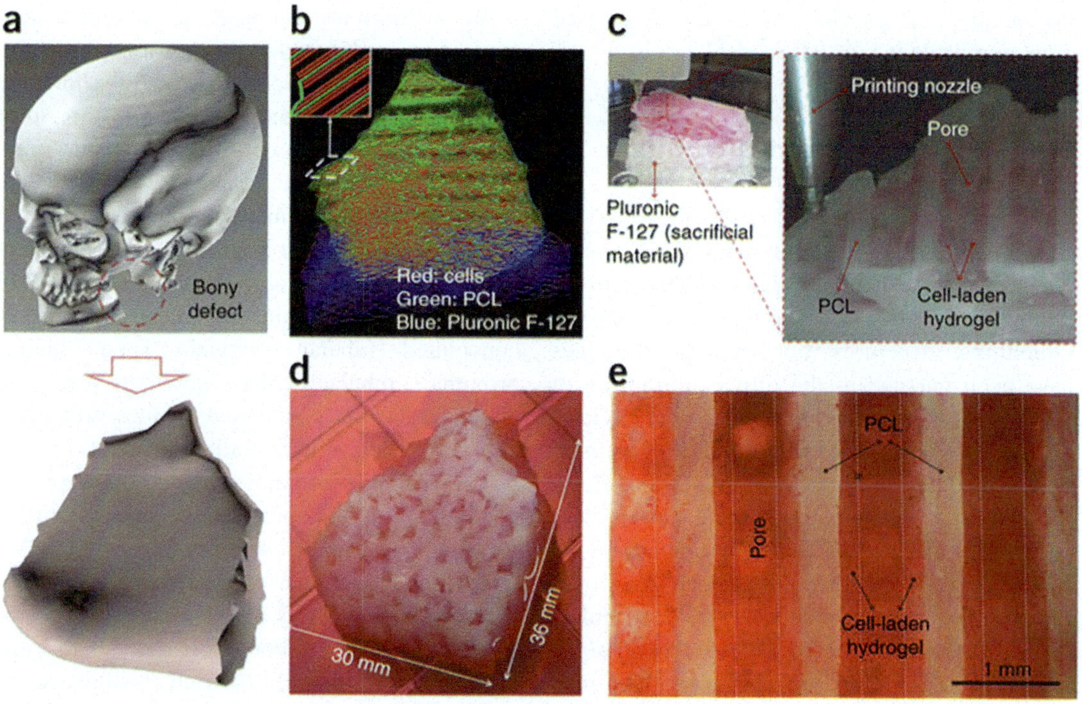

Fig. 1.4 3D bioprinted mandible bone. (**a**) CAD model produced from CT image data; (**b**) visualization of dispensing paths along with the 3D structure; (**c**) picture of 3D printing process; (**d**) picture of 3D printed mandible bone; (**e**) Alizarin Red S staining showing mineral deposits characteristic of osteogenic differentiation. (Reprinted with permission from (Kang et al. 2016) Copyright © 2016, Springer Nature)

bone morphogenic protein and transforming growth factor genes during 14 days, resulting in matrix deposition and mineralization. Upon subcutaneous implantation, they observed higher vascularization and mineralization as compared with constructs without cells. Overall, this gene-activated bioink can be the first step to a totally new generation of bioinks.

Recently, bioprinting caught the attention of scientists to the possibility to produce patient-specific implants emulating the complex tissues' shape and gradient composition (Costa et al. 2018; Datta et al. 2017). One example is the work of Li *et al* (2017). To study a new approach for treating bone and cartilage defects, researchers explored the use of 3D scanning followed by 3D printing. For that, they created different defects in animals (rabbit and mini pig), produced digital models of the defects by high-resolution scanning and accurately printed using photopolymerized hydrogels, demonstrating the feasibility of this strategy. In a different example, Kang et al. (2016) printed cell-laden hydrogels together with biodegradable polymers combine with tricalcium phosphate and sacrificial hydrogels to emulate the shape of a human mandible bone (Fig. 1.4). For this, they used computed tomography (CT) images of the bone and transformed it into a computer model (Fig. 1.4 a, b). Finally, this created model was printed, dispensing cells and materials into de desired shape (Fig. 1.4 c, d). Additionally, they included microchannels along the construct to facilitate nutrients diffusion. After 24 hours of culture, they observed around 90% of viability and after 28 days they were able to detect mineral deposits (Fig. 1.4 e). Although they did not observe the efficacy of this construct once implanted, this was a step forward to the development of patient-specific implants. Despite the advances achieved in engineering bone tissues, most of the strategies do not comprise vascularization, which has hampered its successful maturation and implantation. Yet, the produced tissues can be used as more reliable *in vitro* 3D models.

1.5 3D - Bioprinted Microfluidic in vitro Models

Bioprinted tissues have revolutionized tissue-engineering field but it still has some limitations, as the production of vascularized complex biomimetic structures. For so, new strategies have been pursued as the direct bioprinting of 3D artificial tissues within microfluidic devices, producing a new generation of 3D - bioprinted organ-on-a-chip platforms (Yi et al. 2017). These sophisticated platforms containing tissues with complex microarchitectures capable of humanlike functions, allowed proper investigation of physicochemical stimulus. Although these engineered tissues were obtained by bioprinting, the devices were obtained through the traditional techniques, namely soft lithography. Even so, this was a huge step, since it allowed to obtain bioprinted tissues adapted to the specific designs of microfluidic devices. Although no examples of bioprinted bone tissue-on-a-chip were reported, there are some works worth to mention. One example is the work of Bhise et al. (2016), where the authors printed a construct composed of hepatic spheroids encapsulated inside of a hydrogel inside of a microfluidic device for continuous perfusion. Using this model, they were able to culture spheroids for 1 month and study the effect of acute exposure to acetaminophen, an analgesic and antipyretic drug. In fact, the model allowed the prediction of drug toxicity emulating the *in vivo* environment.

In a different approach, it was created a method to produce a segment of a nephron surrounded by an artificial matrix, as shown in Fig. 1.5 (Homan et al. 2016). For this, they printed a bioink with gel-to-fluid transition features, composed of Pluronic F127 mixed with trombone, inside of a matrix composed of gelatin and fibrinogen (Fig. 1.5 b and c). After cooling the printed construct at 4 °C, the Pluronic F127 mixed with thrombin liquefied, opening a convoluted tubular channel. Then, human proximal tubular cells were perfused to circumscribe the convoluted tubules creating a columnar renal tubular epithelium (Fig. 1.5 d, e and f). With this technique, the authors were able to maintain a perfusable open lumen in culture for long periods of time (≈ 2 months). This interesting model showed to have morphological features and markers similar to native cells. Additionally, the authors also showed that the developed model could be used to study drugs' effect on epithelium's permeability.

In a more complex study, Hyungseok Lee and Dong-Woo Cho (2016) were able to print not only cells but also the microfluidic device in a single procedure. In fact, this interesting work describes the bioprinting of different cell types and materials in a controlled manner to obtain a bioprinted liver-on-a-chip. For that, they printed inside of an empty cavity, human hepatocellular carcinoma cells and human umbilical vein endothelial cells entrapped within a matrix of collagen type 1 and gelatin. Then, they produced the walls of the channel in a layer-by-layer fashion, the top wall and sealed it to avoid leakage. The last step was the production of a connection tube. Overall, the developed liver-on-a-chip platform showed an improved liver

Fig. 1.5 (continued) by an artificial matrix; (**c**) images of the procedure for obtaining a segment of a nephron surrounded by an artificial matrix; (**d**) image of developed convoluted proximal tubule (scale bar = 500 μm); (**e**) higher magnification of area denoted by the white rectangle in (**d**) (scale bar = 200 μm); (**f**) image of cells circumscribing developed convoluted proximal tubule after perfusion (tubulin staining showing primary cilia, scale bar = 50 μm); (**g**) picture of a columnar epithelium; (**h**) human proximal tubular cells after 6 weeks of culture showing a basement membrane composed of laminin and collagen IV (scale bar = 10 μm); (**i, l**) brightfield images of developed convoluted proximal tubule perfused with different concentrations of Cyclosporine A (Cys A) (scale bars = 200 μm); (**j, m**) Images of actin and nuclei staining of developed convoluted proximal tubule perfused with different concentrations of Cys A (scale bars = 200 μm); (**k, n**) high magnification images of (**j, m**) respectively (scale bars = 200 μm). (Adapted with permission from (Homan et al. 2016) Copyright © 2016, Springer Nature)

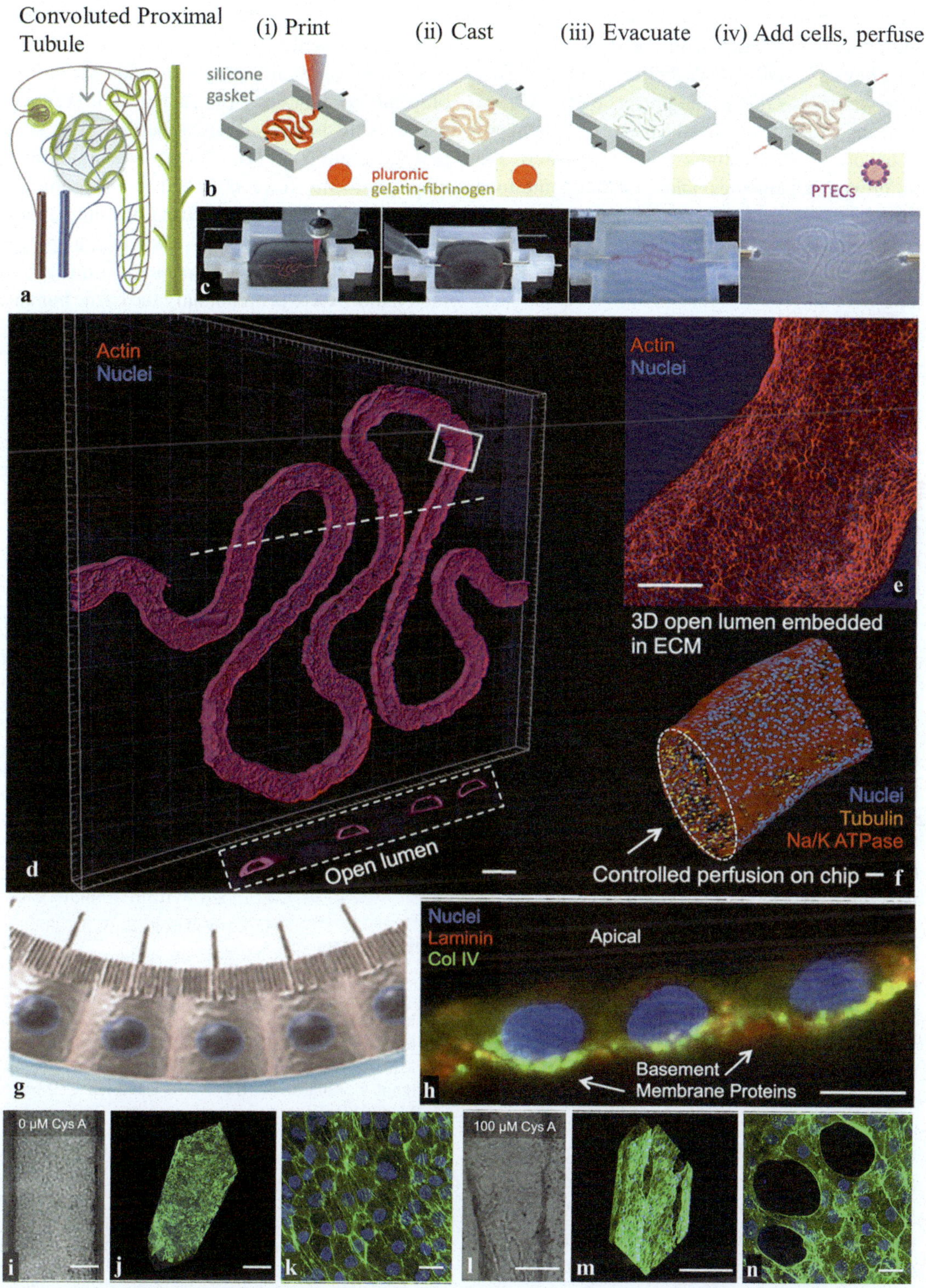

Fig. 1.5 Development of a segment of a nephron surrounded by an artificial matrix. (**a**) Picture of a nephron and convoluted proximal tubule; (**b**) scheme of the procedure for obtaining a segment of a nephron surrounded

function, translated by the secretion of albumin and urea, as compared with the static culture of hepatocytes and endothelial cells. This observation was due to the presence of a flow of medium, which mimicked the native environment. Noteworthy, the authors also compared the culture of hepatocytes under static conditions with the co-culture of hepatocytes with endothelial cells, and showed that the presence of endothelial cells promoted liver functions.

1.6 Conclusions

Bone tissue engineering field has evolved greatly along the years. Even so the development of an *in vitro* model that better mimic the native tissue is still a challenge. Nevertheless, it is clear that it would be an important step to the translation of new drugs into the clinics. The main problem is the complexity presented along with this tissue. Not only due to the cells that compose it, but also due to the different mechanical properties along the bone tissue and the presence of an intricate vasculature network. Given these challenges, new techniques have been pursued. In one hand, the combination of 3D tissue engineering approaches with microfluidic devices allowed to mimic the vasculature naturally present in native tissues. On the other hand, the use of 3D printing has allowed developing spatially controlled structures composed of different materials and/or growth factors and/or cells. In fact, 3D printing allows the fabrication of structures in a more reproducible manner and within small amounts of time and lower costs. Furthermore, it is important to mention that 3D printing has also promoted a new generation of patient-specific implants, boosting personalized medicine. Still, one of the greatest challenges is the development of a bioink for the printing of such structures with properties similar to native bone tissue, maintaining cell viability and promoting cell proliferation and differentiation. Signs of progresses have been made to achieve a proper bioink, such as the reinforcement of mechanical properties of the bioink, the crosslinking used or even the modification with plasmid DNA for control of cells' differentiation. But further optimization will be needed.

Regardless of the progress made so far, 3D bioprinting by itself still struggles to produce tissues with blood vessels structures. For so, a new generation of 3D - bioprinted organ-on-a-chip platforms have emerged as a promising approach. These complex structures are capable of human like functions, allowing a proper investigation of physicochemical stimulus. In fact, it allowed to print not only the tissue but also the microfluidic device in one single step. In this reasoning the result organ-on-a-chip promise to emulate the native tissue as well as the vascularization normally present.

Overall, despite the difficulties associated with 3D printed structures that still need to be overcome, it has been proven to be a valuable tool to accomplish a new generation of 3D bioprinted bone-on-a-chip technology platform.

Acknowledgments The authors thank the funds obtained through the FROnTHERA (NORTE-01-0145-FEDER-0000232) project supported by Norte Portugal Regional Operational Program (NORTE 2020), under the PORTUGAL 2020 Partnership Agreement, through the European Regional Development Fund (ERDF). FRM acknowledges Portuguese Foundation for Science and Technology (FCT) for her post-doc grant (SFRH/BPD/117492/2016), JMO thanks FCT for the distinction attributed under the Investigator FCT program (IF/01285/2015) and VMC acknowledges Investigator FCT program (IF/01214/2014).

References

Adepu S, Dhiman N, Laha A, Sharma CS, Ramakrishna S, Khandelwal M (2017) Three-dimensional bioprinting for bone tissue regeneration. Curr Opin Biomed Eng 2:22–28. https://doi.org/10.1016/j.cobme.2017.03.005

Arrigoni C, Gilardi M, Bersini S, Candrian C, Moretti M (2017) Bioprinting and organ-on-Chip applications towards personalized medicine for bone diseases. Stem Cell Rev 13(3):407–417. https://doi.org/10.1007/s12015-017-9741-5

Aziz AUR, Geng C, Fu M, Yu X, Qin K, Liu B (2017) The role of microfluidics for organ on Chip simulations. Bioengineering 4(2):39. https://doi.org/10.3390/bioengineering4020039

Bendtsen ST, Wei M (2017) In vitro evaluation of 3D bioprinted tri-polymer network scaffolds for bone tissue regeneration. J Biomed Mater Res

A 105(12):3262–3272. https://doi.org/10.1002/jbm.a.36184

Bhattacharjee N, Urrios A, Kang S, Folch A (2016) The upcoming 3D-printing revolution in microfluidics. Lab Chip 16(10):1720–1742. https://doi.org/10.1039/c6lc00163g

Bhise NS, Manoharan V, Massa S, Tamayol A, Ghaderi M, Miscuglio M, Lang Q, Shrike Zhang Y, Shin SR, Calzone G, Annabi N, Shupe TD, Bishop CE, Atala A, Dokmeci MR, Khademhosseini A (2016) A liver-on-a-chip platform with bioprinted hepatic spheroids. Biofabrication 8(1):014101. https://doi.org/10.1088/1758-5090/8/1/014101

Boga JC, Miguel SP, de Melo-Diogo D, Mendonca AG, Louro RO, Correia IJ (2018) In vitro characterization of 3D printed scaffolds aimed at bone tissue regeneration. Colloids Surf B Biointerfaces 165:207–218. https://doi.org/10.1016/j.colsurfb.2018.02.038

Bouet G, Marchat D, Cruel M, Malaval L, Vico L (2015) In vitro three-dimensional bone tissue models: from cells to controlled and dynamic environment. Tissue Eng Part B Rev 21(1):133–156. https://doi.org/10.1089/ten.TEB.2013.0682

Byambaa B, Annabi N, Yue K, Trujillo-de Santiago G, Alvarez MM, Jia W, Kazemzadeh-Narbat M, Shin SR, Tamayol A, Khademhosseini A (2017) Bioprinted osteogenic and vasculogenic patterns for engineering 3D bone tissue. Adv Healthcare Mat 6(16):1700015. https://doi.org/10.1002/adhm.201700015

Carvalho MR, Reis RL, Oliveira JM (2018) Mimicking the 3D biology of osteochondral tissue with microfluidic-based solutions: breakthroughs towards boosting drug testing and discovery. Drug Discov Today 23:711–718. https://doi.org/10.1016/j.drudis.2018.01.008

Costa JB, Silva-Correia J, Oliveira JM, Reis RL (2017) Fast setting silk fibroin bioink for bioprinting of patient-specific memory-shape implants. Adv Healthcare Mat 6(22):n/a-n/a. https://doi.org/10.1002/adhm.201701021

Costa JB, Silva-Correia J, Ribeiro VP, da Silva Morais A, Oliveira JM, Reis RL (2018) Engineering patient-specific bioprinted constructs for treatment of degenerated intervertebral disc. Mat Today Commun doi:https://doi.org/10.1016/j.mtcomm.2018.01.011, 506, 512, 19

Cox SC, Thornby JA, Gibbons GJ, Williams MA, Mallick KK (2015) 3D printing of porous hydroxyapatite scaffolds intended for use in bone tissue engineering applications. Mater Sci Eng C Mater Biol Appl 47:237–247. https://doi.org/10.1016/j.msec.2014.11.024

Cui H, Zhu W, Nowicki M, Zhou X, Khademhosseini A, Zhang LG (2016) Hierarchical fabrication of engineered vascularized bone biphasic constructs via dual 3D bioprinting: integrating regional bioactive factors into architectural design. Adv Healthc Mater 5(17):2174–2181. https://doi.org/10.1002/adhm.201600505

Cunniffe GM, Gonzalez-Fernandez T, Daly A, Sathy BN, Jeon O, Alsberg E, Kelly DJ (2017) Three-dimensional bioprinting of Polycaprolactone reinforced gene activated bioinks for bone tissue engineering. Tissue Eng Part A 23(17–18):891–900. https://doi.org/10.1089/ten.tea.2016.0498

Datta P, Ozbolat V, Ayan B, Dhawan A, Ozbolat IT (2017) Bone tissue bioprinting for craniofacial reconstruction. Biotechnol Bioeng 114(11):2424–2431. https://doi.org/10.1002/bit.26349

Duval K, Grover H, Han L-H, Mou Y, Pegoraro AF, Fredberg J, Chen Z (2017) Modeling physiological events in 2D vs. 3D cell culture. Physiology 32(4):266–277. https://doi.org/10.1152/physiol.00036.2016

Faramarzi N, Yazdi IK, Nabavinia M, Gemma A, Fanelli A, Caizzone A, Ptaszek LM, Sinha I, Khademhosseini A, Ruskin JN, Tamayol A (2018) Patient-specific bioinks for 3D bioprinting of tissue engineering Scaffolds. e1701347. https://doi.org/10.1002/adhm.201701347

Fedorovich NE, De Wijn JR, Verbout AJ, Alblas J, Dhert WJ (2008) Three-dimensional fiber deposition of cell-laden, viable, patterned constructs for bone tissue printing. Tissue Eng A 14(1):127–133. https://doi.org/10.1089/ten.a.2007.0158

Fedorovich NE, Wijnberg HM, Dhert WJ, Alblas J (2011) Distinct tissue formation by heterogeneous printing of osteo- and endothelial progenitor cells. Tissue Eng A 17(15–16):2113–2121. https://doi.org/10.1089/ten.TEA.2011.0019

Florencio-Silva R, GRDS S, Sasso-Cerri E, Simoes MJ, Cerri PS (2015) Biology of bone tissue: structure, function, and factors that influence bone cells. Biomed Res Int 17:1–17. https://doi.org/10.1155/2015/421746

Hao S, Ha L, Cheng G, Wan Y, Xia Y, Sosnoski DM, Mastro AM, Zheng SY (2018) A spontaneous 3D bone-on-a-chip for bone metastasis study of breast cancer cells. Small 14(12):1702787. https://doi.org/10.1002/smll.201702787

He HY, Zhang JY, Mi X, Hu Y, Gu XY (2015) Rapid prototyping for tissue-engineered bone scaffold by 3D printing and biocompatibility study. Int J Clin Exp Med 8(7):11777–11785

Ho CM, Ng SH, Li KH, Yoon YJ (2015) 3D printed microfluidics for biological applications. Lab Chip 15(18):3627–3637. https://doi.org/10.1039/c5lc00685f

Homan KA, Kolesky DB, Skylar-Scott MA, Herrmann J, Obuobi H, Moisan A, Lewis JA (2016) Bioprinting of 3D convoluted renal proximal tubules on Perfusable chips. Sci Rep 6:34845. https://doi.org/10.1038/srep34845

Jariwala SH, Lewis GS, Bushman ZJ, Adair JH, Donahue HJ (2015) 3D printing of personalized artificial bone scaffolds. 3d Print Additive Manuf 2(2):56–64. https://doi.org/10.1089/3dp.2015.0001

Jusoh N, Oh S, Kim S, Kim J, Jeon NL (2015) Microfluidic vascularized bone tissue model with hydroxyapatite-incorporated extracellular matrix. Lab Chip 15(20):3984–3988. https://doi.org/10.1039/C5LC00698H

Kang H-W, Lee SJ, Ko IK, Kengla C, Yoo JJ, Atala A (2016) A 3D bioprinting system to produce human-scale tissue constructs with structural integrity. Nat

Biotechnol 34:312–319. https://doi.org/10.1038/nbt.3413. https://www.nature.com/articles/nbt.3413#supplementary-information

Lee H, Cho D-W (2016) One-step fabrication of an organ-on-a-chip with spatial heterogeneity using a 3D bioprinting technology. Lab Chip 16(14):2618–2625. https://doi.org/10.1039/C6LC00450D

Lee JH, Gu Y, Wang H, Lee WY (2012) Microfluidic 3D bone tissue model for high-throughput evaluation of wound-healing and infection-preventing biomaterials. Biomaterials 33(4):999–1006. https://doi.org/10.1016/j.biomaterials.2011.10.036

Li L, Yu F, Shi J, Shen S, Teng H, Yang J, Wang X, Jiang Q (2017) In situ repair of bone and cartilage defects using 3D scanning and 3D printing. Sci Rep 7(1):9416. https://doi.org/10.1038/s41598-017-10060-3

Lima MJ, Correlo VM, Reis RL (2014a) Micro/nano replication and 3D assembling techniques for scaffold fabrication. Mater Sci Eng C 42:615–621. https://doi.org/10.1016/j.msec.2014.05.064

Lima MJ, Pirraco RP, Sousa RA, Neves NM, Marques AP, Bhattacharya M, Correlo VM, Reis RL (2014b) Bottom-up approach to construct microfabricated multi-layer scaffolds for bone tissue engineering. Biomed Microdevices 16(1):69–78. https://doi.org/10.1007/s10544-013-9806-4

Macdonald NP, Cabot JM, Smejkal P, Guijt RM, Paull B, Breadmore MC (2017) Comparing microfluidic performance of three-dimensional (3D) printing platforms. Anal Chem 89(7):3858–3866. https://doi.org/10.1021/acs.analchem.7b00136

Maia FR, Musson DS, Naot D, da Silva LP, Bastos AR, Costa JB, Oliveira JM, Correlo VM, Reis RL, Cornish J (2018) Differentiation of osteoclast precursors on gellan gum-based spongy-like hydrogels for bone tissue engineering. Biomed Mat (Bristol, England) 13(3):035012. https://doi.org/10.1088/1748-605X/aaaf29

Matic I, Matthews BG, Wang X, Dyment NA, Worthley DL, Rowe DW, Grcevic D, Kalajzic I (2016) Quiescent bone lining cells are a major source of osteoblasts during adulthood. Stem Cells (Dayton, Ohio) 34(12):2930–2942. https://doi.org/10.1002/stem.2474

Rafał W, Krzysztof A (2015) Inkjet 3D printing of microfluidic structures—on the selection of the printer towards printing your own microfluidic chips. J Micromech Microeng 25(8):085013

Salentijn GIJ, Oomen PE, Grajewski M, Verpoorte E (2017) Fused deposition modeling 3D printing for (bio)analytical device fabrication: procedures, materials, and applications. Anal Chem 89(13):7053–7061. https://doi.org/10.1021/acs.analchem.7b00828

Veerla SC, Kim DR, Yang SY (2018) Fabrication of a microfluidic device for studying the in situ drug-loading/release behavior of graphene oxide-encapsulated hydrogel beads. Biomat Res 22(1):7. https://doi.org/10.1186/s40824-018-0119-9

Waheed S, Cabot JM, Macdonald NP, Lewis T, Guijt RM, Paull B, Breadmore MC (2016) 3D printed microfluidic devices: enablers and barriers. Lab Chip 16(11):1993–2013. https://doi.org/10.1039/C6LC00284F

Wang X, Tolba E, Schröder HC, Neufurth M, Feng Q, Diehl-Seifert B, Müller WEG (2014) Effect of bioglass on growth and biomineralization of SaOS-2 cells in hydrogel after 3D cell bioprinting. PLoS One 9(11):e112497. https://doi.org/10.1371/journal.pone.0112497

Wang XF, Song Y, Liu YS, Sun YC, Wang YG, Wang Y, Lyu PJ (2016) Osteogenic differentiation of three-dimensional bioprinted constructs consisting of human adipose-derived stem cells in vitro and in vivo. PLoS One 11(6):e0157214. https://doi.org/10.1371/journal.pone.0157214

Wu Z, Su X, Xu Y, Kong B, Sun W, Mi S (2016) Bioprinting three-dimensional cell-laden tissue constructs with controllable degradation. Sci Rep 6:24474. https://doi.org/10.1038/srep24474

Yi H-G, Lee H, Cho D-W (2017) 3D printing of organs-on-chips. Bioengineering 4(1):10. https://doi.org/10.3390/bioengineering4010010

Microfluidics for Processing of Biomaterials

2

Luca Gasperini, Alexandra P. Marques, and Rui L. Reis

Abstract

Microfluidics techniques can be used to process a wide range of biomaterials, from synthetic to natural origin ones. This chapter describes microfluidic processing of biomaterials, mainly polymeric materials of natural origin, focusing on water-soluble polymers that form non-flowing phases after crosslinking. Some polysaccharides and proteins, including agarose, alginate, chitosan, gellan gum, hyaluronic acid, collagen, gelatin, and silk fibroin are emphasized deu to their relevance in the field. The critical characteristics of these materials are discussed, giving particular consideration to those that directly impact its processability using microfluidics. Furthermore, some microfluidic-based processing techniques are presented, describing their suitability to process materials with different sol-gel transition mechanisms.

Keywords

Microfluidics · Microfabrication · Hydrogels

L. Gasperini (✉)
3B's Research Group, I3Bs – Research Institute on Biomaterials, Biodegradables and Biomimetics, University of Minho, Headquarters of the European Institute of Excellence on Tissue Engineering and Regenerative, Braga, Portugal

ICVS/3B's – PT Government Associate Laboratory, Braga/Guimarães, Portugal
e-mail: luca.gasperini@i3bs.uminho.pt

A. P. Marques · R. L. Reis
3B's Research Group, I3Bs – Research Institute on Biomaterials, Biodegradables and Biomimetics, University of Minho, Headquarters of the European Institute of Excellence on Tissue Engineering and Regenerative, Braga, Portugal

ICVS/3B's – PT Government Associate Laboratory, Braga/Guimarães, Portugal

The Discoveries Centre for Regenerative and Precision Medicine, Headquarters at University of Minho, Guimarães, Portugal

Highlights

- Microfluidics techniques allow a high level of control on fluid flow and can be employed to process materials for microfabrication.
- Based on the nature of flowing the material, different microfluidics set-up of a wide range of complexity can be used.

J. M. Oliveira, R. L. Reis (eds.), *Biomaterials- and Microfluidics-Based Tissue Engineered 3D Models*, Advances in Experimental Medicine and Biology 1230,
https://doi.org/10.1007/978-3-030-36588-2_2

2.1 Biomaterials and Crosslinking Strategies

2.1.1 Thermally Cross-Linked Biomaterials

Some polymeric solutions form a gel upon temperature variations. In these cases, the solvent and solute are miscible within a specific temperature range and up to a critical temperature at which they form a gel. Based on the nature of the polymer and its sol-gel mechanism, the gel is formed either by increasing or decreasing the temperature (Jeong et al. 2002). Some polymeric solutions present a critical temperature below which the polymer and the solution are mixable, lower critical solution temperature (LCST); the gel is formed by increasing the temperature above LCST. On the other hand, the upper critical solution temperature (UCST) indicates the temperature above which polymer and water are miscible, and when the polymeric solution is cooled below the UCST the polymer becomes insoluble and more hydrophobic, forming a gel. This is an enthalpy-driven transition where the polymer-water interactions become unfavorable due to temperature variations. By changing the temperature, the polymer rearranges its conformation to minimize the surface in-contact with water. Many polymeric solutions of natural-origin materials follow this last sol-gel mechanism (Klouda and Mikos 2008; Ward and Georgiou 2011). Agarose, for example, is a polysaccharide derived from red algae formed by alternating units of β-D-galactopyranose and 3,6-anhydro-α-L-galactopyranose that above UCST are organized in a random coil conformation. Below UCST the polymer chains assume a double helix conformation where the helices aggregate and are stabilized by hydrogen bonds (Lahaye and Rochas 1991). The critical temperature depends on the molecular weight and the structure of the polymer and can vary typically within a range of 20–50 °C. While its chemical composition can change with the source, some commercial formulations of agarose are modified to reduce the number of hydrogen bonds, decreasing its critical temperature. As other thermally driven crosslinking, the formation of the gel is reversible, but in the case of agarose it presents a marked thermal hysteresis (Indovina et al. 1979) meaning that after the formation of the gel the liquefaction temperature is much higher than the UCST. The thermal hysteresis allows decreasing the temperature below UCST to form gels that remain stable if reheated above the same temperature. For example, some formulations of agarose can be used to encapsulate cells because they form gels between 25 and 35 °C and the gels remain stables when heated up to the physiological temperature. Gelatin is another common natural-origin biomaterial that forms gel upon cooling. Gelatin is a protein that results from breaking the triple helix of collagen into single strain molecules after acid treatment or hydrolysis (Kuijpers et al. 1999). Gelatin solutions present a critical temperature that depends on its structure and concentration. In some commonly used formulation this critical temperature is very close to physiological temperature. While gelatin, forms gel upon cooling, collagen solutions typically form gels upon heating. Commercial formulations of collagen are commonly found dissolved or soluble in diluted acid (e.g. 0.1% acetic acid). To form a gel, the solution is neutralized with a strong base (usually NaOH) in a concentrated buffer. When heated to physiological temperature collagen undergoes fibrillogenesis through an entropy-driven self-assembly process, forming collagen fibers indistinguishable from the natural ones (Helseth and Veis 1981; Xu et al. 2009).

2.1.2 Polyelectrolytes and Ionically Cross-Linked Biomaterials

Some polymers carry a net charge along their backbone that makes them, in the presence of ions or polyelectrolytes of opposite charge, to form insoluble complexes through those groups responsible for the solubility of the polymer in water(Bekturov and Bimedina 1981). The formed gels can be dissolved in the presence of chelating agents that break down the complex. Some commonly used polymers are negatively charged due to the presence of carboxyl or sulfate groups while positively charged polymers carry amino groups that become protonated in water at low pH (mainly chitosan and polylysine) (Mano et al. 2007).

Alginate is a widely used, negatively charged, polysaccharide derived from brown seaweed. It was one of the first and still one of the most used polymers for cell encapsulation (de Vos et al. 2006; Goh et al. 2012). Alginate forms gels in the presence of multivalent cations that bind adjacent chains creating a 3D hydrophilic network. Alginate is a linear block copolymer made of mannuronic (M) and glucuronic (G) residues being the latter the preferable binding sites for the cations. While the sol-gel transition is quick, the strength of the gel that is formed depends on the structure of the polymer, the relative ratio of G and M blocks and the cations used (Drury et al. 2004; Mancini et al. 1999).

Gellan gum, similarly to alginate, is also a negatively charged polysaccharide that can form hydrogels in the presence of cations but through a different sol-gel mechanism due to the polymer thermos-sensitive character. The polymeric chains of gellan gum are in random coil conformation in solution at high temperature, and upon cooling they rearrange to double-helices that, by further decreasing the temperature, self-assemble in tight structures (Grasdalen and Smidsrød 1987; Oliveira et al. 2010). These structures, necessary for the formation of stable gels, are further stabilized in the presence of positively charged ions that shield the carboxyl side groups of the polymer chain. For this reason, the sol-gel transition can be obtained both by decreasing the temperature of a solution and/or in the presence of ions. Solutions of gellan gum at relatively low concentration (1, 1.5%) do not form strong hydrogels at room temperature, but they present characteristics of microstructured liquids (thixotropic). Differently than alginate where only multivalent cations trigger the transition to gel, in gellan gum any positively charged cation effectively shields the helices increasing the viscosity of the solution or forming a gel. This limits the use of buffers or any solution containing salts to dissolve the polymer. This is particularly relevant if cells are resuspended in the polymeric solution and kept for extended periods of time, as for example prior bioprinting or microfluidics, leading to osmotic shock (Carvalho et al. 2017).

While Alginate and gellan gum are negatively charged, chitosan is one of the few natural positively charged polysaccharides that is being used as a biomaterial. Chitosan derives from the deacetylation of chitin, a structural component of the exoskeleton of some species such as crabs and shrimps (Marguerite Rinaudo 2006). Chitosan contains an amine group that, upon protonation in acidic solutions, renders it soluble and positively charged (Rinaudo et al. 1999). At physiological pH and in the presence of phosphate salts, chitosan is soluble and thermoresponsive, forming a gel when heated up to 37 °C (Couto et al. 2009).

2.1.3 Chemically Cross-Linked Biomaterials

The family of chemically cross-linked biomaterials is composed of those materials that form hydrogels thanks to covalent bonds formed between adjacent polymeric chains after a chemical reaction. Because of this, chemical crosslinking is irreversible, in opposition to the ionic and thermal one that involve reversible mechanisms. Materials that form hydrogels when irradiated by light are particularly interesting in the biomedical field. These crosslinking strategies require the presence of a photoinitiator that forms radicals when irradiated by light. The chemical nature of the photoinitiator dictates the wavelength of the light necessary for the formation of those radicals. The radicals then react with the functional groups of the polymeric backbone forming covalent intermolecular bonds. The attractiveness of crosslink materials just by the action of light has been leading the research on polymer modification to introduce suitable functional groups, such as acrylates, to allow its photocrosslinking (Ferreira et al. 2009; Smeds et al. 2001). One example is the modification of hyaluronic acid, a biodegradable linear anionic polysaccharide present in the extracellular matrix of mammals, that is soluble in water and forms highly viscous solutions at relatively low concentrations but cannot form stable hydrogels (Balazs 2004). The chemical reaction can be optimized to obtain different degrees of substitution and as such to tune the final properties of the gel in terms of porosity and mechanical properties. This type of reaction has been also applied to other biopolymers such as gelatin (Van Den Bulcke et al. 2000) and gellan gum (Coutinho et al. 2010), mentioned above. In these cases, the intrinsic features of the polymers are likely to be changed after the modification of the polymer and thus, the original crosslinking mechanisms might no longer apply.

Another chemical crosslinking method that has been of particular interest is the chemical crosslinking mediated by enzymes. This method is inspired by nature and uses enzymes as catalyzers of the chemical reaction (Moreira Teixeira et al. 2012). By controlling the activity of the enzymes, gels with different properties are formed. One of the most straightforward examples is the combination of fibrinogen and thrombin, naturally present in mammals and involved in vital biological processes. Thrombin catalyzes the cleavage of the fibrinopeptides in the middle of fibrinogen, a water-soluble glycoprotein, to form fibrin monomers. These monomers polymerize into half-staggered oligomers that lengthen into protofibrils, which then aggregate laterally to form fibers. The 3D structure/gel that is formed by branching during polymerization is crosslinked by calcium and the transglutaminase factor XIIIa becoming mechanically stable.

While the above gelling mechanisms are the most common, others of different natures exist. For example a polymer may rearrange forming an insoluble three-dimensional structure as consequence of pH or ionic strength variations of the coagulating bath (Agnello et al. 2017). Some polymers, such as silk fibroin, present complex sol-gel mechanisms influenced by many parameters. The water-soluble silk is metastable and tends to stabilize forming insoluble β-sheets. The sol-gel transition is influenced by mechanical shear stresses, concentration of the protein and salts in solution, and pH and temperature (Kim et al. 2004; Matsumoto et al. 2006).

2.2 Processability

The processability of polymeric solutions for microfluidics is characterized by their resistance to flow. In microfluidics, fluids are forced to flow into microchannels and capillaries by applying pressure to the reservoir where the microchannel is connected. Viscosity is the property of the fluid that represents its resistance to flow and is related to how the molecules in the liquid phase interact. Generically, a highly viscous solution is defined as thick; highly viscous solutions have high resistance to flow and are characterized by high cohesive forces. On the opposite hand, low viscosity solutions are defined as thin and possess a low resistance to flow.

Like other properties of materials, viscosity is a function of different parameters. For Newtonian fluids, viscosity varies with the temperature and viscosity tends to decrease as temperature increases. Polymeric solutions such as the ones that have been discussed in this chapter are generally non-Newtonian. For non-Newtonian fluids, the viscosity not only depends on temperature but also on the shear rate or the shear rate history, being the shear rate the rate of deformation due to shear stresses. In particular, polymeric solutions tend to show a shear-thinning non-Newtonian behavior meaning that their viscosity decreases at higher shear rates. When polymers are forced through a channel they preferentially rearrange in the direction of the flow, minimizing the interaction between the chains and as such decreasing the viscosity of the solution. Thus, the viscosity profile for these materials correlates the shear rate to the value of viscosity. For the majority of the shear-thinning solutions a log-log plot of viscosity vs shear rate allows identifying a linear region corresponding to the power-law distribution that connects two regions where the viscosity is constant, called Newtonian plateaus. The plateau at low shear rates is called zero shear viscosity region while the one at high shear rates is called infinite shear viscosity region. Based on the range of shear rates tested the polymeric solution may present all these regions or only some. Viscosity values are typically measured with a rheometer by placing the polymeric solution between plates, one fixed and the other rotating. A deformation of gradually increasing speed is applied, and the stress necessary to obtain that rotation is recorded by the machine.

The behavior of a non-Newtonian shear thinning solutions can be model using different mathematical models, which leads to their identification with the name of the model (Fig. 2.1). The Carreu and Cross equations can be used to model fluids that present both the power-law regions and the Newtonian plateaus (H. A. Barnes and Walters 1985; Rao 2007). The pressure used in a microfluidic system and the flowrate can be correlated using other equations such as the Hagen-Poiseuille equation, typically presented in the following form (Sutera and Skalak 1993):

$$\Delta P = \frac{8\mu LQ}{\pi R^4}$$

Where ΔP is the difference of pressure between the reservoir and end of the channel, μ is the viscosity, L is the length of the channel, Q is the flow rate, and R is the radius of the channel. This equation is only valid for Newtonian fluids. Other, sometimes called corrections, were developed for non-Newtonian fluids, and contain the parameters obtained after the non-linear regression performed to fit the model to the viscosity data. Examples are the power-law index and the consistency for the Rabinowitsch correction, applicable to power-law fluids.

Other materials that in solution tend to form microstructures, are thixotropic meaning that the measured viscosity depends on the deformation history of the solution (Barnes and Barnes 1997). Those microstructures are usually made of weak forces that are destroyed during deformation. While the viscous forces can rebuild in time, the viscosity will be lower if measured shortly after the deformation. These materials are more challenging to process because their properties, namely viscosity, change while they are processed. This means that the processing parameters should be tuned in function of processing time and adapted to the change of viscosity to obtain consistent results. Furthermore, these materials present an abrupt change of viscosity from the material at rest (off the shelf) and while being processed. Thixotropic materials tend to present apparent yield stress, which is the minimum stress necessary to initiate the flow (Møller et al. 2006).

There are also other properties than viscosity that influence the processability of biomaterials. For example, when working with thermoresponsive materials is essential to define and characterize the temperature range where the sol-gel transition happens. Dynamic tests are performed by applying a sinusoidal shear deformation and recording the stress. This kind of measurement is performed to obtain information on the viscoelastic behavior of the sample. For an ideal elastic sample, the stress is proportional and in phase with the deformation. For purely viscous samples the stress is proportional to the rate of deformation and stress and strain are out of phase. Viscoelastic materials have both contributions,

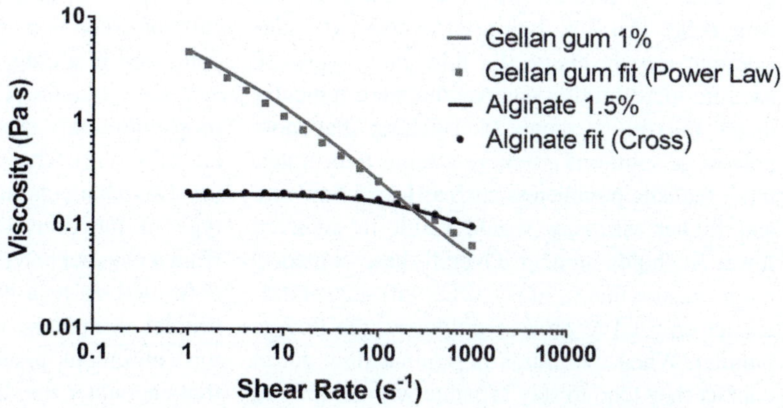

Fig. 2.1 Viscosity profiles of Gellan gum 1% and Alginate 1.5%. The gellan gum profile was fitted with a power-law equation ($R^2 = 0.9875$) while the alginate profile was fitted with a Cross equation ($R^2 = 0.9979$)

and their behavior is described by the storage modulus G' and loss modulus G", which characterize respectively the elastic solid-like and viscous liquid-like contributions. Different experiments can be performed based on the nature of the sample. For example the sol-gel transition temperature of thermoresponsive polymeric solutions can be measured by a temperature sweep. A sinusoidal deformation in the linear viscoelastic region is applied at constant frequency and by gradually changing temperature. G' an G" are recorded, and a gradual increase of G' with temperature is expected close to the gel-point, the point where G' and G" cross.

2.3 Processing Approaches

In microfluidic systems used to process polymers, the flow is characterized by low Reynolds number, or in other terms the flow is laminar. Reynolds number is the ratio between the inertial and the viscous forces of the flow and, for the flow in a circular pipe, is defined as:

$$Re = \frac{\rho v D}{\mu}$$

Where ρ, v and μ are respectively the density, velocity, and viscosity of the fluid, while D is the diameter of the channel. The flow is considered laminar if the Reynolds number is under 2000. To understand why in microfluidic systems the flow is always laminar it's important to provide simple examples. Considering the case of water flowing in a channel of 500 μm of diameter with a flow rate of 20 mL/min the Reynolds number is less than 1000. The flow rate used in this particular example is way above the flow rates typically used in microfluidic systems which are typically in the microliters range, Considering that most polymeric solutions are more viscous than water, more realistic parameters such as lower flow rate and higher viscosity would result in an even lower Reynolds number. Overall these considerations ensure that in most (if not all) microfluidics set-up used to process polymers, the flow is laminar. When two fluids in laminar flow get in contact they tend to stay separate without mixing and this is one crucial feature commonly exploited to process biomaterials. Mixing two different fluids in microfluidic systems can be challenging. This is proved by the amount of literature that can be found related to this subject (Lee et al. 2011; Nguyen and Wu 2005). When mixing is needed, a chip that contains a passive mixer is a simple solution to design. In this configuration, the microchannels are designed to increase the contact area between two or more fluids and thus favoring mixing. Differently, active mixers are miniaturized devices that apply external forces to the fluids.

2.3.1 Droplet-Based

Droplet-based microfluidics is a high throughput method capable of producing highly monodispersed and separated droplets of liquids in an immiscible phase. Droplet-based systems are typically emulsion system, and as such it allows producing oil droplets in water or water-based droplets in oil. The material used for the channels must be selected accordingly, preferable hydrophilic materials for oil in water and hydrophobic ones for water in oil emulsions since the continuous phase has to wet the walls of the channels and to keep the droplets separated avoiding their coalescence after being formed (Jankowski et al. 2013; Saeki et al. 2010). Surfactants (e.g. Tween 20 and Span 80) are commonly added to lower the interfacial energy of the phases facilitating the formation of new interfaces and stabilizing the formed droplet.

Droplets are formed due to a junction, typically between two or more microfluidic channels contained in a chip. The geometry of a junction can vary, but the T and the flow-focusing are most commonly used for the formation of droplets (Fig. 2.2). The T-junction is formed by joining two channels at a 90-degree angle. In this type of junctions the dispersed phase coming from a channel enters the continuous phase and is deformed up to a break by the flow, forming a droplet. Similarly, in a flow-focusing junction, the continuous phase flows from side channels that encounter the droplet-forming solution (dis-

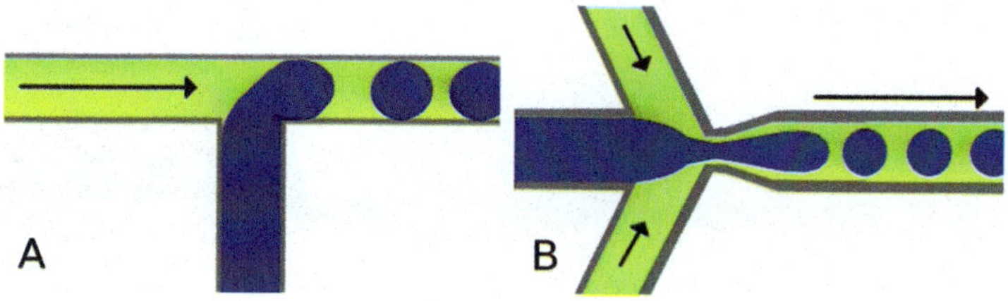

Fig. 2.2 Scheme showing two microfluidic junctions used for the formation of droplets. (**a**) T-junction (**b**) Flow focus. (Gasperini et al. 2014)

perse phase) at the junction forcing it into an orifice present at the beginning of the exit channel that focuses the flow and favors the formation of droplets. In this junction the shearing forces that form the droplet are symmetric. In some set-ups two inlets can be used to carry two different polymeric solutions that intersect at the junction forming a droplet with two separate sections, the Janus beads (Zhao et al. 2009).

In both the T and flow-focusing junctions, the size of the droplet is determined by the nature of the solutions (e.g., viscosity), the size of the channels and the relative flow rate of the two phases. Furthermore, there are three distinct mechanisms of formation of droplets: squeezing, dripping and jetting (De Menech et al. 2008; Nie et al. 2008). Depending on the relative flow rates and the nature of the solutions one of these mechanisms might be favored. In particular, in the dripping mode the formation of the droplet due to the breakup of the fluid happens near the junction. On the other hand, in the jetting mode the flow of the droplet-forming phase is squeezed creating a continuous thread that extends into the outlet channel and breaks far away from the junction. This last regime leads to droplets with a broader size range and is favored at high flowrates (Seemann et al. 2011).

Depending on the nature of the material, the droplets formed by microfluidics can then be crosslinked using different methods. Droplets made of photo cross-linkable and thermally cross-linkable materials are hardened by the mechanisms discussed in the first section of this chapter and inside the channels composing the microfluidic apparatus. For thermally cross-linkable materials, parts of the microfluidic system can be placed in a water bath of controlled temperature guaranteeing the crosslinking before the droplets reaching its end and being collected. Importantly, abrupt expansions of the fluid during this phase should be avoided because they can lead to the formation of local vortexes that destroy the still liquid droplet. For photo cross-linkable materials the section of the microfluidic apparatus where the crosslinking takes place should be made of material that is transparent to the particular light wavelength used. For example, polytetrafluoroethylene is a common choice in many microfluidics setups because the fluorinated polymer has excellent resistance to a wide range of chemicals and good rigidity. It can also easily be found from many suppliers in different sizes, although these pieces are typically semi-transparent or opaque to light. Thus, for photo-crosslinking, other fluorinated polymers, such as fluorinated ethylene propylene, should be chosen. While maintaining excellent chemical resistance this polymer has much better optical properties in the visible lights and close UV range (Šiljegović et al. 2011).

Ionically cross-linkable materials can be processed in the chip or out of the chip. One of the methods that have been followed to form a gel is based on carriers that can be triggered in a controlled manner to release ions. For example, calcium carbonate can release calcium under acidic conditions; calcium carbonate nanoparticles-containing drops were crosslinked when subjected to a continuous phase at low pH coming from an inlet placed upstream the junction due to the release of calcium (Tan and Takeuchi 2007). Other authors have produced droplets by emulsifying an aqueous solution of alginate in a phase containing a dissolved cross-linking agent (Choi et al. 2007; Zhang et al. 2006). In this case the continuous phase containing cross-linking ions is used directly in the junction during the formation of the drop. External crosslinking can also be achieved either by using crosslinking baths con-

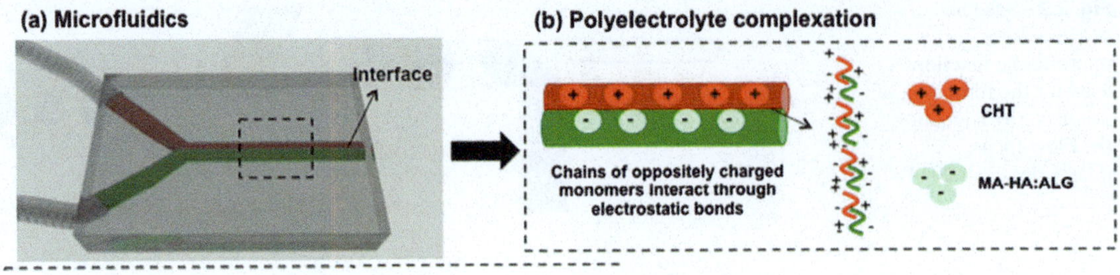

Fig. 2.3 (**a**) Illustrates the low tendency of fluids to mix in a microfluidic device. (**b**) Polyelectrolyte complexation at the interface of the two fluids occurs through electrostatic interactions between the negatively charged solution (green, either MA-HA:ALG or MA-CS:ALG) and the positive counterpart (red, CHT). (Reprinted (adapted) with permission from (Raquel Costa-Almeida et al. 2016). Copyright 2016 American Chemical Society)

taining salts (Chen et al. 2013) or by collecting the droplets in a solid substrate rich in calcium that allows the diffusion of the cations into the droplets (Hirama et al. 2013). By changing the viscosity of the crosslinking bath, during hardening the droplets can gain different shapes, from microgels with a dimple to spherical beads (Hu et al. 2012).

2.3.2 Continuous-Flow

While droplet microfluidics deals with the fabrication and manipulation of discrete droplets of liquids, continuous microfluidics is a technique based on the manipulation of continuous flows (Jun et al. 2014). More straightforward or similar microchip designs to the ones used for the formation of droplets can be used. The simpler design comprising one single channel is used to spin a single component fiber. It is a set up commonly employed for wet-spinning fibers into a coagulating bath (Agnello et al. 2016). In this case the most commonly used materials are the ones that crosslink quickly in the presence of the crosslinking agent present in the coagulating bath (e.g. alginate spun in calcium chloride). This allows the formation of the gel immediately after exiting from the microchip guaranteeing the fabrication of a round fiber.

Other chips may contain junctions similar to the ones of droplet microfluidics to fabricate multicomponent fibers. In this case, instead of using emulsions, the inherent low tendency of fluids to mix is commonly exploited.

As mentioned above, fluids in laminar flow regimes have a low tendency to mix so that two fluid flowing in parallel inside one channel will stay separate (Fig. 2.3). When two polyelectrolytes are used, the interaction at the interface can form a solid phase that aligns in the direction of the flow forming an ordered microstructure (Costa-Almeida et al. 2015). Other geometries of the chip allow obtaining different types of fibers. If it is coaxial, the inner component of the fiber will stay separate from the outer shell. Upon crosslinking the separation is maintained and a core-shell fiber can be fabricated. Some complex geometries also exist that combine in one single chip some of the geometries previously discussed forming multicomponent fibers with accurate and complex structures (Cheng et al. 2014).

2.4 Conclusions

In this chapter, different applications of microfluidics for processing different polymers were presented. The versatility of microfluidics techniques can be exploited using different designs of the chips and different set-ups to target specific applications. Each design has its advantages, and no single device or solution is universal and able to process all materials. Overall there is a wide variety of solutions that can be adopted for processing polymers at the microscale.

Acknowledgments APM acknowledges the European Research Council for Consolidator Grant Project "ECM_INK" ERC-2016-COG-726061.

References

Agnello S, Gasperini L, Reis RL, Mano JF, Pitarresi G, Palumbo FS, Giammona G (2016) Microfluidic production of hyaluronic acid derivative microfibers to control drug release. Mater Lett 182:309–313

Agnello S, Gasperini L, Mano JF, Pitarresi G, Palumbo FS, Reis RL, Giammona G (2017) Synthesis, mechanical and thermal rheological properties of new gellan gum derivatives. Int J Biol Macromol 98:646–653. https://doi.org/10.1016/j.ijbiomac.2017.02.029

Balazs EA (2004) Viscoelastic properties of hyaluronan and its therapeutic use. In: Chemistry and biology of hyaluronan, vol 415. Elsevier, Amsterdam

Barnes HHA, Barnes A (1997) Thixotropy – a review. Journal of Non-Newtonian Fluid Mechanics 70(97):1–33. https://doi.org/10.1016/S0377-0257(97)00004-9

Barnes HA, Walters K (1985) The yield stress myth? Rheol Acta 24(4):323–326. https://doi.org/10.1007/BF01333960

Bekturov EA, Bimedina LA (1981) Interpolymer complexes. Adv Polymer Sci 41:99–147

Carvalho AF, Gasperini L, Ribeiro RS, Marques AP, Reis R l (2017) Control of osmotic pressure to improve cell viability in cell-laden tissue engineering constructs. J Tissue Eng Regen Med 12(2):e1063–e1067. https://doi.org/10.1002/term.2432

Chen W, Kim J-H, Zhang D, Lee K-H, Cangelosi GA, Soelberg SD, Furlong CE, Chung J-H, Shen AQ (2013) Microfluidic one-step synthesis of alginate microspheres immobilized with antibodies. J R Soc, Interface R Soc 10:20130566. https://doi.org/10.1098/rsif.2013.0566

Cheng Y, Zheng F, Lu J, Shang L, Xie Z, Zhao Y, Chen Y, Gu Z (2014) Bioinspired multicompartmental microfibers from microfluidics. Adv Mater 26:5184–5190. https://doi.org/10.1002/adma.201400798

Choi C-H, Jung J-H, Rhee YW, Kim D-P, Shim S-E, Lee C-S (2007) Generation of monodisperse alginate microbeads and in situ encapsulation of cell in microfluidic device. Biomed Microdevices 9(6):855–862. https://doi.org/10.1007/s10544-007-9098-7

Costa-Almeida R, Gasperini L, Rodrigues MT, Babo PS, Mano JF, Reis RL, Gomes ME (2015) Development of biomimetic microengineered hydrogel fibers for tendon regeneration In 2015 4th TERMIS World Congress

Costa-Almeida R, Gasperini L, Borges J, Babo PS, Rodrigues MT, Mano JF, Reis RL, Gomes ME (2016) Microengineered multicomponent hydrogel fibers: combining polyelectrolyte complexation and microfluidics. ACS Biomat Sci Eng 3(7):1322–1331. https://doi.org/10.1021/acsbiomaterials.6b00331

Coutinho DF, Sant SV, Shin H, Oliveira JT, Gomes ME, Neves NM, Khademhosseini A, Reis RL (2010) Modified Gellan gum hydrogels with tunable physical and mechanical properties. Biomaterials 31(29):7494–7502. https://doi.org/10.1016/j.biomaterials.2010.06.035

Couto DS, Hong Z, Mano JF (2009) Development of bioactive and biodegradable chitosan-based injectable systems containing bioactive glass nanoparticles. Acta Biomater 5(1):115–123. https://doi.org/10.1016/j.actbio.2008.08.006

De Menech M, Garstecki P, Jousse F, Stone H a (2008) Transition from squeezing to dripping in a microfluidic T-shaped junction. J Fluid Mech 595:141–161. https://doi.org/10.1017/S002211200700910X

de Vos P, Faas MM, Strand B, Calafiore R (2006) Alginate-based microcapsules for immunoisolation of pancreatic islets. Biomaterials 27(32):5603–5617. https://doi.org/10.1016/j.biomaterials.2006.07.010

Drury JL, Dennis RG, Mooney DJ (2004) The tensile properties of alginate hydrogels. Biomaterials 25(16):3187–3199. https://doi.org/10.1016/j.biomaterials.2003.10.002

Ferreira P, Coelho, JFJ, Almeida JF, Gil MH (2009) Photocrosslinkable Polymers for Biomedical Applications. In R. Fazel (Ed.), Biomedical Engineering - Frontiers and Challenges. InTech, pp. 55–74

Gasperini L, Mano JF, Reis RL (2014) Natural polymers for the microencapsulation of cells. J R Soc Interface 11(100):20140817–20140817. https://doi.org/10.1098/rsif.2014.0817

Goh CH, Heng PWS, Chan LW (2012) Alginates as a useful natural polymer for microencapsulation and therapeutic applications. Carbohydr Polym 88(1):1–12. https://doi.org/10.1016/j.carbpol.2011.11.012

Grasdalen H, Smidsrød O (1987) Gelation of gellan gum. Carbohydr Polym 7(5):371–393. Retrieved from http://www.sciencedirect.com/science/article/pii/014486178790004X

Helseth DL, Veis A (1981) Collagen self-assembly in vitro differentiating specific telopeptide-dependent interactions using selective enzyme modification and the addition of free amino telopeptide. J Biol Chem 256(14):7118–7128. Retrieved from https://pdfs.semanticscholar.org/6d43/701ea790f272f10fdc437bcc8abfc79a38a5.pdf

Hirama H, Kambe T, Aketagawa K, Ota T, Moriguchi H, Torii T (2013) Hyper alginate gel microbead formation by molecular diffusion at the hydrogel/droplet interface. Langmuir: ACS J Surfaces Colloids 29(2):519–524. https://doi.org/10.1021/la303827u

Hu Y, Wang Q, Wang J, Zhu J, Wang H, Yang Y (2012) Shape controllable microgel particles prepared by microfluidic combining external ionic crosslinking. Biomicrofluidics 6(2):26502–265029. https://doi.org/10.1063/1.4720396

Indovina PL, Tettamanti E, Micciancio-Giammarinaro MS, Palma MU (1979) Thermal hysteresis and reversibility of gel-sol transition in agarose-water systems. J Chem Phys 70(6):2841–2847. https://doi.org/10.1063/1.437817

Jankowski P, Ogończyk D, Derzsi L, Lisowski W, Garstecki P (2013) Hydrophilic polycarbonate chips for generation of oil-in-water (O/W) and water-in-oil-in-water (W/O/W) emulsions. Microfluid

Nanofluid 14(5):767–774. https://doi.org/10.1007/s10404-012-1090-8
Jeong B, Kim SW, Bae YH (2002) Thermosensitive sol-gel reversible hydrogels. Adv Drug Deliv Rev 54(1):37–51. Retrieved from http://www.ncbi.nlm.nih.gov/pubmed/11755705
Jun Y, Kang E, Chae S, Lee S-H (2014) Microfluidic spinning of micro- and nano-scale fibers for tissue engineering. Lab Chip 14(13):2145–2160. https://doi.org/10.1039/c3lc51414e
Kim U-J, Park J, Li C, Jin H-J, Valluzzi R, Kaplan DL (2004) Structure and properties of silk hydrogels. Biomacromolecules 5(3):786–792. https://doi.org/10.1021/bm0345460
Klouda L, Mikos AG (2008) Thermoresponsive hydrogels in biomedical applications. Eur J Pharm Biopharm: Off J Arbeitsgemeinschaft Für Pharmazeutische Verfahrenstechnik eV 68(1):34–45. https://doi.org/10.1016/j.ejpb.2007.02.025
Kuijpers AJ, Engbers GHM, Feijen J, De Smedt SC, Meyvis TKL, Demeester J, Krijgsveld J, Zaat SAJ, Dankert J (1999) Characterization of the network structure of Carbodiimide cross-linked gelatin gels. Macromolecules 32(10):3325–3333. https://doi.org/10.1021/ma981929v
Lahaye M, Rochas C (1991) Chemical structure and physico-chemical properties of agar. In: Juanes JA, Santelices B, McLachlan JL (eds) International workshop on Gelidium. Springer, Dordrecht, pp 137–148. https://doi.org/10.1007/978-94-011-3610-5
Lee C, Chang C, Wang Y, Fu L (2011) Microfluidic mixing: a review. Int J Mol Sci 12:3263–3287. https://doi.org/10.3390/ijms12053263
Mancini M, Moresi M, Rancini R (1999) Mechanical properties of alginate gels: empirical characterisation. J Food Eng 39:369–378. Retrieved from http://www.sciencedirect.com/science/article/pii/S0260877499000229
Mano JF, Silva GA, Azevedo HS, Malafaya PB, Sousa RA, Silva SS, Boesel LF, Oliveira JM, Santos TC, Marques AP, Neves NM, Reis RL (2007) Natural origin biodegradable systems in tissue engineering and regenerative medicine: present status and some moving trends. J R Soc Interface 4(17):999–1030. https://doi.org/10.1098/rsif.2007.0220
Matsumoto A, Chen J, Collette AL, Kim U, Altman GH, Cebe P, Kaplan DL (2006) Mechanisms of silk fibroin sol-gel transitions. J Phys Chem B 110(43):21630–21638. https://doi.org/10.1021/jp056350v
Møller PCF, Mewis J, Bonn D (2006) Yield stress and thixotropy: on the difficulty of measuring yield stresses in practice. Soft Matter 2(4):274. https://doi.org/10.1039/b517840a
Moreira Teixeira LS, Feijen J, van Blitterswijk CA, Dijkstra PJ, Karperien M (2012) Enzyme-catalyzed crosslinkable hydrogels: emerging strategies for tissue engineering. Biomaterials 33(5):1281–1290. https://doi.org/10.1016/j.biomaterials.2011.10.067
Nguyen N-T, Wu Z (2005) Micromixers—a review. J Micromech Microeng 15(2):R1–R16. https://doi.org/10.1088/0960-1317/15/2/R01
Nie, Z., Seo, M., Xu, S., Lewis, P. C., Mok, M., Kumacheva, E., … Stone, H. a. (2008). Emulsification in a microfluidic flow-focusing device: effect of the viscosities of the liquids. Microfluid Nanofluid, 5, 585–594. doi:https://doi.org/10.1007/s10404-008-0271-y
Oliveira JT, Martins L, Picciochi R, Malafaya PB, Sousa RA, Neves NM, Mano JF, Reis RL (2010) Gellan gum: a new biomaterial for cartilage tissue engineering applications. J Biomed Mater Res A 93(3):852–863. https://doi.org/10.1002/jbm.a.32574
Rao MA (2007) Flow and functional models for rheological properties of fluid foods. Springer, Boston, pp 27–58. https://doi.org/10.1007/978-0-387-70930-7_2
Rinaudo M (2006) Chitin and chitosan: properties and applications. Prog Polym Sci 31(7):603–632. https://doi.org/10.1016/j.progpolymsci.2006.06.001
Rinaudo M, Pavlov G, Desbrie J (1999) Influence of acetic acid concentration on the solubilization of chitosan. Polymer 40:7029–7032
Saeki D, Sugiura S, Kanamori T, Sato S, Ichikawa S (2010) Microfluidic preparation of water-in-oil-in-water emulsions with an ultra-thin oil phase layer. Lab Chip 10(3):357–362. https://doi.org/10.1039/B916318B
Seemann R, Brinkmann M, Pfohl T, Herminghaus S (2011) Droplet based microfluidics. Rep Prog Phys 75:016601. https://doi.org/10.1088/0034-4885/75/1/016601
Šiljegović M, Kačarević-Popović ZM, Bibić N, Jovanović ZM, Maletić S, Stchakovsky M, Krklješ AN (2011) Optical and dielectric properties of fluorinated ethylene propylene and tetrafluoroethylene–perfluoro(alkoxy vinyl ether) copolymer films modified by low energy N4+ and C4+ ion beams. Radiat Phys Chem 80(12):1378–1385. https://doi.org/10.1016/J.RADPHYSCHEM.2011.08.012
Smeds KA, Grinstaff MW, Pfister-serres A, Miki D, Dastgheib K, Inoue M, Hatchell DL (2001) Photocrosslinkable polysaccharides for in situ hydrogel formation. J Biomed Mater Res 54(1):115–121
Sutera SP, Skalak R (1993) The history of Poiseuille's law. Annu Rev Fluid Mech 25(1):1–20. https://doi.org/10.1146/annurev.fl.25.010193.000245
Tan W-H, Takeuchi S (2007) Monodisperse alginate hydrogel microbeads for cell encapsulation. Adv Mater 19(18):2696–2701. https://doi.org/10.1002/adma.200700433
Van Den Bulcke AI, Bogdanov B, De Rooze N, Schacht EH, Cornelissen M, Berghmans H (2000) Structural and rheological properties of Methacrylamide modified gelatin hydrogels. Biomacromolecules 1(1):31–38. https://doi.org/10.1021/bm990017d
Ward M, Georgiou TK (2011) Thermoresponsive polymers for biomedical applications. Polymers 3(4):1215–1242. https://doi.org/10.3390/polym3031215
Xu T, Molnar P, Gregory C, Das M, Boland T, Hickman JJ (2009) Electrophysiological characterization of embryonic hippocampal neurons cultured in a 3D

collagen hydrogel. Biomaterials 30(26):4377–4383. https://doi.org/10.1016/j.biomaterials.2009.04.047

Zhang H, Tumarkin E, Peerani R, Nie Z, Sullan RMA, Walker GC, Kumacheva E (2006) Microfluidic production of biopolymer microcapsules with controlled morphology. J Am Chem Soc 128(37):12205–12210. https://doi.org/10.1021/ja0635682

Zhao LB, Pan L, Zhang K, Guo SS, Liu W, Wang Y, Chen Y, Zhao XZ, Chan HLW (2009) Generation of Janus alginate hydrogel particles with magnetic anisotropy for cell encapsulation. Lab Chip 9(20):2981–2986. https://doi.org/10.1039/b907478c

Organs-on-a-Chip

3

Lucie A. Low, Margaret Sutherland, Nadya Lumelsky, Seila Selimovic, Martha S. Lundberg, and Danilo A. Tagle

Abstract

Organs-on-chips, also known as "tissue chips" or microphysiological systems (MPS), are bioengineered microsystems capable of recreating aspects of human organ physiology and function and are *in vitro* tools with multiple applications in drug discovery and development. The ability to recapitulate human and animal tissues in physiologically relevant three-dimensional, multi-cellular environments allows applications in the drug development field, including; (1) use in assessing the safety and toxicity testing of potential therapeutics during early-stage preclinical drug development; (2) confirmation of drug/therapeutic efficacy *in vitro*; and (3) disease modeling of human tissues to recapitulate pathophysiology within specific subpopulations and even individuals, thereby advancing precision medicine efforts. This chapter will discuss the development and evolution of three-dimensional organ models over the past decade, and some of the opportunities offered by MPS technology that are not available through current standard two-dimensional cell cultures, or three-dimensional organoid systems. This chapter will outline future avenues of research in the MPS field, how cutting-edge biotechnology advances are expanding the applications for these systems, and discuss the

L. A. Low (✉)
National Center for Advancing Translational Sciences (NCATS), National Institutes of Health, Bethesda, MD, USA
e-mail: lucie.low@nih.gov

M. Sutherland
National Institute for Neurological Disorder and Stroke (NINDS), National Institutes of Health, Bethesda, MD, USA
e-mail: sutherlandm@ninds.nih.gov

N. Lumelsky
National Institute of Dental and Craniofacial Research (NIDCR), National Institutes of Health, Bethesda, MD, USA
e-mail: nadyal@nidcr.nih.gov

S. Selimovic
National Institute of Biomedical Imaging and Bioengineering (NIBIB), National Institutes of Health, Bethesda, MD, USA
e-mail: Seila.selimovic@nih.gov

M. S. Lundberg
National Heart, Lung, and Blood Institute (NHLBI), National Institutes of Health, Bethesda, MD, USA
e-mail: lundberm@nhlbi.nih.gov

D. A. Tagle
National Center for Advancing Translational Sciences (NCATS), National Institutes of Health, Bethesda, MD, USA
e-mail: Danilo.tagle@nih.gov

J. M. Oliveira, R. L. Reis (eds.), *Biomaterials- and Microfluidics-Based Tissue Engineered 3D Models*, Advances in Experimental Medicine and Biology 1230,
https://doi.org/10.1007/978-3-030-36588-2_3

current and future potential and challenges remaining for the field to address.

Keywords

Microphysiological systems · Bioengineering · Microfluidics · Drug development · Disease modeling

Highlights

- Organ-on-chips are three-dimensional tissue platforms that could have utility in a wide range of drug development paradigms
- Platforms could be used for disease modeling, drug screening and to identify patient subgroups, or those who are most likely to benefit from clinical treatments
- Biological and technical challenges remain but validation and commercialization efforts are underway

3.1 Introduction

It is well-known that the current processes for discovering and developing new therapeutic drugs and treatments are hugely costly and complex, take many years, and have extremely high attrition rates, with the simplistic "chevron" depiction of drug development recently being updated to a more realistic and dynamic "map" (Wagner et al. 2017a, b). The failure rate of drugs entering clinical trials is as high as 90%, and recent reports put the timeline and cost of a new drug entering the market as up to 15 years and over $2.6 billion from start to finish (Morgan et al. 2011; Profile Toolkit 2018). During preclinical stages, potential compounds mainly fail due to unacceptable toxicity levels in 2-dimensional cellular and animal models (Roberts et al. 2014; Waring et al. 2015). At later clinical stages, failures are predominantly a result of a lack of drug efficacy (Arrowsmith 2011a, b).

Because many potential therapeutics are lost during development or require years of clinical trials data before approval for broader clinical use, the length of time needed plus high costs of drug development ends up having direct effects on patient quality of life and disease prognoses. Therefore, alternative approaches are desperately needed to speed up the processes, reduce costs and produce safer and more effective drugs. Worldwide efforts are underway to develop, validate and apply alternative methods and paradigms to supplement existing platforms, and to introduce novel assays and methodologies to help speed, streamline and redevelop drug development processes (see Sect. 3.2) (Balls 1995; Casey et al. 2015).

Over the last decade, significant advancements have been realized in the development and utilization of human *in vitro* microphysiological systems (MPS), often termed "organs-on-chip" or "tissue chips", for toxicity analysis, efficacy testing and disease modeling. These microfluidic devices combine materials science and bioengineering approaches with human stem cell, primary and immortalized cell line-based biology to replicate human tissue responses and diseases. 3D cell and tissue models are more complex than standard 2D cell cultures, and often incorporate multiple heterogeneous cell types that result in relatively more faithful recapitulation of human *in vivo* cellular environments by allowing cellular communications in more appropriate contexts (see Sect. 3.2). Additionally, the development of human bioengineered systems at preclinical phases reduces the need for animal testing, which helps eliminate the limitations of testing therapeutics in organisms with different physiologies (e.g. different liver metabolic enzymes (Fashe et al. 2015, Sasahara et al. 2015)) as well as being highly responsive to the 3R principles of animal testing (refinement, reduction, replacement). Growth within this technology sector has been driven by a need to increase success in new drug development pipelines and reduce the escalating R&D costs by offering technical alternatives (Martin et al. 2017), and is projected to experience a 38% growth rate in commercial areas over the next few years (Fig. 3.1).

It is becoming widely acknowledged that MPS could serve as tools with applications across a broad spectrum of drug development, from providing more physiologically-relevant platforms for early target validation of promising com-

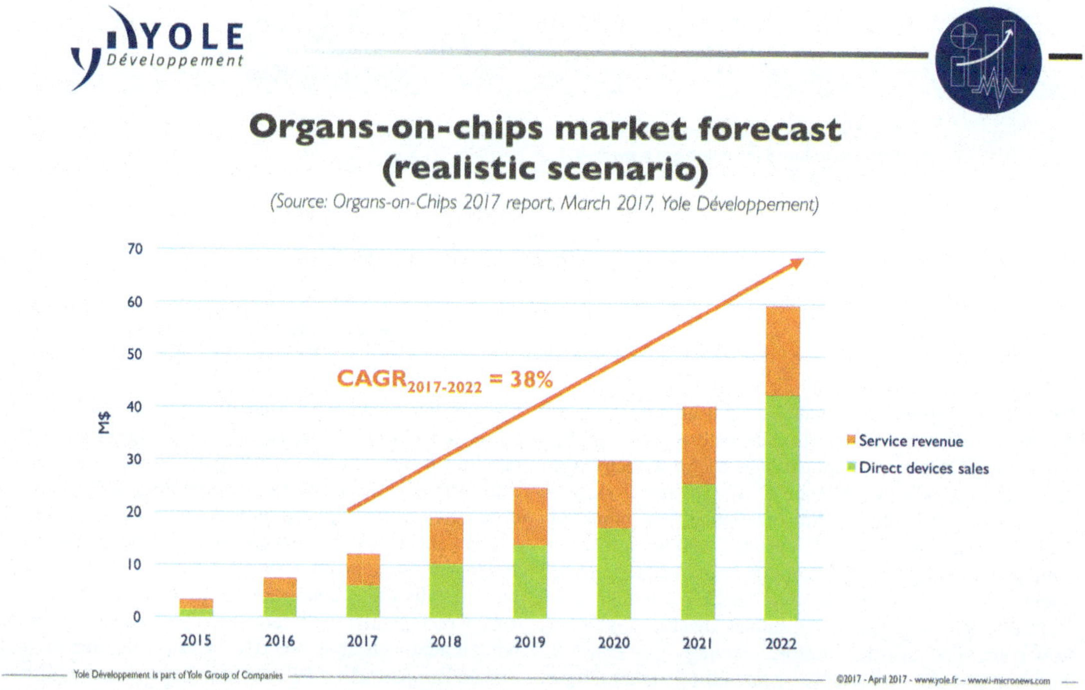

Fig. 3.1 – Market analysis of projected growth of Organs-on-chips market from 2017–2022. CAGR = compound annual growth rate. Courtesy of Yole Développement, from "Organs-on-chips: small market and gigantic promise", Organs-On-Chips 2017 report, April 2017 (http://www.yole.fr/OrgansOnChips_Market.aspx#.W2B8pxpKiV4)

pounds; to disease modeling of both common and rare diseases in order to shed light on pathophysiology and drug responses; to stratification of patient subpopulations or even creating "you-on-a-chip" for truly personalized medicine (Fig. 3.2). This chapter will describe the evolution of MPS over the last decade and compare the pluses and minuses of 2-dimensional (2D) and 3-dimensional (3D) cellular model systems. It will highlight recent key innovations in MPS platform development and finally discuss important opportunities and challenges that remain for the field.

3.2 Comparing 2-Dimensional and 3-Dimensional Tissue Modeling Systems

The recent explosion of activity in the development of 3D *in vitro* tissue systems such as organoids and organs-on-chips for a variety of human tissues signifies recognition that they promise a superior capacity over traditional 2D systems to mimic extracellular matrix (ECM) organization, tissue architecture, specific cell-cell/cell-ECM interactions and sophisticated environmental cues, including biomechanical and biophysical forces (Jackson and Lu 2016). However, 2D and 3D systems each have their own advantages and disadvantages, which should be considered within the context of specific applications (Liu et al. 2018) (Table 3.1). 2D cultures, where cells are grown as monolayers on flat surfaces, have been employed for many years as ubiquitous tools for drug screening, toxicity testing and cellular and molecular analyses of disease mechanisms (Guillouzo 1998). Because of their simplicity, ease of technical manipulation, scalability, and high-throughput capabilities, they remain a widespread model. Indeed, much useful information has been obtained with 2D cultures. For example, over the years they have been widely used for drug toxicity studies (Niu and

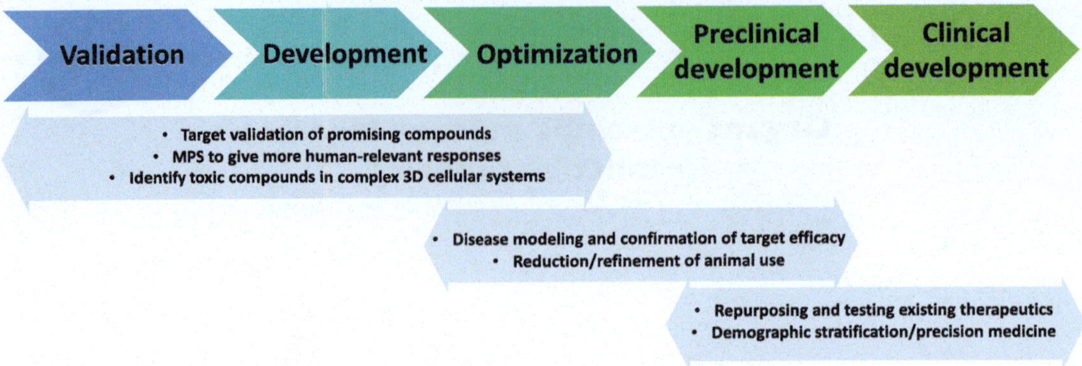

Fig. 3.2 A simplified schematic of the drug development process, excluding post-approval monitoring by regulatory bodies. Organs-on-chips could be useful at each stage of the drug development process. At preclinical stages, they allow screening of hit compounds in complex 3-dimensional multi-cellular systems, allowing earlier elimination of highly toxic compounds and validation and optimization of lead targets more quickly. At later preclinical stages, they can model human diseases and confirm compound efficacy. Before moving into Phase I clinical trials, use of multiple integrated organs-on-chips allows ADME and PK-PB modeling and refinement of dosage guidelines and toxicity profiles for 'first-in-humans' studies. Animal usage is also reduced/refined. At preclinical and clinical stages, organs-on-chips can be created from healthy and diseased populations using primary organ tissues or induced pluripotent stem cell (iPSC) sources and used to test existing and repurposed therapeutics in a risk-free manner for patients. Patient populations, or even individuals, can be modeled on a chip, and personalized therapeutics or treatment regimens tested before human administration. ADME – absorption, distribution, metabolism, excretion. PK-PB modeling – physiologically-based pharmacokinetic modeling

Wang 2015), and for studies of the mechanisms and pathophysiology of many monogenic and polygenic diseases, such as inborn errors of hepatic metabolism (Pournasr and Duncan 2017), Parkinson's disease (Lopes et al. 2017; Singh Dolt et al. 2017), diabetes (Wallet et al. 2017) and others. Gene-editing technologies have been employed for derivation of isogenic induced pluripotent stem cell (iPSC) lines using 2D culture systems, where derived cells for multiple tissues originate from a single donor (Singh Dolt et al. 2017).

3.2.1 Limitations of 2-Dimensional Systems

Despite many useful features, the limitations of 2D systems in mimicking human physiology are obvious. These limitations are primarily related to the loss of normal tissue architecture and biophysical forces, and the heterotypic 3D environment or niche in which cells normally reside *in vivo* that enable dynamic and reciprocal interactions with other cell types, within a highly-organized ECM. Because of these limitations, the differentiated 2D cultures of iPSCs and of primary cells typically exhibit highly immature transcriptional and metabolic profiles of fetal rather than postnatal cells (Jiang et al. 2018).

3.2.2 Adding a Third Dimension to Tissue Culture

Emergence of a new generation of biomaterials with tunable properties as well as the use of micro-fabrication, bioprinting and microfluidic technologies opens wide-ranging possibilities for adding "a third dimension" to traditional 2D cultures. In this regard, organs-on-chips and organoids as well as organoid/chip hybrids are now at the center of much of the activity in the field (Liu et al. 2018) (Table 3.1). Organoids are 3D tissue cultures, often derived from stem cells, which self-organize into simplistic organs and tissues. In MPS (AKA organs-on-chips) however, different cell types and ECMs are placed into pre-

Table 3.1 – Comparisons between 2D and 3D tissue platforms in terms of biological complexity, manufacturing and outputs. Engineered tissues = cells + scaffolds. Adapted from Liu et al. (15)

	Conventional 2D systems	3D systems		
		Engineered tissues	Organoid	Organ-on-chip
Production method and timing	Differentiated, grown on rigid flat surfaces as monolayer; fast	Fabricated with scaffold and casting mold; slow	Embedded in Matrigel to self-organize; slow	Seeded in engineered chambers with perfusion; fast
Maturation	Immature	Improved but still lacking	Improved but still lacking	Improved but still lacking
Cell morphology and type	Unnatural, usually monotype	Size and shape similar to in vivo	Size and shape similar to in vivo	Depends on platform design
ECM	Limited composition and contact with cells	Similar to in vivo	Similar to in vivo	Depends on platform design
Tissue architecture	Absent	Simple	Complex, similar to organ developmental stages	Complexity depends on platform design
Diffusion of signal factors and nutrients	Short distances (through cell membranes); usually receiving supraphysiological doses	Concentration gradients may exist, may be affected by ECM properties	Interior cells may die or lack maturity due to ineffective transport to interior	Precisely controlled temporal and spatial gradients
Vascularization or perfusion?	No	No	No	Yes
High throughput feasibility; controllability; scalability	Present; high; absent	Possibly; low; may develop to macroscale	Absent; very low; may develop to macroscale	Present; very high; microscale
Variability; reproducibility; difficulty of use	Low; high; easy	High; low; difficult	High; low; easy	High; sometimes low; difficult
Possibility for genome editing	Straightforward	Difficult	Difficult	Straightforward
Characterization and analyses	Limited, but easy cell retrieval	Tissue function analyses possible but cell retrieval and phenotypic analyses can be hard.	Tissue function analyses possible but cell retrieval and phenotypic analyses can be hard.	Real-time tissue/ organ function analyses possible, with easy cell retrieval

fabricated chambers and cultured under different conditions in complex bioreactors, with micro-engineered microfluidic channels for fluid flow across the tissue to allow recapitulation of tissue complexity. This design makes it possible to exert fine control over tissue composition and architecture, as well as to pre-program precise tissue oxygenation and nutrient diffusion profiles. These features create opportunities for significant prolongation of the chips' viability and function up to several weeks or even months (Tanataweethum et al. 2018). Because of the sophisticated design features in some systems, they can be equipped with a range of biosensors, can be amenable to high-throughput genetic sequencing, provide context for high-content readouts, and allow real-time monitoring and sample retrieval. Moreover, specific tissue-tissue and cell-cell interactions and biophysical forces such as shear stress, tension, mechanical loading, electrical forces, peristalsis and breathing movements can be applied to MPS in a highly controlled manner to promote cell differentiation, tissue maturation and acquisition of tissue function readouts which are more useful than those obtained with 2D systems.

3.2.3 Organoids vs. Organs-on-Chips and Synergistic Engineering

Unlike the creation of organs-on-chips driven by specific features of their design and fabrication, formation of 3D organoids is primarily controlled by stochastic self-organizing principles of cell-cell and cell-ECM interactions where the cells coalesce to form masses mimicking certain features of normal tissue organization and function (Lou and Leung 2018). Many investigators have used iPSCs to derive organoids for a variety of human tissues, including gut, lung, kidney, heart and brain (Dutta et al. 2017). During organoid formation, as differentiation of the iPSCs proceeds *in vitro* in response to specific differentiation cues, the arising cell types self-assemble to form 3D structures containing different cell types in spatial configurations similar to those found in organs during normal *in vivo* development. While organoids possess complex architecture reminiscent of native tissues, their formation may be much slower compared to that of tissue chips. Additionally, it is difficult to exert precise control over organoid formation, and they are not easily amenable to multiplexing and real-time monitoring; further, sample retrieval to obtain analyzable readouts from these systems can be challenging or impossible. Moreover, lack of vascularization/perfusion in these constructs can lead to inefficient oxygen and nutrient transport into interior cells resulting in a necrotic core; an important limitation that reduces viability and shortens the life of these cultures. Finally, while gene-editing technologies can be easily combined with tissue chip design and derivation, it is difficult to modify genes when employing organoid-based technology as there is little control over the differentiation microenvironment.

Given that tissue chips and organoids are endowed with their own set of advantages and limitations, investigators in the field are now attempting to improve them through so-called synergistic engineering to develop new 3D *in vitro* models at the intersection of organoid and tissue chip technologies (Takebe et al. 2017). A general design principle for such models is to rely on a certain degree of stochasticity, which drives organoid formation, while taking advantage of the precise bioengineering design principles that characterize production of tissue chips. In this regard, capitalizing on the new opportunities offered by sophisticated bioprinting technologies might be promising (Li et al. 2016). It is envisioned that synergistic engineering could be useful for designing multi-tissue composite models, such as vascularized organ buds that can be effectively perfused with culture medium or blood substitutes (Takebe et al. 2015; Zhang et al. 2016), or innervated tissues able to receive neural inputs (Workman et al. 2016).

3.3 The Evolution of Microphysiological Systems Funding and Investment

Building on early versions of organs-on-chips, MPS technology development in the U.S. has been accelerated through an eight-year investment of over $140 M from the National Institutes of Health (NIH), the Food and Drug Administration (FDA), and the Defense Advanced Research Projects Agency (DARPA) (Sutherland et al. 2013). In Europe, a ban on the use of experimental animals in 2012 under the Organisation for Economic Co-operation and Development (OECD) guidelines stimulated the European Union to initiate a €1.4 million, 36-month project to support alternative platform development for drug testing. The core members of the project were InSphero AG, the Bioengineering Laboratory, ETH Zurich, AstraZeneca AB R&D, and F. Hoffmann-La Roche AG (Kanamori et al. 2018). Japanese pharmaceutical companies have also recognized the potential of MPS technology. In July 2017, the Japan Agency for Medical Research and Development (AMED) initiated a five-year research program with a budget of approximately ¥500 million (~$4.5 M) per year with the goal of developing a chip-based MPS that will be used in drug discovery processes (Kanamori et al. 2018). To broaden the access to this technology and support the use for both academic and industry investigators, further development is required to optimize reproducibility, robustness, and device manufacturing. To enable increased adoption of these devices, additional demonstration of utility and cost per chip will need to be considered. Based on advances in academic R&D, several start-up companies have been entering the market in an effort to transform organ-on-chip technology to industrially relevant applications (Zhang and Radisic 2017). To date, broad investment in organ-on-chip technology shows the magnitude of expectations for research related to the use of these devices.

3.3.1 Expansion and Integration of Organ-on-Chip Technology

There are many examples of the rapid development of organ-on-chip technology, but one example of the integration of chip technology with mechanical engineering is epitomized by the lung-on-chip developed by researchers at the Wyss Institute, which integrated relevant biomechanical forces to cultured tissues to illustrate the key roles these forces play in faithfully modeling human tissues (Huh et al. 2010). This lung chip mimics the alveolar-capillary border by using a dual culture system of human alveolar epithelial cells in the upper chamber and pulmonary microvascular endothelial cells in the lower chamber separated by a flexible, microporous membrane. After a sixteen-day incubation period, cyclic changes in pressure are applied to two air chambers running parallel to the cell chambers, thereby stretching and relaxing the membrane to mimic the mechanical forces of breathing. This chip has been used to model pulmonary inflammatory responses to the proinflammatory mediator tumor necrosis factor alpha (TNF-α); show the phagocytic response of neutrophils to foreign *E.Coli* bacteria introduced into the device; escalate the toxic and inflammatory responses of the lung to silica nanoparticles through mechanical stress; and demonstrate the impact of clinically-relevant concentrations of interleukin−2 (IL-2) on microvascular integrity (Huh et al. 2012). The basic design has since been used to model a number of other tissues including gut (Kim et al. 2012), kidney (Jang et al. 2013) and blood-brain barrier (Herland et al. 2016).

Since the publications describing the lung chip and demonstration of its utility, advancements in other organ-on-chip systems has expanded to include liver (Yoon No et al. 2015; Vernetti et al. 2016; Lee-Montiel et al. 2017; Soto-Gutierrez et al. 2017), vasculature (Moya et al. 2013; Vunjak-Novakovic et al. 2013; Fernandez et al. 2016; Sobrino et al. 2016), muscle (cardiac (Agarwal et al. 2013, Wang et al. 2014, Mathur et al. 2015) and skeletal (Madden et al. 2015; Cheng et al. 2016)), kidney (Weber et al. 2016), reproductive tissues including ovary, uterus, cervix and fallopian tube (Laronda et al. 2013; Arslan et al. 2015; Zhu et al. 2016; Xiao et al. 2017), testes, blood-brain barrier (Brown et al. 2015; Brown et al. 2016; Wang et al. 2017), skin (Abaci et al. 2016; Schimek et al. 2018), gut (Foulke-Abel 2016; Shah et al. 2016) and bone (de Peppo et al. 2013), amongst many others (see organ-on-chip reviews (An et al. 2015, Esch et al. 2015, Low and Tagle 2017a, b).

The power of MPS can be seen when the chips are linked together to form integrated systems. Indeed, shared goals of the U.S. government agencies' MPS programs focused on organ-on-chip devices with viabilities of at least a twenty-eight-day period, and the integration of these devices into linked platforms to improve modeling inter-organ interaction. One of the first integrated organ-on-chip devices from 2004 used microfabrication technology to create a three-chamber (lung, liver, "other") microscale mammalian cell culture platform on a one- inch square silicon chip (Sin et al. 2004). The device had a flow rate of 1.76 μL/min and was designed to model an approximated physiological liquid-to-cell ratio plus hydrodynamic shear stress, while maintaining the organ-specific liquid residence time parameters determined through physiologically based pharmacokinetic modeling (PBPK). A dissolved oxygen sensor was integrated into the system, making it possible to integrate real-time sensors into such devices (see Sect. 3.4).

The integration of multiple organs on a common platform provides a wealth of additional research possibilities, including analyses of

Fig. 3.3 – A TissUse (Germany) four-organ chip, incorporating intestine (1), liver (2), skin (3) and kidney (4) tissues. A surrogate blood medium is shown in pink. An excretory flow circuit is shown in yellow. Each microphysiological fluid flow circuit is operated by a separate peristaltic micropump. (Image reproduced under a Creative Commons 3.0 license from Maschmeyer et al.)

organ–organ interactions, pharmacokinetics/pharmacodynamics (PK/PD), hormone/immune/cytokine responses to drug exposure, and physiological reactions to drug metabolites. To address the pharmacokinetics of drug absorption, metabolism and secretion, Maschmeyer et al. (Maschmeyer et al. 2015) developed an integrated microphysiological platform that maintains the functionality of liver, kidney, intestine and skin over 28 days in co-culture (Fig. 3.3). A peristaltic on-chip micropump ensures pulsatile media flow interconnecting the four tissue culture compartments through microfluidic channels. A second microfluidic circuit ensures drainage of the fluid excreted through the kidney epithelial cell layer. This dual microfluidic system enables ADME (absorption/distribution/metabolism/excretion) profiling and repeated dose toxicity testing of drug candidates.

3.3.2 A Marriage of Technologies

The revolution in stem cell technology driven by the discovery that human induced pluripotent stem cells (iPSCs) can be generated from adult cells (Takahashi and Yamanaka 2006; Yu et al. 2007) has led to the prospect of using iPSCs to seed organ-on-chip devices, both as a means to develop integrated multi-organ systems from the same individual and as a possible tool for personalized medicine approaches to numerous diseases. The ability to differentiate iPSCs into multiple cell types eliminates the need for isolation of primary tissues from donors or patients, which is often a difficult and invasive procedure (and is not possible for organ systems like the brain), or may not produce high enough cell yields to be of use. iPSCs also provide a renewable cell source for device seeding, enabling wider reproducibility and therefore utility of the platforms for the research communities. iPSC generation has led to rapid development of, as an example, brain and blood-brain barrier (BBB) organ-on-chip devices, including a vascularized motor neuron model, which demonstrates enhanced neuronal activity, induction of vascular-neural interaction genes, and developmental gene expression profiles that are indicative of more *in vivo*-like signatures when compared to conventional 2D culture systems (Sances et al. 2018).

3.4 How Engineering Advances Have, and Will Continue to, Push the Field Forward

Current approaches to developing organ-on-a-chip platforms are rooted in plate-based cell culture, microfluidics, and 3-dimensional printing. Multi-well cell culture plates gave rise to bioengineered tissues by allowing for the 3D culture of cell suspensions and later organoids, e.g. liver embryoid bodies (Au et al. 2014) and cerebral organoids (Lancaster et al. 2013). Microfluidics then contributed the capability to reduce the culture well size and control with great precision the number of cells, and therefore the organoid size, as well as the quantities of reagents (Du et al. 2016). It also enabled researchers to design chips for larger biomimetic tissue constructs (Tan and Desai 2004) and introduce structural complexity, e.g. permeable membranes for lung (Kniazeva et al. 2011) and intestinal tissue (Kim et al. 2012) engineering. More recently, 3D bioprinting of biopolymers and cell suspensions has provided the flexibility to pattern tissue components on a larger scale (up to 1 mm in height), and with high precision (usually on the order of 50 μm). Most importantly, bioprinting enables easy integration of fluidic channels for nutrient supply and cellular waste removal – effectively vascular channels (Kolesky et al. 2014; He et al. 2016). These developments in engineering, coupled with synergistic engineering to combine organoid and organ-on-chip technology, have moved cell culture studies over the last two decades from two dimensions into the 3D and 4D (temporal control) realms when looking at responses over time. These complex tissues and multi-tissue platforms that are viable and can function for several weeks (Materne et al. 2015; Skardal et al. 2017; Ronaldson-Bouchard et al. 2018) have significant translational potential for drug screening and rare disease research.

One powerful advantage to organ-on-chip technology is the possibility to monitor real-time outcomes of a variety of cell- and tissue-specific responses in non-invasive ways. This, however, requires continuous monitoring of the engineered tissue to ascertain its functionality and response to environmental cues, e.g. the presence of a pathogen, drug, or toxicant. Researchers have begun to incorporate existing cell analysis techniques with microfluidics, such as enzyme-linked immunosorbent assays (ELISA) (Riahi et al. 2016; Luan et al. 2017), trans-epithelial electrical resistance (TEER) (van der Helm et al. 2017; Henry et al. 2017), and multi-electrode arrays (MEA) (Koutsouras et al. 2017; Maoz et al. 2017). Recently, several of these techniques have been simultaneously included into a single organ-on-a-chip platform (Junaid et al. 2017; Zhang et al. 2017). This platform features physical (e.g. temperature, pH detection), chemical (e.g. immunobiosensors to monitor excreted biomarkers or to detect oxygen levels), and optical (e.g. sensors to observe neuro-muscular contractility) sensors, which enable continuous measurements of the cellular environment and behavior. Future work should focus on long-term (e.g. months) stability of the embedded sensors without loss of accuracy or precision; multiplexing sensors for concurrent monitoring of key biomarkers from several engineered tissues; methods for continuous processing of the large sensor and optical data; and feedback loops from the sensors to the environmental controls, including perfusate contents and flow rate. Together, these technical advances promise to increase the utility of MPS in a variety of contexts and create cutting-edge technologies that can give information on cell function in ways that are currently not possible using any other platform.

3.5 Present Considerations and Future Challenges

3.5.1 Cell Sources

For organ-on-chip development, cell sources currently come anywhere from primary tissue biopsies, to commercially available cell lines, to stem cell-derived cell populations. Primary cells derived from some types of tissue biopsies may be grown and maintained in culture for long periods of time and therefore allow for longer term studies than 2D cell culture systems. Comparison of on-chip experiments to fresh tissue collection is important because these collections often best mimic *in vivo* physiology, but they are not widely available from human patients. Other cell sources are induced pluripotent stem cells (iPSCs) from reprogrammed adult somatic cells, and are often used in organ-on-chip platforms. iPSCs may be derived from patients and represent a potentially unlimited cell source, but this technology is still evolving. For instance, although the maturation signals recorded in the vascularized motor neuron-on-chip device are encouraging (Sances et al. 2018), most systems still exhibit embryonic gene expression profiles and hence significant development is still needed to drive iPSC-derived cell types toward an adult phenotype. This is an essential step for application of organ-on-chip devices in modeling of adult-onset diseases, but the recent development of genetic editing techniques offers exciting potential for the creation of multiple tissue types from the same donor (isogenic cell lines), or introduction and deletion of specific genetic sequences to model or correct monogenetic diseases.

3.5.2 Organ-on-Chip Integration: Biological and Technical Challenges

As previously stated, integration of multiple tissue models is of great interest because cells and organ systems communicate with each other by secreting exosomes and soluble factors that are important for peripheral crosstalk. Microfluidic advances have allowed individual organs-on-chips to be connected to one another to mimic the *in vivo* role of the circulation thus enabling the recapitulation of certain aspects of homeostasis. Comprehensive body-on-chip integration aims to more fully mimic the physiology of drug delivery and metabolism in the body. However, several factors need to be considered when designing these systems:

- *Biological challenges:*

These are challenges which require ongoing research in order to understand basic cellular and organ physiology which will require the efforts of researchers in multiple fields of expertise to overcome. Some of these may be related to organ scaling, which is the issue of how to represent the size and/or cell number of linked organ systems in a physiologically relevant manner. For example, the liver is considerably larger than the kidney, so a joint system must scale the tissue constructs accordingly. Additionally, cell metabolism may be different in different organ systems, and with variable cell maturity and time in culture, the basis for normal flow rate for each tissue compartment is not completely understood (Wikswo et al. 2013). Furthermore, to keep the multiple different cell types healthy and functional, multi-organ-on-a-chip platforms require development of a perfusate that maintains physiologically relevant circulation for all included tissues – a fluid that functions either as a blood surrogate or as interstitial fluid (Abaci and Shuler 2015). Such a fluid would contain all nutrients, growth factors, and chemokines required for all tissue types within the system. If attained, this design would greatly simplify the geometry of microfluidic or 3D printed organ-on-a-chip platforms as only one perfusion channel would be needed for all tissues.

- *Technical challenges:*

These are challenges related to the physical linkage and integration of multiple tissue platforms. They include basic challenges such as presence of air bubbles (see Sect. 3.5.3); maintenance of sterility when linking separately cultured platforms (and preventing leakage); and selection of materials for platform fabrication (see Sect. 3.5.3). For physical coupling of microfluidic systems, methods for fluidic integration are categorized as: (1) static, (2) single-pass perfusion, and (3) recirculation (of both common or tissue-specific media). Static microfluidic connections between organ chips depend on physical proximity rather than convective flow, and organ crosstalk is facilitated by culturing all cells and organs in the same well (Li et al. 2012). However, one limitation of such systems is a potential for exponential increase of cell toxicity, which can complicate the interpretation of results. Single-pass perfusion creates culture systems that arrange organ chips in parallel, in series, or both. Here, the drawback is that unidirectionality of the perfusion only enables crosstalk to organs downstream and does not allow for upstream feedback, which occurs *in vivo*. Recirculating microfluidic flow more closely mimics *in vivo* circulation and allows for both downstream and upstream crosstalk (Oleaga et al. 2016). Zhang et al. (2017) created a recirculating microfluidic multi-organ organoid platform which incorporates multiple sensors, thereby allowing real-time information capture that could be employed to create feedback algorithms for evaluating drug responses in near real-time. Other methods of fluidic flow may be driven by gravity and passively-controlled hydraulic resistances (Esch et al. 2016). Additional technical challenges include incorporation into platforms of organ-specific biomimetic stimuli (mechanical, electrical and chemical) to drive the functionality of the organ chips and maintain physiologically-relevant 3D tissues (Huh et al. 2010; Nunes et al. 2013; Sun and Nunes 2016; Marturano-Kruik et al. 2018).

3.5.3 Manufacturing, Materials and Reproducibility

Development of organ-on-chip technology has been hampered by several factors, which include inadequate community standards for manufacturing, poor performance of materials, and reproducibility issues. These factors have limited the high throughput capabilities for organ-on-chip systems. While these difficulties remain, progress is continually being made, especially with technical advances in 3D printing. One of the oldest and most powerful 3D bioprinting techniques is laser-assisted stereolithography (SLA) (Groll et al. 2016). To date, there are many growing companies utilizing SLA technologies, including Organovo, N3D, Bio3D, and LaserPrint. Although a 3D-printed organ-on-a-chip platform is yet to be fully commercialized, it is expected that adoption of these devices will be realized with continued advances in automation and additive manufacturing.

Selecting the right material for microfluidic device fabrication is another challenge for an organ-on-chip platform developer. The chip material should be biocompatible with the chosen tissue and should support a physiologically-relevant environment. Polydimethylsiloxane (PDMS) is commonly used in many academic laboratories because of its low cost, ease of fabrication, optical transparency, lack of cellular toxicity and elastomeric properties that simplify chip bonding (Shirure and George 2017). However, PDMS is hydrophobic and therefore is difficult to surface-functionalize; because of these properties, PDMS absorbs hormones, small molecules and drugs thereby altering their concentrations, and affecting reproducibility of the results (Chen et al. 2016). PDMS is also permeable to gases (Merkel et al. 2000), which leads to spontaneous formation of gas bubbles that impact perfusion flow inside microfluidic devices – the dreaded "gas bubble problem" (Wang et al. 2012). No other currently used materials – glass, thermoplastics, or hydrogels – fully satisfy the needs, hence the search continues for an optimal material – one which is surface "functionalizable", yet non-sticky to cells; gas permeable only under controlled, reversible conditions; and cost-effective and simple to handle.

Finally, although the predictive value and physiological relevance of various organ-on-chip systems have been demonstrated, their potential for medium or high-throughput screenings (HTS) is still to be realized. Adaptations such as massive parallelization and automation are among the biggest challenges for meeting industrial benchmarks of HTS. Further, low-throughput characteristics of perfusion, sampling, and cell injection, as well as limitations of fabrication, scale up, and integration of online analytics/sensors must be addressed before chip platforms could be employed for HTS (Probst et al. 2018).

3.5.4 Standardization and Validation Considerations

As organ-on-chip models functionally evolve, a need for standardized, good manufacturing practice (GMP) and good laboratory practice (GLP) procedures are becoming more important for comparing data across laboratories. Validation of tissue health and functionality assays must be robust, reliable, and reproducible. However, because there is no single 'validation' method or assay that fits all tissue types, standardization of validation methods is problematic. For example, if a system is designed to model and quantify transport across barrier tissues, it becomes important to verify the integrity of tight junctions that comprise the barrier. One approach could be an assessment of tight junction viability through a quantification of tight-junction protein expression (Shah et al. 2016). However, these measures would be less useful in non-barrier tissues, which exemplifies the need to validate tissue functions applicable to the experimental model for its specific context of use. To help address this challenge, in 2017 the National Center for Advancing Translational Sciences (NCATS), part of the NIH in the United States, funded Tissue Chip Testing Centers to independently assess the robustness and reliability of various MPS platforms. In collaboration with the FDA and pharmaceutical industry, these Testing Centers validate the outcomes and assays provided by the platform designers, with continuously evolving validation criteria being developed and published (Low and Tagle 2017a, b; Sakolish et al. 2018). To house this data, and create a standardized, centralized database of MPS data from a variety of users, NCATS also funded the establishment of an MPS Database (MPS-Db) to provide developers and end-users access to a publicly accessible repository of MPS data.

3.5.5 Commercialization and Regulatory/Industry Buy-in of Organ-on-Chip Systems

As noted above, manufacturing advancements are still needed to reach GMP and GLP levels, both for commercialization of organ-on-chip systems and for encouraging buy-in from regulatory bodies that will receive data from these platforms being included in, for example, Investigational New Drug (IND) applications in the United States. Additionally, if organ-on-chip and integrated "body-on-chip" platforms aim to reduce the use or replace animal models, and be qualified through the US FDA validation strategy of Drug Development Tools (DDTs), the validation process of physiologically relevant *in vitro* systems should follow similar, if not the same criteria as outlined in the DDT guidance document (Rebelo et al. 2016). In the US, the current FDA DDT program is voluntary and uses a "fit-for-purpose" qualification. Once an animal model is qualified for a specific context of use as a DDT, the pharmaceutical industry can use the tool for the qualified purpose during product development, and FDA reviewers can be confident in applying the DDT without additional supporting data. Therefore, meeting FDA DDT requirements by the multi-organ or body-on-a-chip platforms should provide a reliable approach for earning industry acceptance of this technology for drug discovery and development. The NIH-funded Tissue Chip Testing Centers and MPS Database in the US and the Organ-on-Chip Development (ORCHID) initiative in Europe are supporting these efforts.

3.6 Conclusions and Future Directions

Although existing challenges need to be addressed to accommodate the use of organ-on-chip technology in a routine lab workflow, the enormous potential of these devices to expand our understanding of human cellular biology, improve toxicity and efficacy testing in drug development and model human disease in the context of precision medicine will continue to accelerate the development of this technology. To give an example of a future avenue of potentially impactful research: Today, chronic inflammatory processes are at the core of disease pathology in seven of the top ten leading causes of mortality in the developed world. Extensive evidence supports the role of inflammation in heart disease (Awan and Genest 2014), cancer (Hanahan and Weinberg 2011), chronic pulmonary disease

(Gan et al. 2004), stroke (Kawabori and Yenari 2015), Alzheimer's disease (Amor et al. 2010), diabetes (Biondi-Zoccai et al. 2003), nephritis (Zheng and Zheng 2016) and Parkinson's disease (Stephenson et al. 2018), as well as holding a crucial role in general injury and healing. A better understanding of the processes that initiate, drive, and terminate inflammation, both acute and chronic, could help to uncover new treatments with impact in wound healing, allergies, and the health of the aging population. Utilization of organ-on-chip technology will enable researchers to study chronic inflammation in multiple organs including lung, gut, brain, and skin, as well as investigate the new roles of immune cells during inflammation associated with cancer.

When considering applications for disease modeling, organ-on-chip technology has also provided new tools for modeling metastatic cancer mechanisms, as these models can recreate cell-cell or cell-ECM interactions, spatiotemporal chemical/physical gradients or the (hydro) dynamic properties of the cellular microenvironment. Over the coming years, the integration of advanced quantitative optical imaging methods, screening and biosensing techniques into on-chip systems together with Big Data analyses is expected to provide clinicians and pharmaceutical companies with a new generation of highly predictive metastasis-on-chip models (Caballero et al. 2017). Regarding future applications of MPS technology, clinical trial readiness studies are beginning to collect peripheral blood mononuclear cells (PBMCs) from blood samples for iPSC generation of many tissue types, thus enabling precision medicine efforts around organ-on-chip technology and generation of "you-on-a-chip" to test individualized treatments. This effort has particularly impactful potential for rare diseases such as Amyotrophic Lateral Sclerosis (ALS) and Frontotemporal Degeneration (FTD), where the use of tissue chip technology to screen therapeutics may lead to the identification of the right drug, for the right patient at the right time.

The organ-on-chip field is fast-developing worldwide, as stakeholders from various sectors are beginning to appreciate the broad potential applications of the technology as it continues to evolve. While challenges remain, the promise of these tools remains transformative, inspiring researchers and clinicians and providing hope for patients.

References

Abaci HE, Shuler ML (2015) Human-on-a-chip design strategies and principles for physiologically based pharmacokinetics/pharmacodynamics modeling. Integr Biol 7(4):383–391

Abaci HE et al (2016) Human skin constructs with spatially controlled vasculature using primary and iPSC-derived endothelial cells. Adv Healthc Mater 5(14):1800–1807

Agarwal A et al (2013) Microfluidic heart on a chip for higher throughput pharmacological studies. Lab Chip 13(18):3599–3608

Amor S et al (2010) Inflammation in neurodegenerative diseases. Immunology 129(2):154–169

An F et al (2015) Organ-on-a-Chip: new platform for biological analysis. Anal Chem Insights 10:39–45

Arrowsmith J (2011a) Trial watch: phase II failures: 2008–2010. Nat Rev Drug Discov 10(5):328–329

Arrowsmith J (2011b) Trial watch: phase III and submission failures: 2007–2010. Nat Rev Drug Discov 10(2):87–87

Arslan SY et al (2015) Novel three dimensional human Endocervix cultures respond to 28-day hormone treatment. Endocrinology 156(4):1602–1609

Au SH et al (2014) Hepatic organoids for microfluidic drug screening. Lab Chip 14(17):3290–3299

Awan Z, Genest J (2014) Inflammation modulation and cardiovascular disease prevention. Eur J Prev Cardiol 22(6):719–733

Balls M (1995) Defining the role of ECVAM in the development, validation and acceptance of alternative tests and testing strategies. Toxicol In Vitro 9(6):863–869

Biondi-Zoccai GGL et al (2003) Atherothrombosis, inflammation, and diabetes. J Am Coll Cardiol 41(7):1071–1077

Brown JA et al (2015) Recreating blood-brain barrier physiology and structure on chip: a novel neurovascular microfluidic bioreactor. Biomicrofluidics 9(5):054124

Brown JA et al (2016) Metabolic consequences of inflammatory disruption of the blood-brain barrier in an organ-on-chip model of the human neurovascular unit. J Neuroinflammation 13(1):306

Caballero D et al (2017) Organ-on-chip models of cancer metastasis for future personalized medicine: from chip to the patient. Biomaterials 149(Supplement C):98–115

Casey W et al (2015) A new path forward: the Interagency Coordinating Committee on the

Validation of Alternative Methods (ICCVAM) and National Toxicology Program's Interagency Center for the Evaluation of Alternative Toxicological Methods (NICEATM). J Am Assoc Lab Anim Sci 54(2):170–173
Chen T et al (2016) A drug-compatible and temperature-controlled microfluidic device for live-cell imaging. Open Biol 6(8):160156
Cheng CS et al (2016) Cell density and joint microRNA-133a and microRNA-696 inhibition enhance differentiation and contractile function of engineered human skeletal muscle tissues. Tissue Eng A 22(7–8):573–583
de Peppo GM et al (2013) Engineering bone tissue substitutes from human induced pluripotent stem cells. Proc Natl Acad Sci 110(21):8680–8685
Du G et al (2016) Microfluidics for cell-based high throughput screening platforms—a review. Anal Chim Acta 903(Supplement C):36–50
Dutta D et al (2017) Disease Modeling in stem cell-derived 3D organoid systems. Trends Mol Med 23(5):393–410
Esch EW et al (2015) Organs-on-chips at the frontiers of drug discovery. Nat Rev Drug Discov 14:248–260
Esch MB et al (2016) Modular, pumpless body-on-a-chip platform for the co-culture of GI tract epithelium and 3D primary liver tissue. Lab Chip 16(14):2719–2729
Fashe MM et al (2015) Species-specific differences in the in vitro metabolism of Lasiocarpine. Chem Res Toxicol 28(10):2034–2044
Fernandez CE et al (2016) Human vascular microphysiological system for in vitro drug screening. Sci Rep 6:21579
Foulke-Abel, J., et al. (2016). Human enteroids as a model of upper small intestinal ion transport physiology and pathophysiology. Gastroenterology 150(3): 638–649. e638
Gan W et al (2004) Association between chronic obstructive pulmonary disease and systemic inflammation: a systematic review and a meta-analysis. Thorax 59(7):574–580
Groll J et al (2016) Biofabrication: reappraising the definition of an evolving field. Biofabrication 8(1):013001
Guillouzo A (1998) Liver cell models in in vitro toxicology. Environ Health Perspect 106(Suppl 2):511–532
Hanahan D, Weinberg RA (2011) Hallmarks of Cancer: the next generation. Cell 144(5):646–674
He Y et al (2016) Developments of 3D printing microfluidics and applications in chemistry and biology: a review. Electroanalysis 28(8):1658–1678
Henry OYF et al (2017) Organs-on-chips with integrated electrodes for trans-epithelial electrical resistance (TEER) measurements of human epithelial barrier function. Lab Chip 17(13):2264–2271
Herland A et al (2016) Distinct contributions of astrocytes and Pericytes to Neuroinflammation identified in a 3D human blood-brain barrier on a Chip. PLoS One 11(3):e0150360
Huh D et al (2010) Reconstituting organ-level lung functions on a Chip. Science 328(5986):1662–1668
Huh D et al (2012) A human disease model of drug toxicity–induced pulmonary Edema in a lung-on-a-Chip microdevice. Sci Trans Med 4(159):159ra147–159ra147
Jackson EL, Lu H (2016) Three-dimensional models for studying development and disease: moving on from organisms to organs-on-a-chip and organoids. Integr Biol Quant Biosci Nano Macro 8(6):672–683
Jang K-J et al (2013) Human kidney proximal tubule-on-a-chip for drug transport and nephrotoxicity assessment. Integr Biol 5(9):1119–1129
Jiang, Y., et al. (2018). Maturation of cardiomyocytes derived from human pluripotent stem cells: current strategies and limitations. Mol Cells 41(7):613–621
Junaid A et al (2017) An end-user perspective on organ-on-a-chip: assays and usability aspects. Curr Opin Biomed Eng 1:15–22
Kanamori T et al (2018) Technical aspects of microphysiological systems (MPS) as a promising wet human-in-vivo simulator. Drug Metab Pharmacokinet 33(1):40–42
Kawabori M, Yenari MA (2015) Inflammatory responses in brain ischemia. Curr Med Chem 22(10):1258–1277
Kim HJ et al (2012) Human gut-on-a-chip inhabited by microbial flora that experiences intestinal peristalsis-like motions and flow. Lab Chip 12(12):2165–2174
Kniazeva T et al (2011) A microfluidic respiratory assist device with high gas permeance for artificial lung applications. Biomed Microdevices 13(2):315–323
Kolesky DB et al (2014) 3D bioprinting of vascularized, heterogeneous cell-laden tissue constructs. Adv Mater 26(19):3124–3130
Koutsouras DA et al (2017) Simultaneous monitoring of single cell and of micro-organ activity by PEDOT:PSS covered multi-electrode arrays. Mater Sci Eng C 81:84–89
Lancaster MA et al (2013) Cerebral organoids model human brain development and microcephaly. Nature 501:373–379
Laronda MM et al (2013) Recreating the female reproductive tract in vitro using iPSC technology in a linked microfluidics environment. Stem Cell Res Ther 4(Suppl 1):S13–S13
Lee-Montiel FT et al (2017) Control of oxygen tension recapitulates zone-specific functions in human liver microphysiology systems. Exp Biol Med 242(16):1617–1632
Li AP et al (2012) Definition of metabolism-dependent xenobiotic toxicity with co-cultures of human hepatocytes and mouse 3T3 fibroblasts in the novel integrated discrete multiple organ co-culture (IdMOC) experimental system: results with model toxicants aflatoxin B1, cyclophosphamide and tamoxifen. Chem Biol Interact 199(1):1–8
Li J et al (2016) Recent advances in bioprinting techniques: approaches, applications and future prospects. J Trans Med 14(1):271
Liu C et al (2018) Modeling human diseases with induced pluripotent stem cells: from 2D to 3D and beyond. Development 145(5):dev156166

Lopes FM et al (2017) Mimicking Parkinson's disease in a dish: merits and pitfalls of the most commonly used dopaminergic in vitro models. NeuroMolecular Med 19(2):241–255

Lou Y-R, Leung AW (2018) Next generation organoids for biomedical research and applications. Biotechnol Adv 36(1):132–149

Low LA, Tagle DA (2017a) Organs-on-chips: Progress, challenges, and future directions. Exp Biol Med (Maywood, NJ) 242(16):1573–1578

Low LA, Tagle DA (2017b) Tissue chips - innovative tools for drug development and disease modeling. Lab Chip 17(18):3026–3036

Luan Q et al (2017) A microfluidic in-line ELISA for measuring secreted protein under perfusion. Biomed Microdevices 19(4):101

Madden L et al (2015) Bioengineered human myobundles mimic clinical responses of skeletal muscle to drugs. elife 4:e04885

Maoz BM et al (2017) Organs-on-chips with combined multi-electrode array and transepithelial electrical resistance measurement capabilities. Lab Chip 17(13):2294–2302

Martin L et al (2017) How much do clinical trials cost? Nat Rev Drug Discov 16:381–382

Marturano-Kruik A et al (2018) Biomechanical regulation of drug sensitivity in an engineered model of human tumor. Biomaterials 150:150–161

Maschmeyer I et al (2015) A four-organ-chip for interconnected long-term co-culture of human intestine, liver, skin and kidney equivalents. Lab Chip 15(12):2688–2699

Materne E-M et al (2015) The multi-organ Chip - a microfluidic platform for long-term multi-tissue Coculture. J Vis Exp: JoVE 98:52526

Mathur A et al (2015) Human iPSC-based cardiac microphysiological system for drug screening applications. Sci Rep 5:8883

Merkel TC et al (2000) Gas sorption, diffusion, and permeation in poly(dimethylsiloxane). J Polym Sci B Polym Phys 38(3):415–434

Morgan S et al (2011) The cost of drug development: a systematic review. Health Policy 100(1):4–17

Moya ML et al (2013) In vitro perfused human capillary networks. Tissue Eng Part C Methods 19(9):730–737

Niu N, Wang L (2015) In vitro human cell line models to predict clinical response to anticancer drugs. Pharmacogenomics 16(3):273–285

Nunes SS et al (2013) Biowire: a platform for maturation of human pluripotent stem cell–derived cardiomyocytes. Nat Methods 10:781–787

Oleaga C et al (2016) Multi-organ toxicity demonstration in a functional human in vitro system composed of four organs. Sci Rep 6:20030

Pournasr B, Duncan SA (2017) Modeling inborn errors of hepatic metabolism using induced pluripotent stem cells. Arterioscler Thromb Vasc Biol 37(11):1994–1999

Probst C et al (2018) High-throughput organ-on-a-chip systems: current status and remaining challenges. Curr Opin Biomed Eng 6:33–41

Profile Toolkit (2018) The Dynamic U.S. Research and Development Ecosystem. from https://http://www.phrma.org/industryprofile/2018/

Rebelo SP et al (2016) Validation of bioreactor and human-on-a-Chip devices for chemical safety assessment. Validation of alternative methods for toxicity testing. C. Eskes and M. Whelan. Springer, Cham, pp 299–316

Riahi R et al (2016) Automated microfluidic platform of bead-based electrochemical immunosensor integrated with bioreactor for continual monitoring of cell secreted biomarkers. Sci Rep 6:24598

Roberts RA et al (2014) Reducing attrition in drug development: smart loading preclinical safety assessment. Drug Discov Today 19(3):341–347

Ronaldson-Bouchard K et al (2018) Advanced maturation of human cardiac tissue grown from pluripotent stem cells. Nature 556(7700):239–243

Sakolish C et al (2018) Technology transfer of the microphysiological systems: a case study of the human proximal tubule tissue Chip. Sci Rep 8(1):14882–14882

Sances S et al (2018) Human iPSC-derived endothelial cells and microengineered organ-Chip enhance neuronal development. Stem Cell Reports 10(4):1222–1236

Sasahara K et al (2015) Pharmacokinetics and metabolism of Delamanid, a novel anti-tuberculosis drug, in animals and humans: importance of albumin metabolism in vivo. Drug Metab Dispos 43(8):1267–1276

Schimek K et al (2018) Bioengineering of a full-thickness skin equivalent in a 96-well insert format for substance permeation studies and organ-on-a-chip applications. Bioengineering 5(2)

Shah P et al (2016) A microfluidics-based in vitro model of the gastrointestinal human–microbe interface. Nat Commun 7:11535

Shirure VS, George SC (2017) Design considerations to minimize the impact of drug absorption in polymer-based organ-on-a-chip platforms. Lab Chip 17:681–690

Sin A et al (2004) The design and fabrication of three-chamber microscale cell culture analog devices with integrated dissolved oxygen sensors. Biotechnol Prog 20(1):338–345

Singh Dolt K et al (2017) Modeling Parkinson's disease with induced pluripotent stem cells harboring α-synuclein mutations. Brain Pathol 27(4):545–551

Skardal A et al (2017) Multi-tissue interactions in an integrated three-tissue organ-on-a-chip platform. Sci Rep 7(1):8837

Sobrino A et al (2016) 3D microtumors in vitro supported by perfused vascular networks. Sci Rep 6:31589

Soto-Gutierrez A et al (2017) Pre-clinical and clinical investigations of metabolic zonation in liver diseases: the potential of microphysiology systems. Exp Biol Med 242(16):1605–1616

Stephenson J et al (2018) Inflammation in CNS neurodegenerative diseases. Immunology 154(2):204–219

Sun X, Nunes SS (2016) Biowire platform for maturation of human pluripotent stem cell-derived cardiomyocytes. Methods 101:21–26

Sutherland ML et al (2013) The national institutes of health microphysiological systems program focuses on a critical challenge in the drug discovery pipeline. Stem Cell Res Ther 4(1):1–5

Takahashi K, Yamanaka S (2006) Induction of pluripotent stem cells from mouse embryonic and adult fibroblast cultures by defined factors. Cell 126(4):663–676

Takebe T et al (2015) Vascularized and complex organ buds from diverse tissues via mesenchymal cell-driven condensation. Cell Stem Cell 16(5):556–565

Takebe T et al (2017) Synergistic engineering: organoids meet organs-on-a-chip. Cell Stem Cell 21(3):297–300

Tan W, Desai TA (2004) Layer-by-layer microfluidics for biomimetic three-dimensional structures. Biomaterials 25(7–8):1355–1364

Tanataweethum N et al (2018) Establishment and characterization of a primary murine adipose tissue-chip. Biotechnol Bioeng 0(0)

van der Helm MW et al (2017) Fabrication and validation of an organ-on-chip system with integrated electrodes to directly quantify transendothelial electrical resistance. JoVE 127:e56334

Vernetti LA et al (2016) A human liver microphysiology platform for investigating physiology, drug safety, and disease models. Exp Biol Med 241(1):101–114

Vunjak-Novakovic G et al (2013) HeLiVa platform: integrated heart-liver-vascular systems for drug testing in human health and disease. Stem Cell Res Ther 4(1):1–6

Wagner J et al (2017a) A dynamic map for learning, communicating, navigating and improving therapeutic development. Nat Rev Drug Discov 17:150

Wagner JA et al (2017b) Application of a dynamic map for learning, communicating, navigating, and improving therapeutic development. Clin Transl Sci 11(2):166–174

Wallet MA et al (2017) Isogenic cellular systems model the impact of genetic risk variants in the pathogenesis of type 1 diabetes. Front Endocrinol 8(276)

Wang Y et al (2012) Systematic prevention of bubble formation and accumulation for long-term culture of pancreatic islet cells in microfluidic device. Biomed Microdevices 14(2):419–426

Wang G et al (2014) Modeling the mitochondrial cardiomyopathy of Barth syndrome with induced pluripotent stem cell and heart-on-chip technologies. Nat Med 20(6):616–623

Wang YI et al (2017) Microfluidic blood–brain barrier model provides in vivo-like barrier properties for drug permeability screening. Biotechnol Bioeng 114(1):184–194

Waring MJ et al (2015) An analysis of the attrition of drug candidates from four major pharmaceutical companies. Nat Rev Drug Discov 14:475–486

Weber EJ et al (2016) Development of a microphysiological model of human kidney proximal tubule function. Kidney Int 90(3):627–637

Wikswo JP et al (2013) Engineering challenges for instrumenting and controlling integrated organ-on-Chip Systems. IEEE Trans Biomed Eng 60(3):682–690

Workman MJ et al (2016) Engineered human pluripotent-stem-cell-derived intestinal tissues with a functional enteric nervous system. Nat Med 23:49

Xiao S et al (2017) A microfluidic culture model of the human reproductive tract and 28-day menstrual cycle. Nat Commun 8:14584

Yoon No D et al (2015) 3D liver models on a microplatform: well-defined culture, engineering of liver tissue and liver-on-a-chip. Lab Chip 15(19):3822–3837

Yu J et al (2007) Induced pluripotent stem cell lines derived from human somatic cells. Science 318(5858):1917–1920

Zhang B, Radisic M (2017) Organ-on-a-chip devices advance to market. Lab Chip 17(14):2395–2420

Zhang YS et al (2016) Bioprinting 3D microfibrous scaffolds for engineering endothelialized myocardium and heart-on-a-chip. Biomaterials 110:45–59

Zhang YS et al (2017) Multisensor-integrated organs-on-chips platform for automated and continual in situ monitoring of organoid behaviors. Proc Natl Acad Sci 114(12):E2293–E2302

Zheng Z, Zheng F (2016) Immune cells and inflammation in diabetic nephropathy. J Diabetes Res 2016:10

Zhu J et al (2016) Human fallopian tube epithelium co-culture with murine ovarian follicles reveals cross-talk in the reproductive cycle. Mol Hum Reprod 22:756–767

Engineering Patient-on-a-Chip Models for Personalized Cancer Medicine

4

David Caballero, Rui L. Reis, and Subhas C. Kundu

Abstract

Traditional *in vitro* and *in vivo* models typically used in cancer research have demonstrated a low predictive power for human response. This leads to high attrition rates of new drugs in clinical trials, which threaten cancer patient prognosis. Tremendous efforts have been directed towards the development of a new generation of highly predictable preclinical models capable to reproduce *in vitro* the biological complexity of the human body. Recent advances in nanotechnology and tissue engineering have enabled the development of predictive organs-on-a-chip models of cancer with advanced capabilities. These models can reproduce *in vitro* the complex three-dimensional physiology and interactions that occur between organs and tissues *in vivo*, offering multiple advantages when compared to traditional models. Importantly, these models can be tailored to the biological complexity of individual cancer patients resulting into biomimetic and personalized *cancer patient-on-a-chip* platforms. The individualized models provide a more accurate and physiological environment to predict tumor progression on patients and their response to drugs. In this chapter, we describe the latest advances in the field of cancer patient-on-a-chip, and discuss about their main applications and current challenges. Overall, we anticipate that this new paradigm in cancer *in vitro* models may open up new avenues in the field of personalized – cancer – medicine, which may allow pharmaceutical companies to develop more efficient drugs, and clinicians to apply patient-specific therapies.

D. Caballero (✉) · S. C. Kundu
3B's Research Group, I3Bs – Research Institute on Biomaterials, Biodegradables and Biomimetics, University of Minho, Headquarters of the European Institute of Excellence on Tissue Engineering and Regenerative Medicine, Barco, Guimarães, Portugal

ICVS 3Bs PT Government Associate Lab, Braga, Guimarães, Portugal
e-mail: dcaballero@i3bs.uminho.pt; kundu@i3bs.uminho.pt

R. L. Reis
3B's Research Group, I3Bs – Research Institute on Biomaterials, Biodegradables and Biomimetics, University of Minho, Headquarters of the European Institute of Excellence on Tissue Engineering and Regenerative Medicine, Guimarães, Portugal

ICVS/3B's PT Government Associate Lab, Braga, Guimarães, Portugal

The Discoveries Centre for Regenerative and Precision Medicine, Headquarters at University of Minho, Guimarães, Portugal
e-mail: rgreis@i3bs.uminho.pt

J. M. Oliveira, R. L. Reis (eds.), *Biomaterials- and Microfluidics-Based Tissue Engineered 3D Models*, Advances in Experimental Medicine and Biology 1230,
https://doi.org/10.1007/978-3-030-36588-2_4

Keywords
Cancer · *In vitro* model · Organ-on-a-chip · Patient-on-a-chip · Personalized medicine

Highlights

- Cancer patient-on-a-chip (CPoC) are highly physiological patient-matched microfluidic models for personalized cancer medicine.
- CPoC allow the study of the mechanistic determinants of tumor growth and progression as well as for drug ADME investigation.
- An active collaboration between all the involved stakeholders is crucial to boost the use of CPoC-based models in the clinics.

4.1 Introduction

Cancer remains a major burden on global health, being the second leading cause of death worldwide after ischemic heart diseases (W.H.O 2018). Despite the significant advances made in understanding and fighting cancer, it is yet not well understood why a large number of patients fail to respond to the treatment, or why some of them respond differently to the same therapy. Current *in vitro* models, such as two- and three-dimensional (2D and 3D, respectively) platforms typically used for drug discovery, development, and screening, are obsolete (Caballero et al. 2017b). They do not recapitulate the complexity of the human body and lack fundamental characteristics of the tumor microenvironment (TME). Similarly, *in vivo* models, despite being more physiological, are not predictive to humans. This lack of accuracy has caused that many anti-cancerous drugs, which have progressed through (pre)clinical trials, being withdrawn from the market due to their unanticipated toxicity in humans. Overall, the *one-size-fits-all* paradigm to treat cancer is not valid anymore. Therefore, more personalized, physiologically-relevant, and predictable tumor models are needed. This new generation of tumor models must reproduce the biological complexity (molecular, structural, and dynamical) of each individual patient, resulting into patient-specific tumor models and therapies. In this regard, recent advances in nanotechnology and tissue engineering have enabled the development of microfluidic-based organs-on-a-chip (OoC) models of human organs and tissues with high predictive power (Bhatia and Ingber 2014; Esch et al. 2015). OoC have been widely employed for the study of human physio-pathological processes, including the numerous events occurring during cancer progression (Caballero et al. 2017b). Nevertheless, the current level of development of OoC have restricted their use mainly for fundamental research applications (Junaid et al. 2017), whereas their clinical and industrial applications still need to be proven. Major challenges that have limited their acceptance by the clinical and industrial market include their technical complexity, lack of standardization, low throughput, and the need for specialized personnel, among others. Indeed, clinical settings for cancer research demand for user-friendly platforms capable to reproduce *in vitro* the functional units and interactions of organs, and their tumors, from patients, and predict their response to treatments. This new generation of tumor models must, therefore, be highly personalized requiring the integration of cells from cancer patients; this results into individualized *cancer patient-on-a-chip* models. This may be critical for cancer therapeutics, where the efficiency (and toxicity) of specific chemo-, immune-, and even radio-therapies can be first tested *in vitro* in a microfluidic model of a cancer patient prior to moving to its human counterpart; this may eventually help physicians in taking decisions. Overall, the development of such a breakthrough in clinical diagnosis may have a tremendous impact not only in the clinics by improving cancer patient prognosis, but also in the biotechnological and pharmaceutical industry where novel products may be evaluated in a more efficient and predictable manners. Finally, this may reduce the associated costs of patient treatment in the healthcare system.

In this chapter, we show the latest advances in the field of patient-on-a-chip models for personalized cancer medicine. We discuss on their scientific, clinical, and industrial applications as well as their main advantages compared to the traditional models currently used in drug discovery, screening, and cancer research. We finally point out on the current limitations and the challenges. This drives us to advance towards the new generation of patient-specific clinical models.

4.2 Why Current Drug Testing Fails?

A recent study has estimated the cost associated to drug discovery and its translation to the market to be around 2.6 billion dollars, with a 145% increase compared to the former decade (DiMasi et al. 2016). Currently, more than 70–80% of cancer research is still based on simplistic *in vitro* models, which oversimplify the *in vivo* context. Indeed, *in vitro* models fail to recapitulate the multicellular content, 3D architecture, vascularization, or fluid exchange of the human body, which is critical to assess correctly human organ(s) response, drug diffusion kinetics, and dose efficiency. Similarly, *in vivo* models are incapable to predict satisfactorily the outcome of drugs on patients due to the absence of the human immune system that is of upmost importance during cancer dissemination and therapeutic response. Altogether, this lack of accuracy causes the failure of about 90% of the drugs (95% for cancer-related drugs) that enter to clinical trials, being the primary causes of failure their efficacy and clinical safety (Hay et al. 2014; Mullard 2016; Seruga et al. 2015). In addition, multiple approved drugs after evaluation using traditional *in vitro* and animal models have been pulled from the market due to their unexpected toxic effects in organs not directly targeted by the drug (Siramshetty et al. 2016). A recent work compiling and analyzing the data on the attrition of drug candidates developed by big pharmaceutical companies have concluded that novel methodologies are indeed needed for an efficient development of new drug (Waring et al. 2015). In this regard, the pharmaceutical and biotechnological industries are currently investing huge amount of resources in the development of such new generation of pre-clinical models.

Overall, drug development would significantly benefit from more advanced and physiological pre-clinical models capable to predict patient outcomes and identify toxicity problems at an early stage of the drug development process. This will univocally improve the efficiency of drug discovery, increase the clinical success rates, and significantly reduce the associated costs to the healthcare system.

4.3 Traditional Platforms for Tumor Modeling and Drug Discovery: An Overview

Tumor models are as diverse as the number and type of cancers. The success or failure of an anticancerous compound is directly correlated to the selected model and its inherent strengths and weaknesses. This selection depends on a large variety of parameters, including simplicity, cost, tumor type or the target under evaluation, among others. The perfect tumor model must accomplish a set of characteristics, such as (i) predictive, (ii) reliable, (iii) reproducible, (iv) fast, (v) affordable, and (vi) simple to use. We next briefly describe about the traditional models typically used for modelling cancer and testing drugs, which include both experimental (*in vitro* and *in vivo*) and computational (*in silico*) approaches, analyzing their advantages and limitations. For a detailed description, the readers may refer to the recent reviews on the topic (Caballero et al. 2017b; Carvalho et al. 2017; Katt et al. 2016; Stock et al. 2016).

4.3.1 In Vitro Tumor Models

4.3.1.1 One-Dimensional Models

One-dimensional (1D) tumor models are widely unknown despite their ability to recapitulate in a simple assay the fibrous architecture of the native extracellular matrix (ECM). Indeed, solid evidences have shown that the cells seeded in 1D environments display morphodynamic features reminiscent of the native scenario. This category include 1D fibers (synthetic or natural-derived), micro-patterned ECM lines, or topographic micro-structures. One-dimensional models have mainly been employed for the study of directional cell motion, a critical phenomenon involved in cancer invasion. As an example, an array of uniaxial lines coated with ECM proteins of 1.5 μm in width was employed to assess the dynamic properties of migrating cancer cells (Doyle et al. 2009). Interestingly, it was found that the cells displayed high velocity regardless of ECM density, similar to cells seeded in 3D native-like matrices and in contrast with 2D surfaces. Additionally, the cells seeded on 1D microstructures displayed phenotypes and cytoskeleton organization similar to those in 3D environments. This suggests that 1D micro-engineered systems can provide a representative model where to investigate signaling pathways, cellular adhesion, and the cytoskeleton organization involved in cancer invasion.

One-dimensional tumor models have also been employed for the study of cell-cell interactions, an event that has been demonstrated to be fundamental in cancer dissemination. As an example, micro-patterned fibronectin (FN) or collagen I lines mimicking the native ECM fibers were used to explore the interaction between tumor cells and tumor-associated macrophages observed *in vivo* (Sharma et al. 2014). Time-lapse microscopy experiments revealed that highly metastatic MTLn3 mammary tumor cells migrated with increased motility and lamellipodia activity upon interacting with bone marrow-derived macrophages. Interestingly, this interaction was perturbed using inhibitors of the involved paracrine signaling resulting in a reduced cell migration dynamics. The obtained results validated the hypothesis that macrophages were mediated by a paracrine loop during cancer cell invasion and metastasis, providing key insights about the important role of macrophages-tumor cells crosstalk in cancer expansion.

Overall, 1D models, albeit simple, may be very useful to assess crucial information of cancer progression. However, they display serious inherent weaknesses, including a limitation in the number and interaction with other cells, and importantly, the 3D architecture of the TME, among others.

4.3.1.2 Two-Dimensional Models

Two-dimensional *in vitro* tumor models have been vastly employed in unravelling the mechanistic determinants of tumor dissemination as well as for assessing the efficiency of anti-cancerous drugs. This is mainly a consequence to their simplicity, easy handling, superior accessibility, high throughput, and in general, reduced cost (Katt et al. 2016). In this type of assay, cells grow on flat surfaces typically made of plastic, such as in cell culture dishes, flasks, or multi-well plates, which are homogenously coated, or patterned, with proteins of the ECM. Besides drug screening applications, and similar to 1D case, 2D tumor models have widely been employed for the study of cancer cell migration and cell-cell/ECM interactions (Doxzen et al. 2013). Recently, this type of strategy was used to separate two isogenic populations of human ovarian cancer cells on micro/nano-patterned surfaces using their different migration dynamics (Shu Fan et al. 2017). Surprisingly, on flat (homogenous) condition, the cells displayed similar migration characteristics, which suggested that on patterned environments cancer cells motility varied in a metastasis-dependent manner.

Two-dimensional tumor models display a high degree of customization that make them ideal candidates to assess the influence of substrate interfaces in cancer dissemination. In this regard, FN-coated surfaces and hyaluronic acid hydrogel were employed to investigate the dissemination of human glioblastoma cells along topological interfaces (Rape and Kumar 2014). The cells were seeded in-between both environments, mimicking the vascular basement membrane of the human brain and the brain parenchyma, respectively. Results showed that the stiffness of the FN surface, together with actomyosin activation, influenced cell migration speed and focal adhesion assembly. This suggested a mechanochemical feedback mechanism governing cell morphodynamics during tumor dissemination. This work provided a simple model to mimic cell migration along vascular interfaces, which could easily be extended to other types of tumors.

It is worth mentioning that 2D models are still the preferred methodology used by pharmaceutical companies for drug discovery and screening due to their intrinsic advantages. However, they also suffer from serious limitations. In particular, 2D tumor models fail to recapitulate the biochemical heterogeneity and 3D architecture of the native TME. Further, the cells grown on these conditions display extreme phenotypes and gene perturbation (Cukierman et al. 2001). They also fail to reproduce organ-organ and tissue-tissue interactions, and dynamic drug/metabolic processes. Consequently, the effect of drugs tested on this type of models cannot be extrapolated to the human scenario.

4.3.1.3 Three-Dimensional Models

Three-dimensional tumor models overcome several of the limitations of their 1D and 2D counterparts by mimicking in a better way the structural, biochemical, and cellular properties of native tissues. Cells seeded in this type of environment display phenotypes and gene expression reminiscent of the *in vivo* scenario, making 3D tumor models a physiologically-relevant platform where to study (cancer) cell-cell/ECM interactions and the actual effect of drugs. Several works have compared the performance of 3D and 2D tumor models for drug discovery, showing that cancer cells seeded in the former respond to drugs in a similar way they do *in vivo* (Breslin and O'Driscoll 2016; Riedl et al. 2017; Stock et al. 2016). This makes 3D tumor models unique candidates to enable the identification of predictive biomarkers associated with tumor origin or with the resistance to a specific therapy, which is more predictive of their outcome in patients.

The types of 3D tumor models available are as diverse as the etiology of the disease. Typical models include standard transwell plates, synthetic or naturally-derived hydrogels, scaffolds, and matrices, micro-carriers, or spheroids. Most of these systems allow a high control on the mechanical properties and cell content (co-culture). They mimic the complexity and permeability of the *in vivo* scenario, preserving the cell phenotype and, in general, gene expression. Importantly, they also reproduce with higher fidelity the stromal reorganization during tumor progression, as recently reported (Brancato et al. 2017). In this work, biodegradable gelatin-based micro-carriers were employed to investigate the desmoplastic reaction activated by the stroma-pancreatic cancer crosstalk. Results showed that pancreatic cancer cells co-cultured with normal fibroblasts within the micro-carriers formed native-like tumor micro-tissues and were able to trigger the cancer-associated fibroblast phenotype, displaying TME markers.

Finally, and despite all the strengths of 3D tumor models, they also display certain limitations, which impacts on their predictive power. Among them, their low throughput, and lack of immune system, perfusion, vasculature, shear stress forces, organ-organ (and tissue-tissue) interactions, and standardization, which threaten the results relevancy and their clinical and industrial translations.

4.3.2 In Vivo Tumor Models

The lack of physiologically-relevant models capable to reproduce *in vitro* the complexity of tumors is one of the main limitations in preclinical drug development. To overcome this, *in vivo* tumor models are typically used before clinical trials on patients (Breyer et al. 2015). They are essential to assess the efficiency and toxicity of anti-cancerous drugs (and other types of compounds) as well as to explore the mechanistic determinants of the disease. Significant progress in the development of drugs has been obtained by using these models, mainly because they provide the biological complexity of a living system and recapitulate the dynamic interplay between organs and tissues. *In vivo* tumor models range from simple model systems typically used for mechanistic investigations, such as *Drosophila*, *C. elegans*, or *Zebrafish*, to more complex animal models, such as rodents or primates, which display higher physiological and genetic similarity to humans.

The screening and validation of potential new anti-cancerous drugs is typically performed using (immuno-compromised) animals either with the tumor implanted sub-cutaneous or in the same organ of origin with cancer cells from patients (patient-derived orthoxenograft model). The latter is the most relevant cancer model system, as it is presumed to faithfully represent the genomic features of primary tumors, as well as the structure, signaling, and metastasis of the human tumor. However, animal models are highly expensive, time-consuming, and allow reduced manipulation. Further, they are in general, very limited in their ability to mimic human carcinogenesis, physiology, and progression; they display serious differences with human tumors, in particular regarding their metabolism and immunology, which impact on their actual effect on humans.

Overall, despite the biological complexity that animal models provide, the above-mentioned limitations represent a major drawback when trying to elucidate new compounds behavior or to understand molecular mechanisms underlying drug response and/or acquired resistance. This is indeed supported by the low percentage of successful translation of drugs from animal models to clinical trials (Mak et al. 2014).

4.3.3 In Silico Tumor Models

The large amount and wide range of experimental data produced in the context of cancer have made *in silico* models of tumors indispensable. Mathematical modelling and computer simulations construct simplified, but representative, models of various factors and events involved in the progression of the disease (Vavourakis et al. 2017). *In silico* modelling span a broad spectrum of models, including statistical, multiscale, continuum, discrete, and agent-based (Edelman et al. 2010). Typically, they are used to complement experimental work since they are capable to extrapolate, or to refine, *in vivo* and *in vitro* experimentation in a rapid and simple way. They also avoid hard-to-measure parameters or compensates non-measurable factors, and do not require expensive and laborious materials and experimentation.

In silico models have been employed to simulate different pathological events ranging from subcellular mechanisms (e.g. gene modulation) to macroscopic interaction between cancer cells and the TME (e.g. tumor invasion). They have also been used for rapidly modelling the pharmacokinetic profile of anti-cancerous compounds, or for predicting the response of a specific therapy to a patient as a function of a large set of parameters, such as the genetic profile of a patient.

Overall, *in silico* tumor models can provide important insights into the mechanistic determinants of the disease and in predicting the efficiency/toxicity of drugs, thus supporting physicians in clinical-decision making. However, they are still based on oversimplified algorithms; their precision depend on a limited number of experimental parameters, which threatens the possibility to patient-personalized prediction. Implementing all the (big) data necessary to develop a personalized computational model and predict patient-specific treatment outcomes would be univocally associated with the need of high computing power. Therefore, a compromise between oversimplification and complexity must be achieved to bring *in silico* modelling to the clinical practice.

4.4 Organ-on-a-Chip Tumor Models: Tumor-on-a-Chip

Recent advances in micro- and nanofabrication technologies and tissue engineering have led to the development of miniaturized organ-on-a-chip (OoC) models of human physio-pathological processes. OoC are microfluidic devices containing human cells, which precisely replicate key features and the 3D architecture of human organs and tissues. OoC offers multiple advantages compared to their *in vitro*, *in vivo*, and *in silico* counterparts. They are 3D, highly physiological, compatible with mass production, and cheap. They also permit the co-culture of different cell types and high control over the physical (geometry, size), hydro/hemodynamic (flow, shear stress, gradient), mechanical (stress, rigidity), and biochemical (functionalization) properties of the model. Their design can be customized, or alternatively, standardized – commercial – chips can be used. Finally, OoC also allow the integration of other tumor models, such as tumor organoids, and are compatible with traditional and advanced microscopy technologies (Caballero et al. 2017b).

Different OoC models of human organs have been described during the last years, including the lung (Huh et al. 2010), spleen (Rigat-Brugarolas et al. 2014), kidney (Kim et al. 2016), brain (Kilic et al. 2016), liver (Lee et al. 2013), gut (Kim et al. 2012), and heart (Grosberg et al. 2011), among many others (Bhatia and Ingber 2014). They all differ in their design and cell content but display a 3D-like arrangement of cells recreating the human scenario. OoC models have widely been employed to investigate pathological processes, and in particular, tumor growth and progression. These OoC models of tumors are typically denoted as *cancer-* or *tumor-on-a-chip* (ToC). ToC models can easily recreate within a micro-engineered chip the (cancer) cell-cell/ECM interactions, the spatiotemporal chemical/physical gradients, or the (hydro)dynamic characteristics of the TME. They may be used to better elucidate the efficiency and toxicity of (new) drugs in the human body (Caballero et al. 2017a), as well as to explore the mechanistic determinants of the disease in a highly physiological, low-cost, and high-throughput manner, solving the limitations of traditional models (Mosig 2017). Different ToC models have been reported reproducing several of the events occurring during cancer progression, including tumor growth and invasion (Hassell et al. 2017), intra/extravasation (Chen et al. 2017; Zervantonakis et al. 2012), or organ specificity (Xu et al. 2016), and others (see Fig. 4.1). Herein, we briefly discuss about some recent examples of ToC platforms highlighting their main applications, strengths, and limitations. For a detailed description, the readers are referred to the recent reviews on the topic and references therein (Caballero et al. 2017b; Katt et al. 2016).

4.4.1 Cancer Invasion and Intravasation Models

The first event in cancer dissemination is the growth of the tumor and the invasion of cancer cells. Several cancer invasion ToC models have been described in the literature (Chen et al. 2016b; Lei et al. 2016; Zhang et al. 2012). These models differ in the architecture of the chip, the type of tumor, and in particular, the composition and structure of the TME. The latter is a very dynamic system exposed to a large variety of mechanical cues, which influence the progression of the tumor. This was recently addressed using an orthotopic human lung ToC model, which recapitulated the growth of lung cancer cells, cytokine secretion profiles, and the clinical responses to therapy previously observed in patients (Hassell et al. 2017). The model contained two channels mimicking the vascular endothelium and the pulmonary epithelium, respectively, which were interconnected through a porous thin membrane. Next, two side chambers were used to apply cyclic stretching mimicking physiological breathing within the chip. Interestingly, the results showed that the induced cyclic motion inhibited tumor growth and invasion, which was mediated by changes in the epidermal growth factor receptor and the MET protein kinase. This work demonstrated the high capabilities of ToC models in recapitulating human physiology and in unravelling the mechanism at work governing pathological processes.

During the last few years, intravasation-on-a-chip models have been employed to investigate the invasion of tumor cells into the vascular and lymphatic systems (see Fig. 4.1) (Lee et al. 2014; Zervantonakis et al. 2012). Typically, these models are based on simple micro-engineered blood vessel-like channels and chambers mimicking the TME. These models differ on their complexity, ranging from simple hollow polydimethylsiloxane (PDMS) channels coated with endothelial cells, to more complex systems where the native

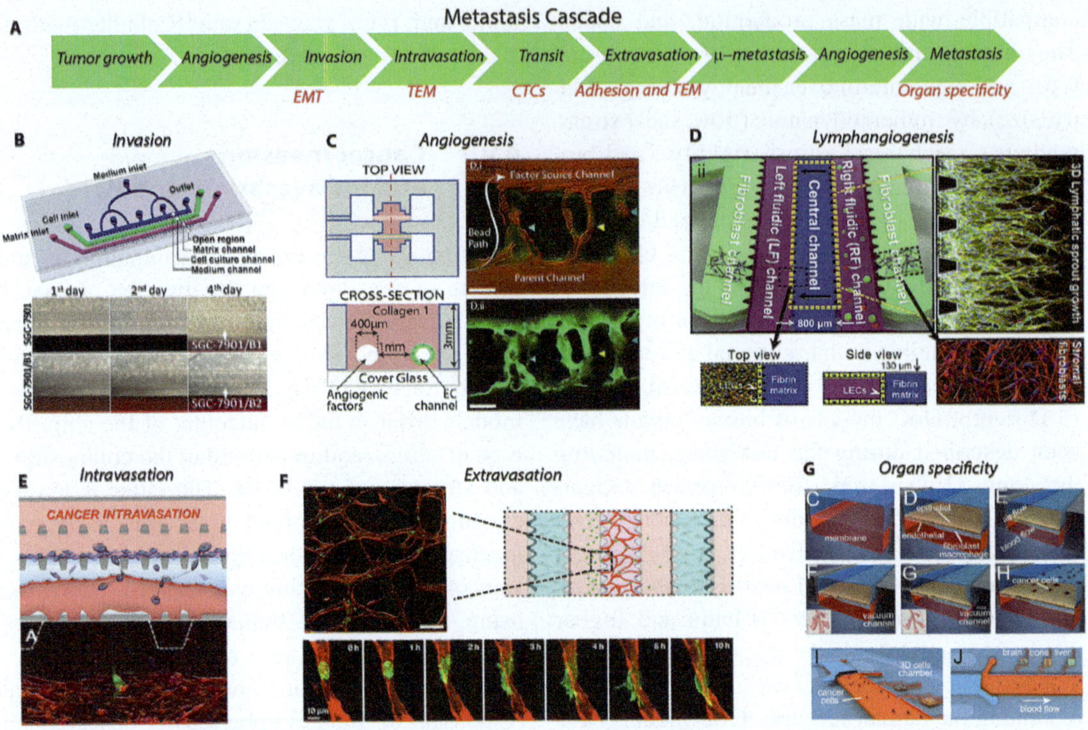

Fig. 4.1 **Tumor-on-a-chip models.** (**a**) Scheme showing the different events occurring during cancer metastasis. ToC models of (**b**) tumor invasion, (**c**) angiogenesis, (**d**) lymphangiogenesis, (**e**) intravasation, (**f**) extravasation, and (**g**) organ specificity. Reproduced with permission from (Caballero et al. 2017b). Copyright © Elsevier

TME is mimicked employing hydrogel-based ECMs encapsulating cancerous cells. Recently, a 3D ToC device made of concentric three-layer cell-laden hydrogels reproducing the tumor cells, the stroma, and the vascular network, was used to assess the breast cancer cell intravasation (Nagaraju et al. 2018). This model showed that the breast cancer cells displayed different invasion capabilities depending on their metastatic capability and on the presence of the vascular layer. In particular, MDA-MB-231 breast cancer cells displayed enhanced invasion and intravasation in the presence of HUVEC cells, in contrast to MCF7 cancer cells. Similarly, a 3-channel microfluidic chip containing endothelial and cancer cells interconnected via a 3D ECM hydrogel was employed to investigate the mechanism of tumor cells-endothelium interactions. Results showed that the secretion of TNF-α by macrophages increased cancer cells invasion capability, endothelial permeability, and intravasation rate (Zervantonakis et al. 2012).

Overall, even though ToC models of cancer invasion and intravasation can provide key insights about their mechanism at work, they need some refinement in terms of throughput and compatibility with standard analytical and characterization technologies to boost their clinical and industrial applicability.

4.4.2 Angiogenesis Models

One of the main events in the cascade of tumor dissemination is angiogenesis, which is defined as the formation of a new vascular network from pre-existing blood vessels. Angiogenesis is regulated by an extensive variety of factors including mechanical and biochemical cues, which trigger the invasion of cancer cells into the microcirculation system. Many ToC models of tumor angiogenesis have been reported during the last few years. They have provided valuable information about the regulators of cancer cell invasion, including the important role of fluid flow and shear stress, the interaction of cells with the TME, or the influence of pro-angiogenic factors in the invasion of endothelial cells (see Fig. 4.1) (Bischel et al. 2013; Theberge et al. 2015). As an example, a ToC model recreating a single – parent – human blood vessel was developed to monitor vascular sprouting (Pauty et al. 2018). In this work, vascular endothelial growth factor-A was added to stimulate the growth of sprouts on the vessel, indicators of the nascent formation of capillaries and mimicking *in vivo* sprouting angiogenesis. The model, albeit simple, provided a good platform to examine the effect of anticancerous drugs. In particular, two Food and Drug Administration (FDA)-approved antiangiogenic drugs (sorafenib and sunitinib) were used. Interestingly, these drugs inhibited the growth of new vessels even upon vascular endothelial growth factor-A stimulation, thus validating the ToC model for drug screening applications. Similarly, a cylindrical collagen-based microfluidic channel was used to seed endothelial cells that self-organized forming an artificial vessel. *In vivo*-like angiogenesis was mimicked by injecting a gradient of different pro-angiogenic factors. Results showed that a combination of growth factors and not an individual one was needed for new vessel formation. Altogether, the above-described approaches show promising clinical applications even though they still need to be proven.

4.4.3 Extravasation Models

Extravasation is the event where cancer cells, after invading and circulating along the vascular and lymphatic systems, escape out of the vessels and invade the neighboring tissue to establish a secondary tumor site. Similar to intravasation, this transendothelial invasion is a rare event, which is difficult to capture and analyze *in vivo*. Several extravasation-on-a-chip devices have been reported to investigate the mechanism at work of cancer cell extravasation (see Fig. 4.1) (Chen et al. 2016a, 2017; Cui et al. 2017; Ma et al. 2018). Most of these models are based on chips containing three microfluidic channels separated by micro-pores or slits. Typically, one of the channels is coated with endothelial cells mimicking the blood vessel; a second channel is used to recreate the TME, and typically, a third channel is employed to flow drugs or to create gradients of biochemical compounds. As an example, this strategy was employed to study the extravasation of human metastatic breast cancer cells towards specific secondary bone-mimicking microenvironment (Jeon et al. 2015). The side channels were used to flow biochemical factors and breast cancer cells. The middle channel contained a fibrin gel mimicking the bone microenvironment and embedding human umbilical vein endothelial cells, human bone marrow-derived mesenchymal stem cells, and osteoblast-differentiated cells. Importantly, a functional native-like microvascular network was generated allowing the lateral flow of media across the gel and the arrest, and extravasation, of MDA-MB-231 breast cancer cells. To demonstrate the specificity of breast cancer cells towards bone, two additional microenvironments were generated, including muscle and acellular collagen. Finally, the role of physiological fluid shear stress was addressed finding that the extravasation rate was larger in static condition, most likely a consequence of the observed decrease in microvessel permeability, and highlighting the utility of this approach. Altogether, this work highlights the potential of ToC models to investigate the mechanistic determinants of extravasation. This may be of particular interest for the pharmaceutical industry to develop and evaluate the efficiency of new drugs targeting this pathological event.

To summarize, all the above-mentioned works demonstrate that ToC models can be an extremely useful tool to reproduce the complex events occurring during tumor dissemination, providing key insights about the mechanism at work of the disease. This is of high interest for clinical applications, where new simple and reliable tumor models that overcome the limitations of current *in vitro* and *in vivo* models, are needed. However,

ToC models still lack fundamental elements of the human physiology, such as organ-organ and tissue-tissue interconnection, which are critical for cancer metastasis studies as well as for drug discovery, screening, and toxicology applications.

4.5 Multi-Organ-on-a-Chip Models

Organs and tissues in the human body are not remained in isolation, but are integrated into a highly dynamic and interconnected microenvironment with a continuous exchange of biochemical and mechanical signals. Indeed, an action targeting one specific organ or tissue, such as the addition of a drug, may affect others. In this regard, standard OoC models mimicking individual organs and their functional units are incapable to investigate secondary organ/tissue effects. To address this limitation, highly integrated OoC devices combining different organs and tissues on a single chip have been developed. These devices have been denoted as *multi-organ* (or tissue*) on-a-chip* (multi-OoC) and provide a highly physiological and dynamic microenvironment where to recapitulate human organ-organ and tissue-tissue interactions. Multi-OoC have, therefore, an enormous potential for disease modelling as well as for drug discovery and screening applications. This allows a comprehensive testing in the drug absorption, distribution, metabolism, and excretion (ADME) process. Similarly, they offer large possibilities for the prediction of drug efficiency and toxicity to secondary organs by assessing the passage of a drug throughout the human body.

Multi-OoC can be classified as static, semi-static, and flexible depending on their configuration (Rogal et al. 2017). It may be noted that alternative categories can also be attributed depending on their hydrodynamic characteristics, namely static (diffusion-driven), single-pass (open loop), pump-driven (closed-loop), and pump-less (gravity-driven) recirculating platforms (Wang et al. 2018). Traditional multi-OoC platforms follow a static configuration, where the individual organ and tissue models are interconnected following a pre-defined order. Typically, a single-route microfluidic channel and a common – universal – medium are used. In this case, the flow is unidirectional where only upstream organs interact with the downstream ones. In contrast, semi-static systems interconnect several organ modules using a microfluidic channel platform and Boyden-based chambers containing the individual organ models (Edington et al. 2018). Typically, semi-static systems allow the recirculation of culture media mimicking systemic blood flow. Finally, flexible multi-OoC models interconnect several – individual – OoC devices through microfluidic tubing allowing multiple and modular configurations. In this case, the number, type, and interconnection order can be freely chosen.

Recent works have reported the use of multi-OoC devices for disease modelling and drug discovery/screening applications (Hwan and Hwan 2018; Wang et al. 2018) (see Table 4.1); these models differ in their designs and complexities. As an example, a multi-OoC platform was recently developed comprising the liver, lung, and heart, and used to study inter-organ response to drug administration (Skardal et al. 2017). The model interconnected three independent OoC models by means of a central fluid routing breadboard in a *plug & play* configuration, and used a standard peristaltic pump to perfuse the drug-containing media. Interestingly, drug response was found to depend on inter-organ and tissue interactions, highlighting the importance of this approach.

Multi-OoC have also been applied to cancer research to assess organ specificity in cancer metastasis (Kong et al. 2016; Skardal et al. 2016a; Xu et al. 2016). This is critically important because the dissemination of tumor cells from the primary tumor to secondary organs is responsible for more than 90% of cancer-related deaths (Valastyan and Weinberg 2011). Recently, a bone perivascular niche-on-a-chip device was employed to investigate the metastatic colonization of breast cancer cells to bone (Marturano-

Table 4.1 Recent works on multi-organ and patient-on-a-chip devices

Organs modelled	Main applications	Advantages	Limitations	References
Liver Heart	Toxicity	Multiple real-time sensors integrated Platform compatible with existing OoC models Mini-microscope integrated	Low throughput Reduced number of organs Large size Scaling between organs not addressed	Zhang et al. (2017)
Liver Heart	Toxicity	Recirculation of media Parallelization Integrated micro-valves Common media	Reduced number of organs Scaling between organs not addressed	Kam(ei et al. 2017)
Gut Liver	Metastasis Drug screening	Simple Customizable Vascular network mimicking	No control on tumor structure or growth Lack of both endothelial and stromal cells No alternative metastatic sites Low throughput Reduced number of organs	Skardal et al. (2016a)
Gut Liver	Inflammation Pharmaco-kinetics	Long-term cultures Recirculation Scaling of organs	Reduced number of organs Low throughput	Chen Wen et al. (2017), Tsamandouras et al. (2017)
Lung Liver	Drug discovery	Programmable micro-pump integrated Precise fluid flow control Reconfigurable Non-sorption materials	Reduced number of organs Scaling between organs not addressed Low throughput	Coppeta et al. (2017)
Liver Heart Lung	Drug screening Toxicology	Independent or interconnected response Integrated biosensors Use of 3D organoids Modular configuration	Low throughput No pharmaco-kinetic studies Reduced number of organs	Skardal et al. (2017)
Intestine Liver Skin Kidney	ADME of drugs	Long period co-culture stability and tissue functionality Reproducible 4-tissue homeostasis Dynamic perfusion	Low throughput. Lack of both endothelial and stromal cells	Maschmeyer et al. (2015)
Cardiac Muscle Neuronal Liver	Toxicology Drug discovery	Multiple organs and drugs tested Measurement of electrical, mechanical, and metabolic response Modular, reconfigurable and pumpless system Common serum-free medium	Important organs for ADME applications missing Scaling between organs not addressed For some organs, cell lines were used	Oleaga et al. (2016)
Lung Brain Bone Liver	Metastasis	Multiple metastatic sites Multiple features of the *in vivo* lung TME, μ-vascular endothelium, immune cells, fibroblasts and distant organs	No assessment of shear stress and vascular targeting for drug delivery Low throughput	Xu et al. (2016)

(continued)

Table 4.1 (continued)

Organs modelled	Main applications	Advantages	Limitations	References
GI tract Lung Liver Bone marrow Kidney	ADME of drugs.	Multiple organs Pumpless (gravity-driven) Scalable Scaling arguments considered	Complex Use of cell lines Low term culture	Miller and Shuler (2016)
Intestine Liver Kidney proximal tubule Blood-brain barrier Skeletal muscle	ADME of drugs. Drug organ-specificity. Toxicology	Multiple organs *In vivo*-like response Dynamic perfusion	Complex Low throughput Scaling between organs not addressed No vascularization	Vernetti et al. (2017)

Kruik et al. 2018). The chip incorporated a native decellularised 3D bone matrix embedding human endothelial and bone marrow-derived mesenchymal stem cells. This *in vivo*-like composition allowed the formation of a stable microvasculature in controlled flow conditions. MDA-MB-231 breast cancer cells were introduced into the device both under static and interstitial flow conditions. Surprisingly, results showed a higher four-fold proliferation rate of cancer cells in static conditions, suggesting that physiologically-relevant interstitial flow activated a slow-proliferative state in cancer cells colonizing the niche. Next, in a similar work, the specificity of human breast cancer cells to bone was studied using a 3-channel multi-OoC device (Bersini et al. 2014). In this work, MDA-MB-231 human breast cancer cells, human umbilical vascular endothelial cells, and human osteo-differentiated bone marrow-derived mesenchymal stem cells, were employed to replicate the invasive malignant tumor cells, the vascular network, and the metastatic site of bones, respectively. Results showed that highly metastatic breast cancer cells introduced in the microvascular network extravasated in larger amounts towards the pre-conditioned bone microenvironment compared to control conditions.

It may be noted that the above-mentioned works albeit informative, mainly address the metastasis of cancer cells into a specific secondary tissues, limiting the understanding of the key players involved in organs specificity during metastasis. To address this issue, multi-OoC devices displaying multiple metastatic sites have also been reported. As an example, a multi-organ lung-on-a-chip device was recently employed to replicate the specificity of metastasis from the lung to the brain, bone, and liver (Xu et al. 2016). The model contained the upstream native-like lung cancer microenvironment, the microvascular endothelium, immune cells, fibroblasts, and the three downstream organs. Results showed that the cells from the respective metastatic sites overexpressed specific proteins characteristic of metastasis, and importantly, similar to the *in vivo* scenario.

Overall, the advanced capabilities of multi-OoC platforms provide them with unique capabilities for the development of highly predictive cancer models, where inter-organs and tissues interactions can be easily assessed. Therefore, multi-OoC may find solid applications in drug discovery, screening, and toxicology, with an enormous potential in the field of personalized cancer medicine (Esch et al. 2014). However, improvements must also be addressed in using primary cells (or iPS-derived cells) and in integrating personalized biomimetic 3D microenvironments if these models are intended to be adopted by the clinical market.

4.6 Multi-Organ-on-a-Chip Models of Cancer Patients: Cancer Patient-on-a-Chip

The combination of multi-OoC technology and tissue engineering tools with primary cells from patients, has led to the development of clinical-oriented *patient-on-a-chip* models (Abaci and Shuler 2015; Hampton 2017; Low and Tagle 2018; Skardal et al. 2016b; Williamson et al. 2013) (see Fig. 4.2). In these models, the different organs and tissues are interconnected following a physiologically-relevant arrangement to mimic the systemic organization and diversity of events occurring inside the human body (see Fig. 4.2). The number of organs required to develop a patient-on-a-chip are not fixed, which may depend on the biological characteristics of each individual patient and the clinical goal of the model. Typically, for drug delivery and screening studies, the model may include the gut, liver, kidney, blood vessel, and the targeted organ of interest. Additional organs may also be included to assess off-target side effect of drugs, such as the heart, to investigate cardiotoxicity, or the brain to examine blood-brain barrier (BBB) permeability.

A special category of patient-on-a-chip models is *cancer patient-on-a-chip* (CPoC), where cells from cancer patients are used. These models display a tremendous potential in personalized medicine by reproducing in a multi-OoC the tumor of an individual patient. This may permit to (i) unravel the specific mechanism at work of patient tumor dissemination and organ-specificity during metastasis, (ii) reproduce organ-organ and tissue-tissue interactions, (iii) provide real time data about treatment efficiency and side effects, (iv) assess the ADME and toxicity of chemo- and immune-therapeutic drugs, or radiotherapeutics treatments, (v) investigate drug pharmacokinetics and pharmacodynamics, and (vi) test, and predict, personalized therapies in a high-throughput and controlled microenvironment (Zhang et al. 2017). Importantly, since most

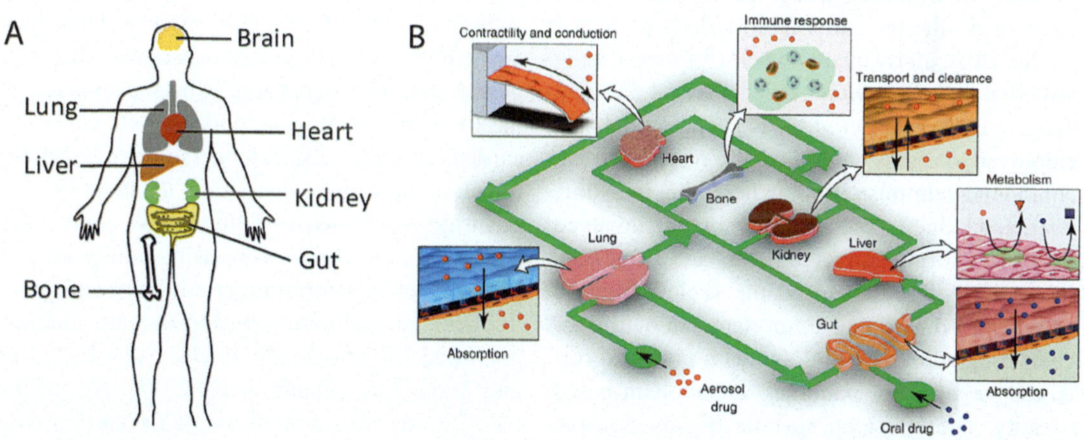

Fig. 4.2 The patient-on-a-chip concept. Schemes showing (**a**) several organs in the human body, and (**b**) the patient-on-a-chip model representing some of them, namely the lung, gut, liver, kidney, bone, and heart. The different organs are interconnected following a physiologically-relevant configuration to account for organ/tissue-organ/tissue interactions capable to recapitulate systemic drug absorption, distribution, metabolism, and excretion. In this case, drug administration can be performed through the lung (inhaled) or through the gut (orally administrated) to study their metabolism in the liver, their transport and clearance in the kidney, their immune response in e.g., the bone, and their toxicity in e.g., the heart. Panel (**b**) is reproduced with permission from (Huh et al. 2011). Copyright © Elsevier

cancer patient fail to respond equally to the same treatment, this precision medicine approach may also help physicians in taking decisions on therapy selection and identify treatment specific regimes or unexpected risk factors. Finally, CPoC may also be employed to model genetic variation or to identify population sub-groups that display a higher risk from adverse side effects. However, and despite the enormous potential of CPoC, the amount of truly CPoC models reported in the literature are scarce. This is due to the technical and biological complexity required, which makes that most of the reported works are based on traditional multi-OoC platforms (see Table 4.1).

One of the main advantages of CPoC models is their ability to investigate fundamental aspects of the human – cancer – physiology by linking multiple organs and tissues together, such as the sequential metabolism of individual (or combinations of) drug administration in the human body (Ronaldson-Bouchard and Vunjak-Novakovic 2018). This may provide valuable information about their toxicities and pharmacokinetics, which is critical during the process of drug development, thus increasing the efficiency of the therapy and reducing the risk of drug toxicity. Recently, a CPoC model was developed by sequentially interconnecting several organ models, namely the intestine, liver, kidney proximal tubule, BBB, and skeletal muscle, following an *in vivo*-like configuration (see Fig. 4.3a and Table 4.1) (Vernetti et al. 2017). The model was employed to assess organ-specific response upon compound administration, where the intestine, liver, and kidney represented the organs involved in drug absorption, metabolism, and excretion, respectively. The BBB and the skeletal muscle were added due to their importance in neurovascular drug administration and as a potential off-target organ to examine metabolism-based toxicity. The effect on specific organs of three known compounds, namely terfenadine, trimethylamine-N-oxide, and vitamin D3, was investigated. Results showed that the compounds were successfully absorbed, metabolized, and transported, and importantly, in agreement with reported clinical data. Similarly, a CPoC approach interconnecting functional heart, muscle, brain, and liver models was used to assess the systemic toxicity of five drugs with known side effects during long-term time periods and continuous recirculation flows (see Fig. 4.3b and Table 4.1) (Oleaga et al. 2016). Interestingly, the model integrated metabolic, electrical, and mechanical readouts to analyze organ functionality in real time. In general, the obtained results were consistent with those reported in the literature, namely a decrease in cell functionality and viability upon drug treatment. In particular, doxorubicin, a well-known chemotherapeutic drug, showed acute cardio- and hepatotoxicity. Doxorubicin cardiotoxicity was also reported in a different work by using a simple CPoC model, which interconnected a heart- and a cancer-liver-on-a-chip (see Fig. 4.3c and Table 4.1) (Kamei et al. 2017). Results showed that cardiotoxicity was a result of the metabolic production of doxorubicinol by HepG2 cells. Similarly, the interconnection of heart, liver, and lung was used to assess the toxicity and organ specificity of bleomycin, a drug typically used for treating some cancers (see Fig. 4.3d and Table 4.1) (Skardal et al. 2017). The addition of the drug exhibited an unanticipated cardiotoxicity, where cardiac cells ceased beating. Surprisingly, the addition of the drug using an individual heart-on-a-chip model did not show any effect; cells did not stop beating. This suggested that bleomycin might be metabolized in the liver and produce secondary toxic factors to cardiac cells. Unfortunately, the device did not integrate sensing capabilities to identify and quantify such toxic compounds. This was addressed in a similar work using a fully-automatized CPoC platform integrating physical, electrochemical, and optical biosensors into a human heart-and-liver-on-a-chip device (see Fig. 4.3e and Table 4.1) (Zhang et al. 2017). Again, the CPoC model was used to assess the cardiotoxicity of common chemotherapeutic drugs towards hepatocellular carcinoma cells. It was found that liver cancer organoids shrunk upon drug treatment as demonstrated by the measured reduction in the amount of secreted albumin and in the increase of characteristic hepatotoxic biomark-

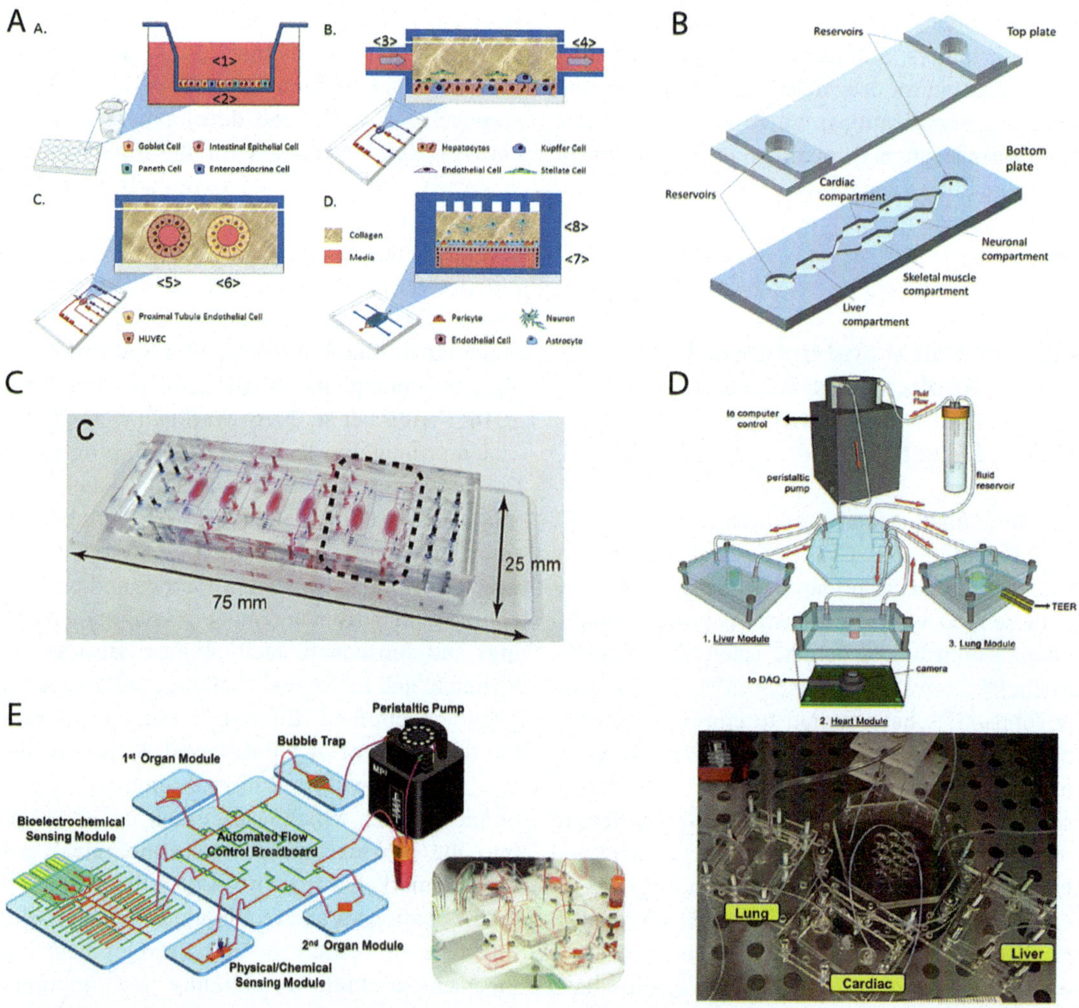

Fig. 4.3 **Cancer patient-on-a-chip models.** (**a**) Five-organ CPoC (intestine, liver, kidney proximal tubule, BBB, and skeletal muscle) used to assess organ-specific response upon compound administration. The scheme shows four of the organs systems used for functional coupling, *A)* intestine, *B)* liver, *C)* kidney, and *D)* BBB (Vernetti et al. 2017) © Nature Publishing Group. (**b**) Four-organ CPoC (heart, muscle, brain, and liver) model used to assess the systemic toxicity of five drugs with known side effects. The scheme shows the different cell compartments (Oleaga et al. 2016) © Nature Publishing Group. (**c**) Two-organ CPoC model (heart and liver) used to assess doxorubicin cardiotoxicity. The image shows the actual chip containing three sets of co-culture systems (Kamei et al. 2017) © Royal Society of Chemistry. (**d**) Three-organ CPoC model (heart, liver, and lung) used to assess the toxicity and organ specificity of bleomycin. The top and bottom images show a scheme and an actual photograph of the entire system set up, respectively (Skardal et al. 2017) © Nature Publishing Group. (**e**) Two-organ CPoC model (heart and liver) used to assess the cardiotoxicity of common chemotherapeutic drugs. The scheme shows the entire system set up, and the inset, an actual picture of the two-organs model (Zhang et al. 2017) © National Academy of Sciences. Figures are reproduced with the permission from the references

ers. Similarly, the addition of doxorubicin induced pronounced cardiotoxicity as demonstrated by the high levels of CK-MB, a cardiac biomarker characteristic of heart damage.

Overall, CPoC represent a new generation of patient-matched models capable to evaluate and predict the outcome of individualized therapies. This is a consequence of their unique ability to replicate into a microfluidic platform the biological multi-organ complexity of individual patients and their tumors. Similarly, CPoC models offer a great potential for drug discovery, screening, and toxicology applications, where biotechnological and pharmaceutical companies can develop drugs in a more efficient and fast manner. This will univocally reduce the current high attrition rates of drugs.

4.7 Clinical and Industrial Applications: Current Challenges and New Directions

Besides fundamental research, where ToC, (multi)-OoC, and CPoC have been widely employed in cancer research applications, the use of these models by the industry is just beginning. Pharmaceutical companies, such as Pfizer™, Roche™, Merck™, AstraZeneca™, and Johnson & Johnson™, have started to employ this technology to accelerate the screening and development of novel therapeutic agents. In addition, several OoC-based companies have started to receive public funds to explore the clinical and industrial capabilities of their technology. As an example, 45 M$ have recently been invested in Emulate™ (US) to boost the application of OoC technology for drug discovery. The chips have been employed by the FDA to assess the toxicology of food supplements and additives (Fitzpatrick 2017). Interestingly, collaborations with hospitals have also been initiated to establish *Patient-on-a-Chip* clinical programs aiming at developing personalized CPoC-based models of patient diseases using reprogrammed stem cells from patients.

Many OoC companies have emerged during the last few years offering their chips for a diverse varieties of biomedical applications, including pharmacological and toxicology studies, drug discovery, and others. Other industrial and clinical applications of CPoC include the screening of cancer biomarkers to assess the properties of metastatic tumor. In this regard, the identification in peripheral blood of circulating tumor cells (CTCs), DNA/RNA, exosomes or antibodies emerge as potential biomarkers of cancer metastasis with high clinical value. This liquid biopsy approach has led to the development of *liquid biopsy-on-a-chip* devices for the individualized diagnosis, prognosis, and optimized treatments on cancer patients (Tadimety et al. 2017). Indeed, several CPoC-based systems have already been reported for the detection of CTCs from different tumor origins, which highlights the enormous potential of these platforms for industrial and clinical applications (Alix-Panabieres and Pantel 2014; Ferreira et al. 2016; Warkiani et al. 2016).

Unfortunately, despite the latest progress in the development of ToC, (multi)-OoC, and CPoC models, their clinical use is still very limited. The future generation of these type of models may need to display advanced properties and be compatible with current and modern imaging, screening, and biosensing technologies. Besides the technical and biological challenges related to the interconnection of different organs and tissues, the development of CPoC-based devices is also accompanied with considerable number of economic barriers, which restrict their incorporation into the pharmaceutical, biotechnological, and clinical market. A recent set of articles reported on the challenges and requirements of employing this type of models in drug development and mechanistic studies, including the important visions of researchers, clinicians, industry, and policy makers (Watson et al. 2017). The topics discussed included the optimal volume of samples to be employed, scaling arguments, model cost, and market needs, among others, highlighting the challenges to be considered when developing predictable CPoC-based models to obtain a systemic-like response. In the following, we briefly discuss about some of these challenges. For a detailed information, the readers are referred to specialized reviews on the topic.

4.7.1 Technological Challenges

Technological challenges can be categorized as basic and complex. The former include standard bubble formation or sterility maintenance. The need for new non-sorptive materials for chip fabrication instead of standard PDMS, or the lack of an integrated perfusion system capable to provide physiological fluid flows for the different tissues (speed and shear stress) to account for the mechanical stimuli/response of cells, are also included in this category. Next, new biomaterials mimicking better the native environment will need to be created. In this regard, the use of cells obtained from patients capable to build the TME are anticipated to be a major revolution. Complex challenges include the lack of standardization, which limits the direct comparison of the data between laboratories. To address this, several generic microfluidic devices have been developed and are commercially available (Junaid et al. 2017). Alternatively, the use of modular, plug & play, pre-fabricated, and universal Lego™-like microfluidic units may standardize the fabrication of CPoC-based systems facilitating the comparison of results between labs (Bhargava et al. 2014; Loskill et al. 2015). Each individual unit may mimic a specific body function, and integrate technical, sensing, and pumping functions. Next, CPoC-based models are more complex compared to static models and, in general, cells are more difficult to collect for molecular analysis. Other technical challenges include the development of models compatible with traditional or advanced imaging (e.g., super-resolution microscopy), screening (e.g., multi-OMICS), and biosensing technologies (e.g., single-molecule biosensing). The latter is of upmost importance for monitoring physiological parameters (e.g. O_2, H_2O_2, pO_2, pH, and others) in real-time. Similarly, CPoC-based systems must also be compatible with automatized (robotized) platforms for an efficient and fast handling. However, all these technological improvements may be associated with an increase in complexity, which may limit the reproducibility of results and manipulation speed/simplicity. Therefore, future CPoC may need to find a balance between technological simplicity, biological relevance, and adaptability.

4.7.2 Biological Challenges

Biological challenges include the integration of a TME with controlled mechanochemical and biological properties, where the cells are capable to recapitulate the phenotype observed *in vivo* over long culture periods. In this regard, the culture of different cells types, such as tumor cells, endothelial, macrophages, or fibroblasts, is still a major limitation in most models. In addition, keeping the cells and tissues alive for long-term periods (>1 month) in physiological-like conditions, together with a systemic liquid-to-cell volume ratio, is a major challenge. Similarly, gathering a sufficient number of mature cells from patients to seed inside the chips, as well as the time required to build a functional CPoC, are major bottlenecks. The use of iPSCs may provide unlimited number of cells and the creation of specific patient and disease-specific cells (Singh et al. 2015). Next, organ/tissue functionality must be ensured by supplying the different organs and tissues with specific culture media, or alternatively, a universal one. Further, medium recirculation and functional vascularization is crucial for linking the different organs and tissues to the vasculature as well as for mimicking systemic drug administration (Osaki et al. 2018). Importantly, scaling arguments must be considered. This accounts for designing the different tissues and organs, which are physiologically scaled relatively to each other as well as in terms of tissue-media ratios. Next, the role of dormant cancer cells or cancer stem cells in resuming tumor proliferation must be assessed to analyze chemotherapeutic resistance in cancer patients. In this regard, primary cells from cancer patients (or even reprogrammed iPSCs from adult somatic cells) may be used instead of standard cell lines, which behavior may differ between labs as well as their response to drugs compared to *in vivo* scenario. Even more complex, having (and demonstrating) that the tissue(s)/organ(s) model display a gene expression pattern characteristic of the *in vivo* scenario is challenging, but fundamental for clinical applications. Similarly, the integration of several and personalized organs and tissues on chip is critical to assess for metastatic-site influence. This is of upmost importance for the development of personalized treatments since not all cancer patients respond equally to a specific treatment. Next, the integration of human microvasculature, containing both the blood and lymphatic vasculature is crucial to elucidate the mechanistic determinants of tumors dissemination, to investigate how drugs transit along the vasculature and are administrated to the TME, and to assess how the tumor respond to drugs injected into the "human" vasculature. Finally, the impact of toxic compounds exchange between organs must be explored (Frey et al. 2014). This is highly important for investigating the efficiency, toxicity, and side effects of drugs in secondary organs. This means that the number of organs and tissue models shall be as high as possible. However, an increase in the number of integrated organ/tissue models is associated with a significant increase in the complexity related to chip fabrication, biological content, handling, and cost, among others.

4.7.3 Economical and Market Challenges

Economical and market barriers include the high manufacturing cost of typical CPoC systems. A decrease in the fabrication cost must be accomplished to facilitate their adoption by the pharmaceutical, biotechnological, and clinical market. To this aim, the chips must be simple, miniaturized, mass produced, and integrate bioelectronic components for sample monitoring and screening. Next, clinical-oriented CPoC models will need to be developed under GMP (good manufacturing practice) conditions using biomaterials (and cells) displaying rigid requirements in terms of cost, biocompatibility, and durability. In this regard, PDMS, typically used for the fabrication of chips due to its intrinsic properties (e.g. optical transparency, rigidity and biocompatibility), is not compatible with mass industrial and commercial production for its elevated cost and fabrication methodology. Other materials displaying alternative physico-chemical properties, such as synthetic or natural-derived hydrogels, and fabrication methods, such as 3D (bio)printing, offer an enormous potential for the low-cost fabrication of organ-on-a-chip devices (Caballero et al. 2017a), thus offering better market opportunities. Finally, important market limitations include the lack of specialized industry and personnel capable to manipulate the chips and interpret the results, which makes their adoption by the clinical market in the near future unrealistic.

Overall, the current level of development of CPoC-based devices, and the technological, biological and economic barriers that they still challenge, limit their current use into the clinical practice. The collaboration between different communities, namely academia, clinics, and industry, as well as the involvement of policy makers, research organizations, societies, patients, and other stakeholders, is fundamental to boost their development and use in the clinics. However, the future looks promising and several European and International-funded organizations and initiatives have been created to consolidate personalized and patient-centred technologies in healthcare treatment.

4.8 Conclusions

Organ-on-a-chip models of cancer patients, or cancer patient-on-a-chip, provide a large plethora of advantages compared to their *in vitro*, *in vivo*, and *in silico* counterparts. This new generation of highly personalized cancer models displays biomimetic compositions and 3D architectures capable to reproduce in a single microfluidic chip the biological, structural, and dynamical complexity of human organs and tissues, and their interactions, of individual cancer patients. Indeed, cancer-patient-on-a-chip (CPoC) models are a viable physiologically-relevant platform for personalized medicine, drug discovery and screening, as well as to explore the mechanistic determinants of cancer progression. By using CPoC models, patient-matched therapies can be tested by physicians to select the most adequate treatment for each individual patient. Importantly, by combining different organs and tissues models, the – metabolic – interaction between different organs can be modelled and investigated. This is of particular importance for metastasis studies and to assess the side effect of drugs into secondary organs. Altogether, CPoC are expected to replace standard *in vitro* static assays, bridging the gap between pre-clinical (animal) models and humans. To become this a reality, an active collaboration between the different stakeholders is fundamental to boost their translation finally towards the biotechnological and clinical market.

Acknowledgements D.C. acknowledges the financial support from the Portuguese Foundation for Science and Technology (FCT) under the program CEEC Individual 2017 (CEECIND/00352/2017). All the authors acknowledge the financial support from the European Union Framework Programme for Research and Innovation Horizon 2020 on Forefront Research in 3D Disease Cancer Models as *in vitro* Screening Technologies (FoReCaST) under grant agreement no. 668983. D.C. and S.C.K also acknowledge the support from the FCT under the scope of the project Modelling Cancer Metastasis into the Human Microcirculation System using a Multiorgan-on-a-Chip Approach (2MATCH) (02/SAICT/2017 – n° 028070) funded by the Programa Operacional Regional do Norte supported by FEDER. Conflicts of interest: none.

References

Abaci HE, Shuler ML (2015) Human-on-a-chip design strategies and principles for physiologically based pharmocokinetics/pharmacodynamics modeling. Integr Biol 7:383–391. https://doi.org/10.1039/c4ib00292j

Alix-Panabieres C, Pantel K (2014) Technologies for detection of circulating tumor cells: facts and vision. Lab Chip 14:57–62. https://doi.org/10.1039/c3lc50644d

Bersini S et al (2014) A microfluidic 3D in vitro model for specificity of breast cancer metastasis to bone. Biomaterials 35:2454–2461. https://doi.org/10.1016/j.biomaterials.2013.11.050

Bhargava KC, Thompson B, Malmstadt N (2014) Discrete elements for 3D microfluidics. Proc Natl Acad Sci U S A 111:15013–15018. https://doi.org/10.1073/pnas.1414764111

Bhatia SN, Ingber DE (2014) Microfluidic organs-on-chips. Nat Biotechnol 32:760. https://doi.org/10.1038/nbt.2989

Bischel LL, Young EWK, Mader BR, Beebe DJ (2013) Tubeless microfluidic angiogenesis assay with three-dimensional endothelial-lined microvessels. Biomaterials 34:1471–1477. https://doi.org/10.1016/j.biomaterials.2012.11.005

Brancato V et al (2017) Bioengineered tumoral micro-tissues recapitulate desmoplastic reaction of pancreatic cancer. Acta Biomater 49:152–166. https://doi.org/10.1016/j.actbio.2016.11.072

Breslin S, O'Driscoll L (2016) The relevance of using 3D cell cultures, in addition to 2D monolayer cultures, when evaluating breast cancer drug sensitivity and resistance. Oncotarget 7:45745–45756. https://doi.org/10.18632/oncotarget.9935

Breyer MD, Look AT, Cifra A (2015) From bench to patient: model systems in drug discovery. Dis Model Mech 8:1171–1174. https://doi.org/10.1242/dmm.023036

Caballero D, Blackburn SM, de Pablo M, Samitier J, Albertazzi L (2017a) Tumour-vessel-on-a-chip models for drug delivery. Lab Chip 17:3760–3771. https://doi.org/10.1039/c7lc00574a

Caballero D, Kaushik S, Correlo VM, Oliveira JM, Reis RL, Kundu SC (2017b) Organ-on-chip models of cancer metastasis for future personalized medicine: from chip to the patient. Biomaterials 149:98–115. https://doi.org/10.1016/j.biomaterials.2017.10.005

Carvalho MR, Lima D, Reis RL, Oliveira JM, Correlo VM (2017) Anti-cancer drug validation: the contribution of tissue engineered models. Stem Cell Rev Rep 13:347–363. https://doi.org/10.1007/s12015-017-9720-x

Chen Wen LK et al (2017) Integrated gut/liver microphysiological systems elucidates inflammatory inter-tissue crosstalk. Biotechnol Bioeng 114:2648–2659. https://doi.org/10.1002/bit.26370

Chen MB, Lamar JM, Li R, Hynes RO, Kamm RD (2016a) Elucidation of the roles of tumor integrin β1 in the extravasation stage of the metastasis cascade. Cancer Res 76:2513–2524. https://doi.org/10.1158/0008-5472.can-15-1325

Chen Z-z et al (2016b) Establishment of a gastric cancer subline with high metastatic potential using a novel microfluidic system. Sci Rep 6:38376. https://doi.org/10.1038/srep38376

Chen M, Whisler J, Fröse J, Yu C, Shin Y, Kamm R (2017) On-chip human microvasculature assay for visualization and quantification of tumor cell extravasation dynamics. Nat Protoc 12:865–880

Coppeta JR et al (2017) A portable and reconfigurable multi-organ platform for drug development with onboard microfluidic flow control. Lab Chip 17:134–144. https://doi.org/10.1039/c6lc01236a

Cui X, Guo W, Sun Y, Sun B, Hu S, Sun D, Lam RHW (2017) A microfluidic device for isolation and characterization of transendothelial migrating cancer cells. Biomicrofluidics 11:014105. https://doi.org/10.1063/1.4974012

Cukierman E, Pankov R, Stevens DR, Yamada KM (2001) Taking cell-matrix adhesions to the third dimension.

Science 294:1708–1712. https://doi.org/10.1126/science.1064829

DiMasi JA, Grabowski HG, Hansen RW (2016) Innovation in the pharmaceutical industry: new estimates of R&D costs. J Health Econ 47:20–33. https://doi.org/10.1016/j.jhealeco.2016.01.012

Doxzen K et al (2013) Guidance of collective cell migration by substrate geometry. Integr Biol 5:1026–1035. https://doi.org/10.1039/c3ib40054a

Doyle AD, Wang FW, Matsumoto K, Yamada KM (2009) One-dimensional topography underlies three-dimensional fibrillar cell migration. J Cell Biol 184:481–490. https://doi.org/10.1083/jcb.200810041

Edelman LB, Eddy JA, Price ND (2010) In silico models of cancer Wiley interdisciplinary. Rev Syst Biol Med 2:438–459. https://doi.org/10.1002/wsbm.75

Edington CD et al (2018) Interconnected microphysiological systems for quantitative biology and pharmacology studies. Sci Rep 8:4530. https://doi.org/10.1038/s41598-018-22749-0

Esch MB, Smith AS, Prot JM, Oleaga C, Hickman JJ, Shuler ML (2014) How multi-organ micro-devices can help foster drug development. Adv Drug Deliv Rev 69–70:158–169

Esch EW, Bahinski A, Huh D (2015) Organs-on-chips at the frontiers of drug discovery. Nat Rev Drug Discov 14:248. https://doi.org/10.1038/nrd4539

Ferreira MM, Ramani VC, Jeffrey SS (2016) Circulating tumor cell technologies. Mol Oncol 10:374–394. https://doi.org/10.1016/j.molonc.2016.01.007

Fitzpatrick S (2017) 'Organs-on-chips' technology: FDA testing groundbreaking science. U.S. Food & Drug Administration, Silver Spring

Frey O, Misun PM, Fluri DA, Hengstler JG, Hierlemann A (2014) Reconfigurable microfluidic hanging drop network for multitissue interaction and analysis. Nat Commun 30:4250. https://doi.org/10.1038/ncomms5250

Grosberg A, Alford PW, McCain ML, Parker KK (2011) Ensembles of engineered cardiac tissues for physiological and pharmacological study: heart on a chip. Lab Chip 11:4165–4173. https://doi.org/10.1039/c1lc20557a

Hampton T (2017) Body-on-a-chip reveals multitissue interactions during drug exposure. JAMA 318:2069–2070. https://doi.org/10.1001/jama.2017.17869

Hassell BA, Goyal G, Lee E, Sontheimer-Phelps A, Levy O, Chen CS, Ingber DE (2017) Human organ chip models recapitulate orthotopic lung cancer growth, therapeutic responses, and tumor dormancy in vitro. Cell Rep 21:508–516. https://doi.org/10.1016/j.celrep.2017.09.043

Hay M, Thomas DW, Craighead JL, Economides C, Rosenthal J (2014) Clinical development success rates for investigational drugs. Nat Biotechnol 32:40–51. https://doi.org/10.1038/nbt.2786

Huh D, Matthews BD, Mammoto A, Montoya-Zavala M, Hsin HY, Ingber DE (2010) Reconstituting organ-level lung functions on a chip. Science 328:1662–1668. https://doi.org/10.1126/science.1188302

Huh D, Hamilton GA, Ingber DE (2011) From 3D cell culture to organs-on-chips. Trends Cell Biol 21:745–754. https://doi.org/10.1016/j.tcb.2011.09.005

Hwan LS, Hwan SJ (2018) Organ-on-a-chip technology for reproducing multiorgan physiology. Adv Healthc Mater 7:1700419. https://doi.org/10.1002/adhm.201700419

Jeon JS, Bersini S, Gilardi M, Dubini G, Charest JL, Moretti M, Kamm RD (2015) Human 3D vascularized organotypic microfluidic assays to study breast cancer cell extravasation. Proc Natl Acad Sci U S A 112:214–219. https://doi.org/10.1073/pnas.1417115112

Junaid A, Mashaghi A, Hankemeier T, Vulto P (2017) An end-user perspective on organ-on-a-chip: assays and usability aspects Curr Op. Biomed Eng 1:15–22. https://doi.org/10.1016/j.cobme.2017.02.002

Kamei K-i et al (2017) Integrated heart/cancer on a chip to reproduce the side effects of anti-cancer drugs in vitro. RSC Adv 7:36777–36786. https://doi.org/10.1039/c7ra07716e

Katt ME, Placone AL, Wong AD, Xu ZS, Searson PC (2016) In vitro tumor models: advantages, disadvantages, variables, and selecting the right platform. Front Bioeng Biotechnol 4. https://doi.org/10.3389/fbioe.2016.00012

Kilic O et al (2016) Brain-on-a-chip model enables analysis of human neuronal differentiation and chemotaxis. Lab Chip 16:4152–4162. https://doi.org/10.1039/c6lc00946h

Kim HJ, Huh D, Hamilton G, Ingber DE (2012) Human gut-on-a-chip inhabited by microbial flora that experiences intestinal peristalsis-like motions and flow. Lab Chip 12:2165–2174. https://doi.org/10.1039/c2lc40074j

Kim S, LesherPerez S, Kim B, Yamanishi C, Labuz J, Leung B, Takayama S (2016) Pharmacokinetic profile that reduces nephrotoxicity of gentamicin in a perfused kidney-on-a-chip. Biofabrication 8:015021. https://doi.org/10.1088/1758-5090/8/1/015021

Kong J et al (2016) A novel microfluidic model can mimic organ-specific metastasis of circulating tumor cells. Oncotarget 7:78421–78432. https://doi.org/10.18632/oncotarget.9382.

Lee S-A, No DY, Kang E, Ju J, Kim D-S, Lee S-H (2013) Spheroid-based three-dimensional liver-on-a-chip to investigate hepatocyte-hepatic stellate cell interactions and flow effects. Lab Chip 13:3529–3537. https://doi.org/10.1039/c3lc50197c

Lee H, Park W, Ryu H, Jeon NL (2014) A microfluidic platform for quantitative analysis of cancer angiogenesis and intravasation. Biomicrofluidics 8:054102. https://doi.org/10.1063/1.4894595

Lei KF, Tseng H-P, Lee C-Y, Tsang N-M (2016) Quantitative study of cell invasion process under extracellular stimulation of cytokine in a microfluidic device. Sci Rep 6:25557. https://doi.org/10.1038/srep25557

Loskill P, Marcus SG, Mathur A, Reese WM, Healy KE (2015) μOrgano: a lego®-like plug & play system for modular multi-organ-chips. PLoS One 10:e0139587. https://doi.org/10.1371/journal.pone.0139587

Low LA, Tagle DA (2018) 'You-on-a-chip' for precision medicine. Expert Rev Precis Med Drug Dev 3:137–146. https://doi.org/10.1080/23808993.2018.1456333

Ma Y-HV, Middleton K, You L, Sun Y (2018) A review of microfluidic approaches for investigating cancer extravasation during metastasis. Microsys Nanoeng 4:17104. https://doi.org/10.1038/micronano.2017.104

Mak IWY, Evaniew N, Ghert M (2014) Lost in translation: animal models and clinical trials in cancer treatment. Am J Transl Res 6:114–118

Marturano-Kruik A et al (2018) Human bone perivascular niche-on-a-chip for studying metastatic colonization. Proc Natl Acad Sci U S A. https://doi.org/10.1073/pnas.1714282115

Maschmeyer I et al (2015) A four-organ-chip for interconnected long-term co-culture of human intestine, liver, skin and kidney equivalents. Lab Chip 15:2688–2699. https://doi.org/10.1039/c5lc00392j

Miller PG, Shuler ML (2016) Design and demonstration of a pumpless 14 compartment microphysiological system. Biotechnol Bioeng 113:2213–2227. https://doi.org/10.1002/bit.25989

Mosig AS (2017) Organ-on-chip models: new opportunities for biomedical research. Future Sci OA 3:FSO130. https://doi.org/10.4155/fsoa-2016-0038

Mullard A (2016) Parsing clinical success rates. Nat Rev Drug Discov 15:447. https://doi.org/10.1038/nrd.2016.136

Nagaraju S, Truong D, Mouneimne G, Nikkhah M (2018) Microfluidic tumor–vascular model to study breast cancer cell invasion and intravasation. Adv Healthc Mater 0:1701257. https://doi.org/10.1002/adhm.201701257

Oleaga C et al (2016) Multi-organ toxicity demonstration in a functional human in vitro system composed of four organs. Sci Rep 6:20030. https://doi.org/10.1038/srep20030

Osaki T, Sivathanu V, Kamm RD (2018) Vascularized microfluidic organ-chips for drug screening, disease models and tissue engineering. Curr Opin Biotechnol 52:116–123. https://doi.org/10.1016/j.copbio.2018.03.011

Pauty J et al (2018) A vascular endothelial growth factor-dependent sprouting angiogenesis assay based on an in vitro human blood vessel model for the study of anti-angiogenic drugs. EBioMedicine 27:225–236. https://doi.org/10.1016/j.ebiom.2017.12.014

Rape AD, Kumar S (2014) A composite hydrogel platform for the dissection of tumor cell migration at tissue interfaces. Biomaterials 35:8846–8853. https://doi.org/10.1016/j.biomaterials.2014.07.003

Riedl A et al (2017) Comparison of cancer cells in 2D vs 3D culture reveals differences in AKT–mTOR–S6K signaling and drug responses. J Cell Sci 130:203–218. https://doi.org/10.1242/jcs.188102

Rigat-Brugarolas LG et al (2014) A functional micro-engineered model of the human splenon-on-a-chip. Lab Chip 14:1715–1724. https://doi.org/10.1039/c3lc51449h

Rogal J, Probst C, Loskill P (2017) Integration concepts for multi-organ chips: how to maintain flexibility?! Future Sci OA 3:FSO180. https://doi.org/10.4155/fsoa-2016-0092

Ronaldson-Bouchard K, Vunjak-Novakovic G (2018) Organs-on-a-chip: a fast track for engineered human tissues in drug development. Cell Stem Cell 22:310–324. https://doi.org/10.1016/j.stem.2018.02.011

Seruga B, Ocana A, Amir E, Tannock IF (2015) Failures in phase III: causes and consequences. Clin Cancer Res 21:4552–4560. https://doi.org/10.1158/1078-0432.ccr-15-0124

Sharma VP, Beaty BT, Cox D, Condeelis JS, Eddy RJ (2014) An in vitro one-dimensional assay to study growth factor-regulated tumor cell–macrophage interaction. Methods Mol Biol Clifton, NJ 1172:115–123. https://doi.org/10.1007/978-1-4939-0928-5_10

Shu Fan Z, Singaram G, Yuan Hao X, Sally KYT, Alice STW, Stella WP, Yun Wah L (2017) Substrates with patterned topography reveal metastasis of human cancer cells. Biomed Mater 12:055001. https://doi.org/10.1088/1748-605X/aa785d

Singh VK, Kalsan M, Kumar N, Saini A, Chandra R (2015) Induced pluripotent stem cells: applications in regenerative medicine, disease modeling, and drug discovery. Front Cell Dev Biol 3:2. https://doi.org/10.3389/fcell.2015.00002

Siramshetty VB, Nickel J, Omieczynski C, Gohlke B-O, Drwal MN, Preissner R (2016) WITHDRAWN – a resource for withdrawn and discontinued drugs. Nucleic Acids Res 44:D1080–D1086

Skardal A, Devarasetty M, Forsythe S, Atala A, Soker S (2016a) A reductionist metastasis-on-a-chip platform for in vitro tumor progression modeling and drug screening. Biotechnol Bioeng 113:2020–2032. https://doi.org/10.1002/bit.25950

Skardal A, Shupe T, Atala A (2016b) Organoid-on-a-chip and body-on-a-chip systems for drug screening and disease modeling. Drug Discov Today 21:1399–1411. https://doi.org/10.1016/j.drudis.2016.07.003

Skardal A et al (2017) Multi-tissue interactions in an integrated three-tissue organ-on-a-chip platform. Sci Rep 7:8837. https://doi.org/10.1038/s41598-017-08879-x

Stock K et al (2016) Capturing tumor complexity in vitro: comparative analysis of 2D and 3D tumor models for drug discovery. Sci Rep 6:28951. https://doi.org/10.1038/srep28951

Tadimety A, Syed A, Nie Y, Long CR, Kready KM, Zhang JXJ (2017) Liquid biopsy on chip: a paradigm shift towards the understanding of cancer metastasis. Integr Biol 9:22–49. https://doi.org/10.1039/c6ib00202a

Theberge AB, Yu J, Young EWK, Ricke WA, Bushman W, Beebe DJ (2015) Microfluidic multiculture assay to analyze biomolecular signaling in angiogenesis. Anal Chem 87:3239–3246. https://doi.org/10.1021/ac503700f

Tsamandouras N, Chen WLK, Edington CD, Stokes CL, Griffith LG, Cirit M (2017) Integrated gut and liver

microphysiological systems for quantitative in vitro pharmacokinetic studies. AAPS J 19:1499–1512. https://doi.org/10.1208/s12248-017-0122-4
Valastyan S, Weinberg RA (2011) Tumor metastasis: molecular insights and evolving paradigms. Cell 147:275–292. https://doi.org/10.1016/j.cell.2011.09.024
Vavourakis V, Wijeratne PA, Shipley R, Loizidou M, Stylianopoulos T, Hawkes DJ (2017) A validated multiscale in-silico model for mechano-sensitive tumour angiogenesis and growth. PLoS Comput Biol 13:e1005259. https://doi.org/10.1371/journal.pcbi.1005259
Vernetti L et al (2017) Functional coupling of human microphysiology systems: intestine, liver, kidney proximal tubule, blood-brain barrier and skeletal muscle. Sci Rep 7:42296. https://doi.org/10.1038/srep42296
W.H.O (2018) Cancer. http://www.who.int/cancer/en/
Wang YI, Carmona C, Hickman JJ, Shuler ML (2018) Multiorgan microphysiological systems for drug development: strategies, advances, and challenges. Adv Healthc Mater 7:1701000. https://doi.org/10.1002/adhm.201701000
Waring MJ et al (2015) An analysis of the attrition of drug candidates from four major pharmaceutical companies. Nat Rev Drug Discov 14:475. https://doi.org/10.1038/nrd4609
Warkiani ME, Khoo BL, Wu L, Tay AKP, Bhagat AAS, Han J, Lim CT (2016) Ultra-fast, label-free isolation of circulating tumor cells from blood using spiral microfluidics. Nat Protoc 11:134–148. https://doi.org/10.1038/nprot.2016.003
Watson DE, Hunziker R, Wikswo JP (2017) Fitting tissue chips and microphysiological systems into the grand scheme of medicine, biology, pharmacology, and toxicology. Exp Biol Med 242:1559–1572. https://doi.org/10.1177/1535370217732765
Williamson A, Singh S, Fernekorn U, Schober A (2013) The future of the patient-specific body-on-a-chip. Lab Chip 13:3471–3480. https://doi.org/10.1039/c3lc50237f
Xu Z et al (2016) Design and construction of a multi-organ microfluidic chip mimicking the in vivo microenvironment of lung cancer metastasis. ACS Appl Mater Interfaces 8:25840–25847. https://doi.org/10.1021/acsami.6b08746
Zervantonakis IK, Hughes-Alford SK, Charest JL, Condeelis JS, Gertler FB, Kamm RD (2012) Three-dimensional microfluidic model for tumor cell intravasation and endothelial barrier function. Proc Natl Acad Sci U S A 109:13515–13520. https://doi.org/10.1073/pnas.1210182109
Zhang Q, Liu T, Qin J (2012) A microfluidic-based device for study of transendothelial invasion of tumor aggregates in realtime. Lab Chip 12:2837–2842. https://doi.org/10.1039/c2lc00030j
Zhang YS et al (2017) Multisensor-integrated organs-on-chips platform for automated and continual in situ monitoring of organoid behaviors. Proc Natl Acad Sci U S A 114:E2293–E2302. https://doi.org/10.1073/pnas.1612906114

5 Biomaterials and Microfluidics for Liver Models

Alain da Silva Morais, Joaquim M. Oliveira, and Rui L. Reis

Abstract

Over the past years, important progresses have been made in the field of tissue engineering. Many of the early trials to improve the development of an engineered tissue construct were centered on the concept of seeding cells onto biomaterial scaffold. By means of innovative manufacturing machineries, the conception of a preformed scaffold became possible. Nowadays, several tissue engineering challenges are associated with applying this scaffold technology to one vital organ construct: liver. The development of microscale tissue ("micro-tissue") constructs to mimic partially the complex structure–function interactions of liver parenchyma have been obtained through the engineering of sophisticated biomaterial scaffolds, liver-cell sources, and in vitro culture techniques. For in vitro applications, micro-tissue constructs are being upgraded into cell-based assays for testing acute, chronic and idiosyncratic toxicities of drugs or pathogens. The present chapter will focus on the biomaterials currently used for the development of in vitro liver constructs as well as the description of the microfluidic-based models that show great promise for liver regenerative medicine approaches.

Keywords

Biomaterials · Decellularized matrices · Liver models · Microfluidics · Natural polymers · Hydrogels

A. da Silva Morais (✉)
3B's Research Group, I3Bs – Research Institute on Biomaterials, Biodegradables and Biomimetics, University of Minho, Headquarters of the European Institute of Excellence on Tissue Engineering and Regenerative Medicine, AvePark, Parque de Ciência e Tecnologia, Zona Industrial da Gandra, Barco, Guimarães, Portugal

ICVS/3B's–PT Government Associate Laboratory, Braga, Guimarães, Portugal
e-mail: alain.morais@i3bs.uminho.pt

J. M. Oliveira · R. L. Reis
3B's Research Group, I3Bs – Research Institute on Biomaterials, Biodegradables and Biomimetics, University of Minho, Headquarters of the European Institute of Excellence on Tissue Engineering and Regenerative Medicine, Guimarães, Portugal

ICVS/3B's PT Government Associate Lab, Braga, Guimarães, Portugal

The Discoveries Centre for Regenerative and Precision Medicine, Headquarters at University of Minho, Guimarães, Portugal

Highlights

- Micro-technology approaches, combining specific biomaterials and microfluidic

J. M. Oliveira, R. L. Reis (eds.), *Biomaterials- and Microfluidics-Based Tissue Engineered 3D Models*, Advances in Experimental Medicine and Biology 1230,
https://doi.org/10.1007/978-3-030-36588-2_5

platforms, present a high potential in the creation of liver models.
- Most promising approaches, some already applied *in vivo*, are highlighted.

5.1 Introduction

Over the past years, important progresses have been made in the field of tissue engineering. Many of the early trials to improve the development of an engineered tissue construct were centered on the concept of seeding cells onto biomaterial scaffold. By means of innovative manufacturing machineries, the conception of a preformed scaffold became possible.

The liver, the largest solid organ of the body, plays a central role in the maintenance of metabolic homeostasis and the biotransformation and elimination of xenobiotic substances from the body, associated with endocrine and exocrine activities. Despite his highly regenerative potential, damages to hepatocytes by drug and viral infections or toxins, resulted in a significant reduction of functionality and regeneration capability (Protzer et al. 2012). Liver failure can be divided into acute liver failure (ALF), in the absence of pre-existing liver disease, chronic decompensation of an end-stage liver disease, or Acute-on-chronic liver failure (ACLF) of known or unknown underlying chronic liver disease (Kim and Kim 2013). To understand the burden of liver disease in Europe, the incidence and prevalence of two end-stage of liver pathology have been studied, cirrhosis and primary liver cancer. Liver diseases have been found to contribute significantly to the global burden of mortality and morbidity (Collaborators 2015) causing approximately 2% of all deaths. World Health Organization (WHO) estimate that hepatocellular carcinoma (HCC) accounts 47,000 deaths in Europe with a growing incidence over the next 10–15 years, and represents more than 80% of primary liver cancers (Siegel et al. 2012). The most common causes of HCC and cirrhosis in Europe are alcohol, viral hepatitis B and C and metabolic syndromes related to overweight and obesity (Singal and Anand 2013; Poh et al. 2015).

Nowadays, orthotopic liver transplantation (OLT) is the only clinical accepted therapeutic option for patients with acute or advanced liver disease (Burra and Freeman 2012; Dutkowski et al. 2015). However this surgical procedure is limited due to low organ availability and the requirements for donor livers (Adam et al. 2012), resulting in a continuing need for alternative strategies for liver transplantation. Among these, hepatocytes transplantation (Hughes et al. 2012), extracorporeal liver support (Stange 2011; Stutchfield et al. 2011) and stem cell therapy (Drosos and Kolios 2013) are being used in order to extend the life of patient or to regenerate the hepatic tissue until availability of a donor liver (see Fig. 5.1).

Tissue engineering (TE) is based on the involvement of cells, signals (physically from a bioreactor or chemically from growth factors), and biocompatible scaffolds, providing an ideal environment for cell growth and differentiation with the goal to develop constructs relevant both *in vitro* and *in vivo* (Bajaj et al. 2014). Liver tissue engineering aimed to develop extracorporeal liver-support systems or hepatic regeneration based on cell transplantation or implantable devices (Ananthanarayanan et al. 2011). These strategies planned to integrate these structures of liver function and mimic hepatic structural design array from keeping liver cells in biocompatible and biodegradable scaffolds to a more bottom-up method that involves the elaboration of the microscopic subunit of the whole tissue. The use of biocompatible and biodegradable materials is relevant for TE applications. Those biomaterials play an important role by providing a defined and tunable 3D environment that is required for complex cellular organization and organoid growth *in vitro* or *in vivo*. For liver tissue engineering, the development of new artificial matrices needed to satisfy general requirements of biomaterials, including cell and tissue compatibility. Moreover, the chemical and physical properties of the materials are often customized to encourage cell functions, including liver cells (Liu and Chen 2005). For example, to mimic the hepatic native extracellular matrix (ECM), the stiffness of a developed biomaterial-based matrix is required to be

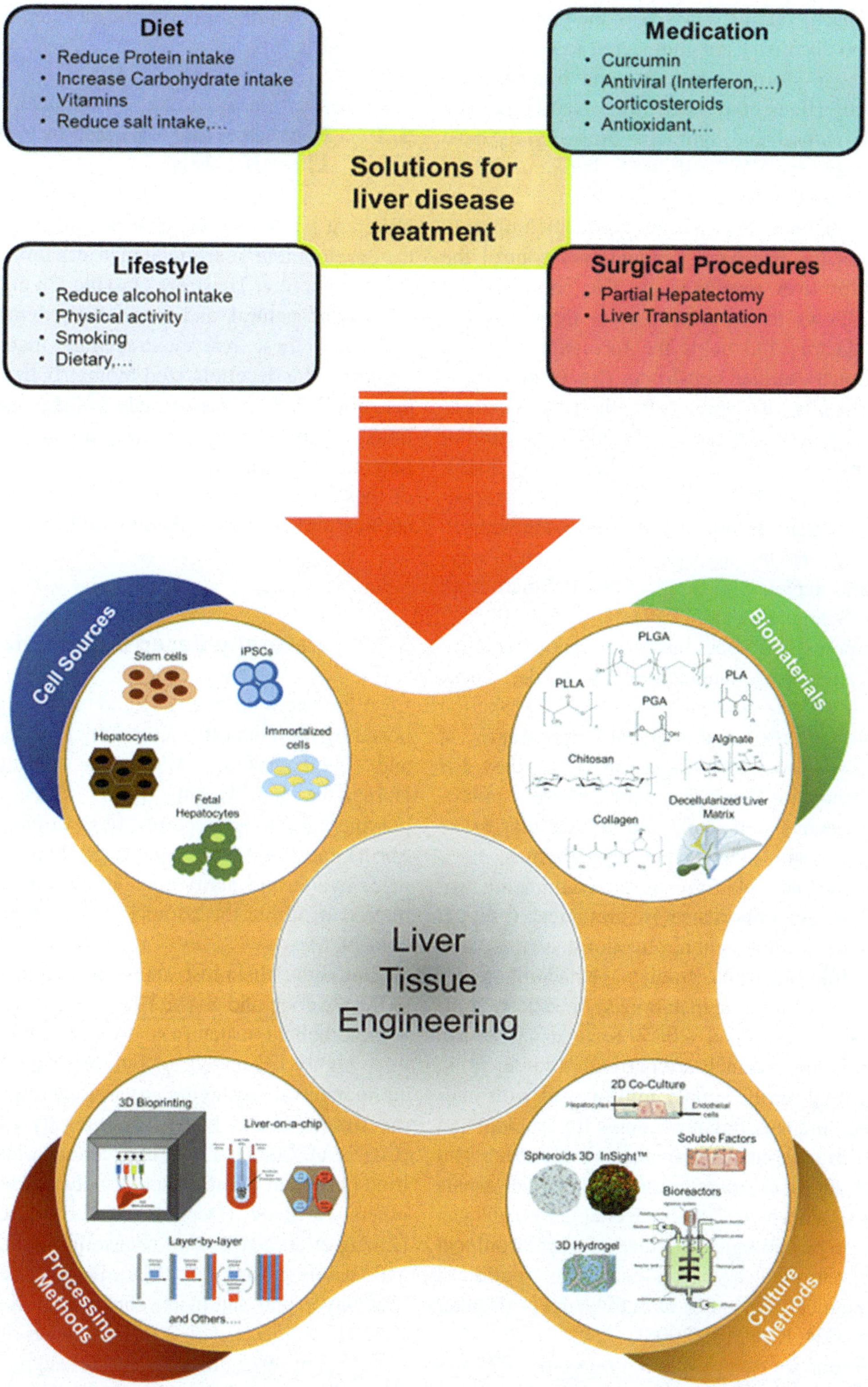

Fig. 5.1 **Rationale Scheme.** Once the conventional therapeutic methods are unsuccessful to treat decompensated liver diseases, there is a continuing need for advanced approaches resulting in the development of liver tissue engineering strategies

soft (Discher et al. 2005). Also, the matrix topology and chemistry are engineered to guide hepatocytes to form spheroids and differentiate correctly (Jiang et al. 2010). For liver TE, a variety of natural and synthetic biomaterials have been assessed (Willerth and Sakiyama-Elbert 2008; Ananthanarayanan et al. 2011).

In combination with biomaterial, the improvement of the microenvironment will control the cells for liver regeneration. The co-culture of cells allows the communication through paracrine factors and helps the formation of more functional cellular constructs (Su et al. 2017). Additionally, the culture of cells in 3D environments allows them to reorganize in a way similar as in native tissue (Knight and Przyborski 2015). Mimic *in vitro* the *in vivo* condition remains possible through the use of bioreactors allowing to finely control the supply and circulation of oxygen and depleted nutrients (Ebrahimkhani et al. 2014; Schuerlein et al. 2017).

Important advances have been made in microfabrication and nanotechnology fields. These aimed to precisely control the spatial distribution of biomolecules and substrate topography at micro- to nanometer resolution. Then, they can be used to engineer extracellular microenvironments, cell shape and intercellular tissue structures with sophisticated topographies.

Microfluidic platforms for cell culture are rapidly gaining importance in biological research and drug development applications, such as stem cell differentiation studies (Khademhosseini et al. 2006) and drug metabolism or toxicity studies (Sivaraman et al. 2005; Kane et al. 2006). Microfluidic channels are particularly interesting since they can be simply multiplexed with integrated fluid handling procedures for efficient and high throughput cellular analysis (Dittrich and Manz 2006), *in situ* monitoring of cellular events (Tourovskaia et al. 2005; Lee et al. 2006) and can mimic a physiological cellular microenvironment with controlled shear stresses and distribution of biochemical molecules at cellular level (Walker et al. 2004; Kim et al. 2006).

Herein, we overview and discuss the advances on the biomaterials as well as on the microfluidic process currently used for the development of liver models.

5.2 Biomaterials – Based *In Vitro* Liver Models

Different types of polymers have been used for the development of artificial matrices for liver tissue engineering. These are classified in main categories of natural and synthetic polymers. In addition to these, native extracellular matrices in the form of a decellularized biomatrix have been also applied. The biomaterials and their applications in liver tissue engineering are summarized herein. The advantages and drawbacks of some of these biomaterials in liver tissue engineering applications were described in Table 5.1.

5.2.1 Naturally-Based Biomaterials

Liver extracellular matrix (ECM) is mainly composed by protein collagen, mostly type I and III, and glycoprotein fibronectin (Martinez-Hernandez and Amenta 1993). Thus, natural polymers such as proteins (e.g. collagen and fibrin) and polysaccharides (e.g. chitosan, alginate, hyaluronic acid) may allow the artificial matrix to mimic the composition of the natural liver ECM.

Collagen, the most abundant component of ECM (Aycock and Seyer 1989), is composed by a triple-helical structure to form a 3D architecture, highly organized, to interrelate with other components then playing an essential role to keep the structural and biological integrity of ECM (Cen et al. 2008). Therefore, collagen has been used in several TE approaches including between others cartilage (Deponti et al. 2013), bone (Zhang et al. 2018), skin (Pensalfini et al. 2017) and vessels (Chan et al. 2016). Despite the use of monolayer and sandwich forms of collagen in liver TE (Zeigerer et al. 2017), collagen alone did not present enough stiffness to support cells (Vasanthan et al. 2012). Moreover, the degradation of collagen is much faster than synthetic

Table 5.1 Biomaterials applied for fabrication of in vitro liver models

Biomaterials		Advantages	Drawbacks	Some liver models (Refs.)
Natural	Collagen	Native component of the ECM liver matrix, biodegradable, biocompatible, presented several cell binding motifs	Lacks mechanical properties, costly	Aycock and Seyer (1989), Cen et al. (2008), Lyu and Untereker (2009), Vasanthan et al. (2012), Chan et al. (2016) and Zeigerer et al. (2017)
	Fibrin	Promotes vascularization, hydrolytically biodegradable	Fast degradation, poor mechanical strength, able to induce immune responses	Bruns et al. (2005), Ahmed et al. (2008), Banihashemi et al. (2015) and Stevens et al. (2017)
	Chitosan	Glycosoaminoglycan-like polysaccharide, supports spheroid formation, biocompatible, promotes high cellular levels of liver specificity	Low mechanical strength, able to induce immune responses	Rinaudo (2006), Seo et al. (2006), Wang et al. (2016) and Fan and Yang (2017)
	Alginate	Hydrophilic, allows cell seeding and microencapsulation, supports spheroid formation, easy handled	Reduced adherence, able to induce immune responses	Dvir-Ginzberg et al. (2003), Lee and Mooney (2012), Alan et al. (2015) and Kim et al. (2017)
	Gelatin	Derived from natural polymer collagen, excellent biocompatibility, non-immunogenic	Dissolved at temperatures at or above 37 °C, made gels at lower temperatures around room temperature	Mariod and Fadul (2013) and Mari et al. (2017)
	Hyaluronic acid	Natural constituent of the liver biomatrix, biocompatible, allows coculture of hepatocytes	Highly viscous, reduced mechanical strength, quickly degraded	Collins and Birkinshaw (2013) and Hemshekhar et al. (2016)
	Silk	Low cost purification, biocompatible, ease of chemical modification	Possibility of minimal inflammatory responses	Naresh and Utpal (2012) and Kundu et al. (2013)
	Elastin	Accurate control over scaffold composition, biodegradable, biologically relevant functionalization, stimuli responsive	Need functionalization to improve cell adhesion	Kanta (2016) and Miranda-Nieves and Chaikof (2017)
	Laminin	Promote hepatogenic differentiation	Expensive	Khalaj et al. (2016), Watanabe et al. (2016) and Garg (2017)

(continued)

Table 5.1 (continued)

Biomaterials		Advantages	Drawbacks	Some liver models (Refs.)
Synthetic	PLA	Allowed hepatocyte aggregation, biocompatible	Acidic degradation product, initiate peptide degradation and inflammation, lack cell recognition signals	Shah Mohammadi et al. (2014) and Perez et al. (2017)
	PLLA	Preserved hepatic function for up to 4 weeks, promote hepatogenic differentiation		Török et al. (2010) and Wang et al. (2013)
	PGA	Good mechanical properties, degradation rate		Kazemnejad (2009) and Knight and Payne (2013)
	PLGA	Biocompatible, biodegradable, can be modulated by changes in ratio of PLLA:PLG, promote stem cell differentiation		Hammond et al. (2006), Makadia and Siegel (2011), Mironov Anton et al. (2016) and Caddeo et al. (2017)
	PCL	Inert, biocompatible, biodegradable	Highly hydrophobic, slow degradation rate	Semnani et al. (2016) and Malikmammadov et al. (2018)
	PDMS	Oxygen permissible membrane, allow high cell seeding density	Highly hydrophobic, permeable to water vapours	Powers et al. (2002) and Sia and Whitesides (2003)
	PEG	Hydrophilic, resistant to protein adsorption, responsive to chemical modification, can be polymerized in presence of cells	Highly hydrophobic, slow degradation rate	Stevens et al. (2015)
	PVA	Inert hydrophilic matrix, preserved functions of cryopreserved hepatocytes	Deficient of cell binding motifs	Bidault et al. (2013), Jain et al. (2015) and Kumar and Han (2017)
	Polyacrylamide	Promotes development of spheroid with controlled size and high yield	Non-biodegradable	Cozzolino et al. (2016)
Other	Decellularized matrix	Whole structural and functional constituents of the native liver matrix stimulates effective cell functions in vitro	Transplanted graft survival ≤8 days, starts clotting in vivo due to exposed collagen	Lee and Cho (2012), Skardal et al. (2012)

PLA Polylactic acid, *PGA* Polyglycolic acid, *PLLA* Poly (L-lactic acid), *PLGA* Poly(lactic-co-glycolic acid), *PCL* Poly (ε-caprolactone), *PDMS* Poly (dimethylsiloxane), *PEG* Polyethylene glycol, *PVA* Polyvinyl alcohol. Source: Adapted with permission from (Jain et al. 2014) Copyright © 2013, Asian Pacific Association for the Study of the Liver

polymers (Lyu and Untereker 2009). These weaknesses limited the use of collagen alone in liver TE.

Fibrin is a soluble protein made and secreted by the liver. Since it can be easily obtained from blood plasma of the own patient, reducing then the risk of immunological reaction or infections, it has been widely used TE applications (Ahmed et al. 2008). Despite the ability of the cells to proliferate and produce ECM within the fibrin scaffold (Banihashemi et al. 2015), the main disadvantages of fibrin are the reduced mechanical properties and fast enzymatic degradation (de la Puente and Ludena 2014). Current approaches for liver TE are based on the use of fibrin gel as matrix (Bruns et al. 2005; Lee et al. 2010a). Recently Stevens et al. (Stevens et al. 2017) used fibrin hydrogel to encapsulate endothelial cell cords and hepatic cellular aggregates in order to create "liver tissue seeds". These liver seed grafts, appropriate for ectopic implantation, were able to significantly increase cell expansion in response of tissue injury (Stevens et al. 2017).

Chitosan is a unique biopolymer obtained from N-deacetylation of the chitin (Rinaudo 2006; Elieh-Ali-Komi and Hamblin 2016) which showed significant analgesic and antibacterial proprieties as well as mucoadhesive capacity (Croisier and Jérôme 2013). Up to now, different chitosan-based architectures were developed including gels, sponges, fibers and films (Croisier and Jérôme 2013) and their potential for tissue engineering application has been assessed (Kim et al. 2008b; Sultana et al. 2015; Agrawal and Pramanik 2016; Dash et al. 2017; Silva et al. 2017). In liver tissue engineering approaches, due to its low bioactivity toward tissue cells and the absence of cell bonding domains, chitosan is frequently modified with functional molecules or combined with other materials (Seo et al. 2006; Cheng et al. 2016; Fan and Yang 2017). As example, the modification of chitosan nanofibrous scaffold with surface-galactose ligands can both reduce the time of degradation of the scaffold and improves the bioactivity of primary hepatocytes *in vitro* (Feng et al. 2009). Also, functionalization of chitosan with lactose moieties showed to improve cell adhesion and that this effect is dependent of the increase of the substitution degree (Wang et al. 2016). However the authors observed that moderate substitution degree of chitosan with lactose moieties is related with better mechanical stability and biocompatibility of the complex, suggesting it as a promising material for liver tissue engineering (Wang et al. 2016). Moreover, Teotia et al. (2015) modified chitosan with porous polysulfone (Psf) and polysulfone-tocopheryl polyethylene glycol succinate (Psf-TPGS). The composite membrane developed improved the adhesion and expansion of HepG2 cells on the external surface, maintaining the internal surface with a decent blood compatibility. Then the authors suggested the use of this hollow fiber to vascularize the bioartificial liver (Teotia et al. 2015).

Alginate is a natural polysaccharide widely used in tissue engineering (Dvir-Ginzberg et al. 2003; Lee and Mooney 2012; Venkatesan et al. 2015) thanks to its source abundance, low price and biocompatibility. *In vitro* studies have assessed the encapsulation of different cell types into alginate scaffolds (e.g. hydrogel, microspheres, gel and bioink). Human induced pluripotent stem cells (hiPSCs) and human embryonic stem cells (hESCs) were printed into alginate hydrogels and, after hepatocyte differentiation, demonstrated good viability as well as liver morphology and functions (Alan et al. 2015). The same results have been observed when encapsulation of mouse primary hepatocytes or hepatocytes spheroids within alginate scaffolds (Miranda et al. 2010; Kim et al. 2017). The impact of the functionalization of alginate has also been assessed. Capsules using galactosylated alginate have been developed for hepatocytes entrapment (Yang et al. 2002; Lou et al. 2017). The authors observed that the encapsulated hepatocytes exhibited higher cell viability and maintained liver-specific functions.

Gelatin and its derivatives is a natural biopolymer derived from animal collagen and is divided in two types: type A obtained from partial acid hydrolysis of animal collagen and type B obtained from alkaline hydrolysis (Mariod and Fadul 2013). However this polymer presented a huge limitation since it is soluble in aqueous solutions. A recent review focused on the development of gelatin-based composites in order to surpass this limitations and modulate the properties of the formulations for tissue engineering applications (Mari et al. 2017). For example, Choi et al. (2014) developed a human gelatin tissue-adhesive hydrogel with haemostatic potential assessed using a haemorrhaging liver rat model. Recently, functionalization of gelatin into a photo-crosslinkable gelatin methacryloyl (GelMA) has been used to create protein-based 3D scaffolds with uniform pore interconnectivity, structural stability and tailorable degradation properties (Lee et al. 2017). The scaffolds developed have been shown to improve *in vitro* hepatocytes attachment and intercellular interaction compared to 2D culture system.

Hyaluronic acid (HA) is the principal component of perisinusoidal space in liver (Nallagangula et al. 2018). Then this non-sulfated glycosaminoglycan represented an excellent choice for application in various tissue engineering strategies (Collins and Birkinshaw 2013; Hemshekhar et al. 2016) including its use as hydrogel for liver application. Catapano et al. used non-woven fabrics of HA esters to culture liver cells (Catapano et al. 2001). These non-wave fabrics were later enriched with components of ECM by Zavan et al. (2005). The authors demonstrated an increase of cell survival of the hepatocytes up to 14 days *in vitro* and up to 35 days after implantation of hepatocytes aggregates (Zavan et al. 2005). Since transdifferentiation of hepatic stellate cells (Ito cells) into myofibroblasts is a key event in hepatic fibrogenesis (Higashi et al. 2017) and the stiffness of the environment has been shown to influence the progression of the fibrosis (Olsen et al. 2011), the methacrylation of HA (MeHA) hydrogel has also been assessed. Studies have demonstrated that MeHA "stiff" hydrogels affected the morphology and differentiation potential of HSC into myofibroblasts (Guvendiren et al. 2014; Caliari et al. 2016).

Other natural-based biomaterials have also been investigated in order to develop new models for liver tissue engineering applications. **Silk** fibroins, which can be obtained from several sources such as silkworms, spiders and flies, have demonstrated huge potential for tissue regeneration (Kundu et al. 2013) including the development of a biomimetic artificial liver extracellular matrix (Naresh and Utpal 2012). **Elastin**, one of the most stable protein found in the ECM of elastic organs like arteries, lung and liver. Liver fibrosis development has been associated to a dysregulation of elastin synthesis and metabolism (Kanta 2016). However, although the presence of elastin in hepatic tissues, the use of elastin-based biomaterials in liver tissue engineering is limited and mostly focused on improving hepatocytes 2D cultures and spheroids formation (Miranda-Nieves and Chaikof 2017). **Laminin**, a structural protein of the ECM of skeletal muscle fibers with a crucial role in tissue development, maintenance and function regulation, was also involved in the development of scaffolds for tissue engineering applications (Garg 2017). Recent studies have focused on the use of laminin-based matrices for the maintenance of primary human hepatocytes (Watanabe et al. 2016), hepatogenic differentiation of human bone marrow mesenchymal stem cells (Khalaj et al. 2016) and hepatic and hepatic specification of human pluripotent stem cells (Kanninen et al. 2016).

A single biomaterial can barely meet all the requirements of hepatic regeneration, thus blends materials appeared to take advantages of each component. Several studies have been based in the use of these composite materials for liver tissue engineering applications. For examples: (i) Lee et al. developed a highly bioactive ink consisting in a mixture of collagen/extracellular matrix (ECM) and alginate (Lee Hyeong et al. 2015), (ii) Shang et al. prepared a galactosylated chitosan (GCs)/HA porous hybrid sponge (Shang et al. 2014), (iii) Tripathi and Melo created a sponge-like biocomposite agarose–chitosan scaffold (Tripathi and Melo 2015), (iv) Deng et al. used an alginate/xyloglucan scaffold (Deng et al. 2015), (v) Nakamura et al. prepared a heparin-conjugated gelatin scaffold able to hold vascular endothelial growth factor (VEGF) (Nakamura et al. 2013), (vi) Arisaka et al. modified a thermo-responsive surface with heparin-binding epidermal growth factor-like growth factor (HB-EGF) (Arisaka et al. 2016), (vii) Yang et al. characterized *in vitro* and *in vivo* silk fibroin (SF)/gelatin composite scaffolds (Zhao et al. 2011), (viii) Kasoju and Bora fabricated SF and GCs nanofibrous mesh (Naresh and Utpal 2012), (ix) Gong et al. developed chitosan/gelatin scaffolds with fine three-dimensional channels (Gong et al. 2014), and (x) Xia et al. developed a bioreactor-based bioartificial liver (BAL) with stacked sandwich culture plates containing hepatocytes together with porous and non-porous collagen-coated PET films (Xia et al. 2012). Some examples of naturally-based *in vitro* models were presented in Fig. 5.2.

5.2.2 Synthetic Biomaterials

While naturally-based biomaterials presented several advantages for liver tissue engineering applications, they still present many deficiencies

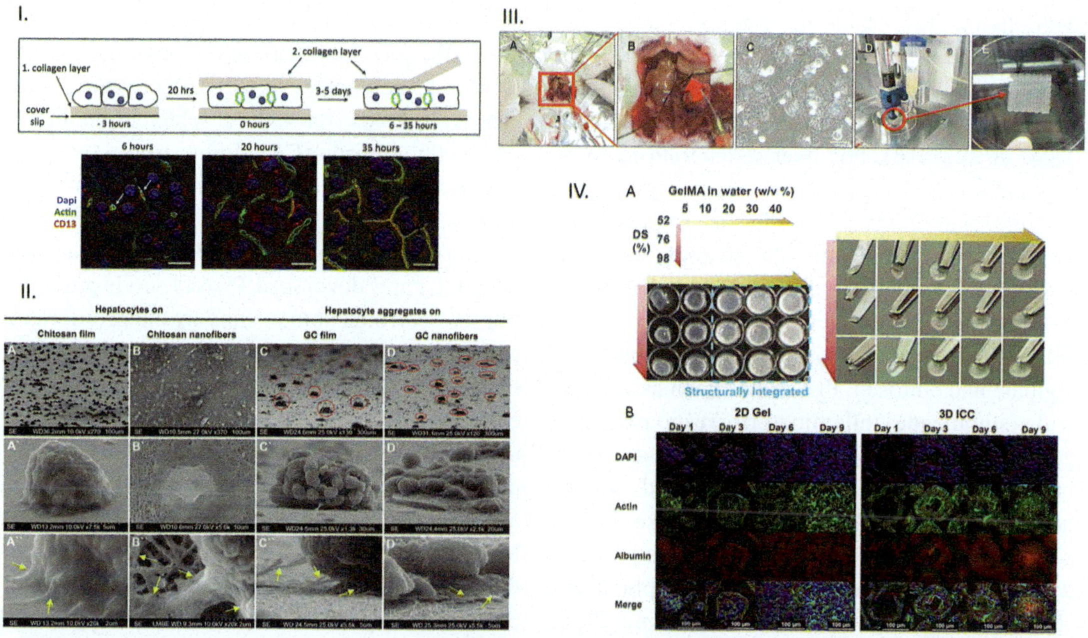

Fig. 5.2 Naturally-based in vitro models for liver tissue engineering applications. (I) Development of cell polarity in collagen sandwich hepatocytes (Reprinted with permission (Zeigerer et al. 2017) © 2016 The Authors. Published by Elsevier Inc.). (II) SEM images of rat primary hepatocytes and hepatocyte aggregates cultured on chitosan and galactosylated chitosan (GC) films and nanofibers after 7 days (Reprinted with permission (Feng et al. 2009) © 2009 Elsevier Ltd. All rights reserved). (III) Isolation of mouse primary hepatocytes and 3-dimensional (3D) bioprinting of alginate scaffold (Reprinted with permission (Kim et al. 2017) © 2017, the Korean Surgical Society. https://creativecommons.org/licenses/by-nc/4.0/). (IV) A. Gelatin methacryloyl-based (GelMA) inverted colloidal crystal (ICC) hydrogels. B. Evaluation of liver-specific functions of Huh7.5 cell constructs in 3D GelMA ICC scaffolds and on 2D GelMA substrates. (Reprinted with permission (Watanabe et al. 2016) © 2018 Springer Nature Limited. All rights reserved. https://creativecommons.org/licenses/by/4.0/)

as for example reduced mechanical properties, limited availability, and variable degradation rate. Thus, as alternative, synthetic biomaterials with tunable degradability and mechanical properties have been developed and assessed. To improve cell adhesion, the synthetic biomaterials were functionalized with polysaccharides, proteins or polypeptides (Vasanthan et al. 2012). The key synthetic biomaterials used for liver tissue engineering applications are aliphatic polyesters (e.g. PLA, PGA, PCL…), polyethers (e.g. PEG) and polyamides (e.g. PVA).

Aliphatic polyesters are synthetic materials which possess ester bonds vulnerable to hydrolysis and which can offer a microenvironment to orient tissue regeneration through improvement of material degradability and mechanical support. **Polylactic acid** (PLA) is a biocompatible and biodegradable aliphatic polyester (Perez et al. 2017). Several studies have applied PLA-based scaffolds for different tissue regeneration applications. For example PLA scaffolds, produced with different architecture, have been widely used for tissue engineering (Shah Mohammadi et al. 2014; Magiera et al. 2017; Xu et al. 2017). Regarding liver applications, Bierwolf et al. demonstrated that PLA nanofibers, created by electrospinning technique, improved proliferation and expression of liver markers by primary rat hepatocytes (Bierwolf et al. 2010). The biodegradable **poly(L-lactic acid)** (PLLA) aliphatic polymer has demonstrated potential to improve hepatocyte expansion and function *in vitro* (Wang et al. 2013) as well as it represented an optimized method for primary hepatocyte transplantation *in vivo* (Török et al. 2010). **Polyglycolic acid** (PGA) is a poly(α-hydroxyl) ester which can be tuned to improve degradation rate and mechanical properties. Studies have been focused on the use of PGA for different tissue engineering applications (Cao and Kuboyama 2010; Knight and Payne 2013; Toosi et al. 2016). For liver regeneration, some encouraging results have been presented (Kazemnejad 2009). In order to improve the mechanical property and the degradation rate, PLA and PGA are frequently blended into the co-polymer **poly(lactic-co-glycolic acid)** (PLGA). Compared to the biocompatibility of alginate and chitosan, PLGA has been associated with significant inflammatory response (Lee et al. 2012). Despite that, PLGA-based scaffolds have been frequently investigated and studies have been focused on the use of these PLGA scaffolds for the development of models for drug delivery and tissue engineering applications (Makadia and Siegel 2011; Pan and Ding 2012; Gentile et al. 2014; Mironov Anton et al. 2016). PLGA has also been assessed from a while in liver studies (Hammond et al. 2006; Caddeo et al. 2017). As example, Liu et al. (2016) developed a PLGA scaffold allowing the co-culture of hepatocytes and mesenchymal stem cells (MSCs). They observed that the differentiation of MSCs and the survival of hepatocytes within the scaffold is dependent of the ratio hepatocytes/MSCs, with the ratio 5:1 respectively showed to be the most efficient. Moreover, they showed that the direct transplantation of the construct could improve the rejection response and the restoration of liver functions in mice model. As for other biomaterials, composites with different PLGA-based compositions and forms have been created such as PLGA combined with collagen, gelatin or plasma treated (Hasirci et al. 2001), and PLGA-based porous scaffolds (Hammond et al. 2011) or microspheres incorporating growth factors (Zhu Xin et al. 2007). **Poly (ε-caprolactone)** (PCL), and the PLC-based biomaterials, are aliphatic polyester used for several biomedical applications based on the good processability and biocompatibility of PCL (Malikmammadov et al. 2018). For liver regeneration purpose applications, several studies have been realized. Zhang et al. (2014) functionalized PCL with galactose to improve PCL' wettability and prepared galactosylated PCL (GPCL) membranes which maintained the physiological functions of hepatocytes *in vitro*. Semnani et al. (2016) created nanofibrous scaffolds through co-electrospinning of PCL with chitosan (CS) aiming to use this composite complex as matrix to enhance liver regeneration. Bishi et al. (2013) synthetized **poly (L-lactic acid)-co-poly(ε-caprolactone)** (PLACL) and developed PLACL/collagen nanofibrous scaffolds able to support the proliferation and differentiation potential of human MSCs into hepatocyte-like cells. Shim et al. (2013) fabricated PCL/PLGA/collagen scaffolds. The authors observed that encapsulated rat primary hepatocytes within the composite scaffolds are viable and able to secrete albumin to up to 10 days. Moreover, PLC has also been used to develop delivery systems of specific proteins. For example, Yu et al. (2014) developed a biodegradable PCL nanofiber matrix able to release the vascular endothelial growth factor (VEGF) which improved, through a concentration-dependent effect, liver regeneration in rats submitted to 70% partial hepatectomy.

A possible drawback for the use of aliphatic polyesters is based on their degradation products which can alter the microenvironment of the cells as well as their natural hydrophobic properties, limiting their application in tissue engineering (Hammond et al. 2006). Therefore others synthetic biomaterials have been considered such as polyethers and polyamides. **Polyethylene glycol** (PEG) is a synthetic polyether which solubility in aqueous solutions and organic solvents contributed to its biocompatibility and processability (Ivanova et al. 2014). Therefore PEG-based hydrogels were widely used in tissue engineering applications (Burdick and Stevens 2005). For liver tissue engineering approaches, a degradable hydrogel has been developed (Stevens et al. 2015). This model was based on PEG-diacrylamide (PEGDAAm) grafted with matrix-metalloproteinase sensitive peptides which improved 3D cell migration while the hydrogel is degraded by cell action. Stevens et al. showed that, after ectopic implantation in nude mice, the hepatic PEGDAAm-based tissue was still present and preserve functionality for

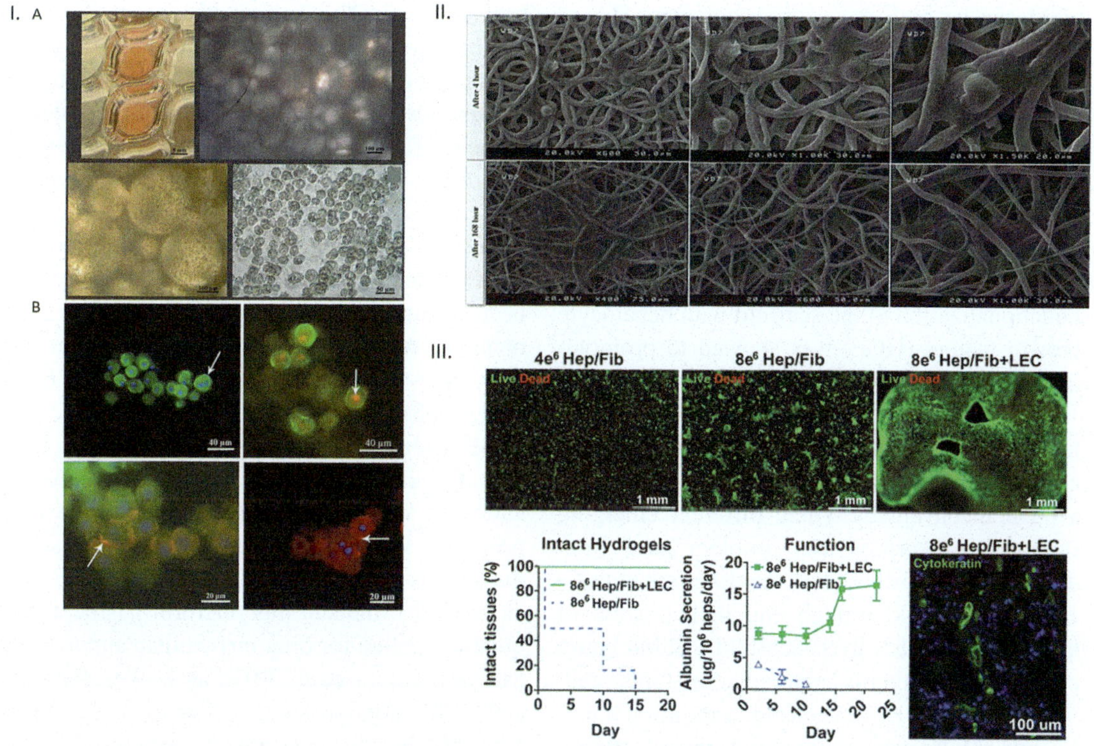

Fig. 5.3 Synthetic-based in vitro models for liver tissue engineering applications. (I) A. Nanofibrous scaffolds seeded with rat primary hepatocytes. B. Immunofluorescent staining of hepatocyte-specific factors (Reprinted with permission (Bierwolf et al. 2010) © 2010, John Wiley and Sons). (II) Scanning electron microscope (SEM) results of Hepa 1–6 epithelial liver mouse cells cultured on PCL/chitosan nanofibers after 4 and 168 h (Reprinted with permission (Semnani et al. 2016) © 2017 Taylor & Francis). (III) Co-culture of rat primary hepatocytes with liver endothelial cells in PEG-diacrylamide hydrogel improves engineered tissue longevity (Reprinted with permission (Stevens et al. 2015) Copyright © 2015 The Authors. Journal of Biomedical Materials Research Part A Published by Wiley Periodicals, Inc.)

more than 3 weeks. **Polyvinyl alcohol** (PVA), a water-soluble polymer, has been widely used for tissue engineering applications (Kumar and Han 2017). For liver regeneration models, different approaches have been investigated since the development of self-supported fibrin-PVA interpenetrating polymer networks (IPN) (Bidault et al. 2013) to macroporous hydrogel consisting of acrylonitrile and N-vinyl-2-pyrrolidone copolymer (PANNVP) and IPN together with chitosan as hepatocyte carriers (Jain et al. 2015). Moreover, application of **polyacrylamide** has also been studied in liver regeneration. Cozzolino et al. (2016) optimized the *in vitro* conditions for hepatic stem cells and differentiated hepatocytes through the modulation of polyacrylamide hydrogels' stiffness. Some examples of synthetic *in vitro* models were presented in Fig. 5.3.

5.2.3 Decellularized Matrices

Decellularization protocol is based on the removal of all cells of the tissue resulting in an extracellular matrix (ECM) scaffold of the native tissue (Lee and Cho 2012). Decellularized matrices, with defined chemical composition and architecture mimicking the targeting tissue but deprived of immune responses, are widely applied in tissue engineering strategies. For example, Skardal et al. showed that extracts from whole fresh liver tissue and from acellular liver

ECM can be both mixed with collagen, HA or heparin-conjugated HA in order to create hydrogels able to maintain primary human hepatocytes functions *in vitro* (Skardal et al. 2012). The authors proposed that this 3D model could be applied for cell therapy and drug and toxicology screening purposes. A similar approach has been applied to develop a liver ECM (L-ECM) scaffold (Nakamura and Ijima 2013). Functionalization of the scaffold with hepatocyte growth factors (HGF) was showed to promote-liver specific functions of hepatocytes.

Even if L-ECM scaffolds are interesting, the lack of a vascular network which is essential for nutrient and metabolites distribution, represented a huge inconvenient. Then, different emerging strategies to surpass this problem have been developed. One is based on the development of a vascular network through endothelialization of the decellularized liver scaffold (Shirakigawa et al. 2013), another is centered in the functionalization of the L-ECM scaffold, presenting a persevered native vasculature, with heparin to reduce the thrombogenicity without affecting the re-cellularization of the scaffold neither the hepatic functions of the cells *in vitro* (Bruinsma et al. 2015). Promising results have also been obtained by repopulation of L-ECM scaffolds with hepatocytes and endothelial progenitor cells (Zhou et al. 2015a) (Fig. 5.4).

Despite the decellularized scaffolds presented huge advantages in their composition and structure compared to other biomaterial-based scaffolds, some challenges persist to be overcome including the development of optimized decellularization protocols in order to keep the native ECM structure and allowing the complete removal of the native cells as well as the improvement of the re-cellularization strategies.

5.3 Microfluidics-Based Liver Models

Microfabrication, and more specifically nanotechnology domain, have recognized important advances aiming to determine precisely, up to the nanometer resolution, the spatial dispersal of the biomolecules (Nikkhah et al. 2012). So, this technology represented a sophisticated tool for the engineering of intercellular tissue structures and extracellular micro-environments. The development of microfluidic technologies has engendered huge interest for a large variety of domains due to the capacity to create highly structured systems to evaluate the physical importance of fluid mechanisms and transport. Then, there has been a huge increase in microfluidics development in terms of improvement of fabrication process and strategies for surface alterations (Kim et al. 2008a).

Microfluidic or lab-on-a-chip platforms have quickly gain significance in biological research and drug development applications, once they propose a valuable technology able to be use for the development of new biomaterials (Barata et al. 2016). Indeed, this platform has been frequently applied for drug metabolism and toxicity assays (Kimura et al. 2018) as well as for stem cell differentiation studies (Zhang et al. 2017). Microfluidic channels are of particular interest once, in order to promote high throughput and useful cellular assay, they can be multiplexed with integrated fluid handling procedures (Lin et al. 2016). Moreover, this procedures have also been applied for the evaluation of cellular distribution of biochemical molecules (Banerjee et al. 2017), to reproduce a physiological cellular microenvironment with controlled shear stresses (Bissoyi et al. 2016), and the monitoring of cellular events *in situ* (Kwapiszewska et al. 2014).

Manufacturing of microfluidics is based on different approaches. Bhattacharjee et al. (2016) have reviewed the following strategies: injection molding, soft lithography, 3D printing and paper microfluidics. After comparing them to each other, the authors concluded that the 3D printing strategy, due to its ability to quickly offer a rapid prototype of a physical model, is the most relevant and then represented a more appropriate technology for specific microfluidic models (Macdonald et al. 2017). Besides the conventional microfluidics models, centrifugal microfluidic or lab-on-a-disc platforms have emerged since the past few years (Tang et al. 2016). These platforms presented several advantages com-

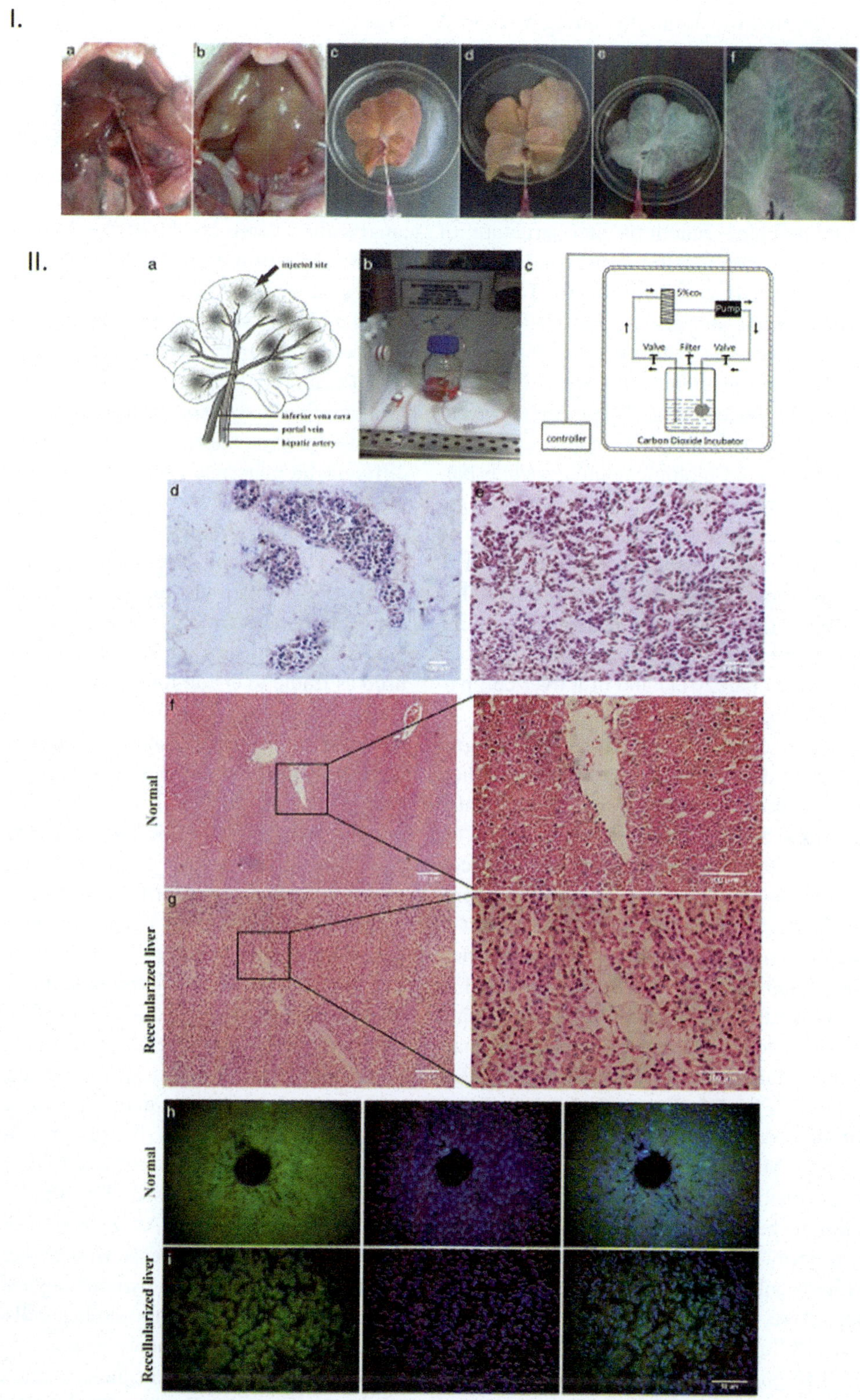

Fig. 5.4 Decellularization and recellularization of rat livers with hepatocytes and endothelial progenitor cells. (I) Sequential whole-organ decellularization progress (a–f). (II) Recellularization process through parenchymal injection method. (Reprinted with permission (Zhou et al. 2015a) © 2015 International Center for Artificial Organs and Transplantation and Wiley Periodicals, Inc.)

pared to the traditional-one including a reduced amount of instrumentation, the proficient removal of residual volumes or unwanted bubbles, as well as a density-based sample separation and transportation (Strohmeier et al. 2015). Microfluidic chip-based organs-on-chips technology has been highlighted due to its capacity to mimic and recreate the complex functions and structure of human organs through the cells inside microfabricated channels (Sun et al. 2016). The use of these microfluidic devices for cell-cell interactions studies and as platforms for theranostic applications has been recently reviewed (Choi et al. 2017; Rothbauer et al. 2018). Another technology recently described and mainly applied for the manipulation of biological units, such as particle separation in microfluidic models, designing, cell electroporation, and isolation of dead versus live cells, is dielectrophoresis (DEP) (Chandrasekaran et al. 2016). This technique has been applied in the development of an artificial microchip composed by liver sinusoid-like structures (Schütte et al. 2011) and of a liver-cell patterning lab chip system (Ho et al. 2013).

The liver, one of the most complex organ, plays a central role in the development of new drugs through the fabrication of 3D bioprinted liver devices. Several studies have described the development of appropriate drugs and gene carriers associated with the rise of different microfluidic devices allowing to assess the impact of drugs under conditions mimicking the biological system (Damiati et al. 2018). As depicted in Fig. 5.5, in the past decade, different hepatic models were developed from microscale primary rat and human hepatocyte culture chips (Lee et al. 2007) (Fig. 5.5-II) to a multi-walled liver microfluidic device (Khetani and Bhatia 2007) (Fig. 5.5-I). More recently, Yu et al. (2017) created a perfusion-incubator-liver-chip (PIC) able to maintain hepatocyte viability and functions for long-term 3D culture. This model could be used for chronic hepatotoxicity testing (Fig. 5.5-II). Therefore, liver-on-a-chip devices were frequently developed to test the acute drug toxicity impact through assessing pharmacokinetic parameters such as bioavailability, cell permeability, and hepatic clearance. Nowadays, the liver-on-a-chip constructs were usually developed to improve the hepatic functions in comparison to the conventional culture models (Harink et al. 2013). Stimulation of primary rat hepatocytes with hepatocyte growth factor (HGF), diffusing through microfluidic channels, improved cell phenotype compared to unstimulated cells (Patel et al. 2015). Moreover, the development of a microfluidic device with liver microsomes encapsulated within a 3D PEG-based hydrogel showed that this model was able to screen the impact of liver metabolism on different compounds (Lee et al. 2013).

Microfluidic technology has also been applied to develop liver microtissues or scaffolds using different biomaterials (Ahadian et al. 2017). In the beginning of the twenty-first century, Powers et al. developed a microfabricated array bioreactor based on a silicon sheet scaffold (Powers et al. 2002). Poly(dimethylsiloxane) (PDMS), a synthetic polymer widely used in the field of biotechnology (Sia and Whitesides 2003), was used as frame to create for example a microfabricated bioreactor (Ostrovidov et al. 2004), a chitosan-based bioartificial liver (Lee et al. 2010b) as well as an extended bile canalicular structure (Goral et al. 2010). These models showed improvement of liver specific functions by the cells used for the assays. Alginate, a natural biomaterial previously described, was also used in combination with PDMS to develop an "organ in a droplet" device (Chen et al. 2016) through the use of a droplet-based microfluidic method. That technique permitted the co-culture of fibroblasts and hepatocytes cell lines in a spatially defined organization resulting in a 3D liver construct.

Nevertheless, the evaluation of the influence of hepatic disorders on hepatocytes functions remained essential. Thus, microfluidic devices have been recently used for this purpose. Alcoholic liver disease (ALD), one of the major cause of chronic liver disease worldwide, is characterized between others by hepatocytes apoptosis, activation of the hepatic stellate cells (HSCs) and alterations in the innate immunity, leading the liver to fibrosis and in most of the cases to cirrhosis (Gao and Bataller 2011). Zhou et al. developed, by the microfluidic co-culture

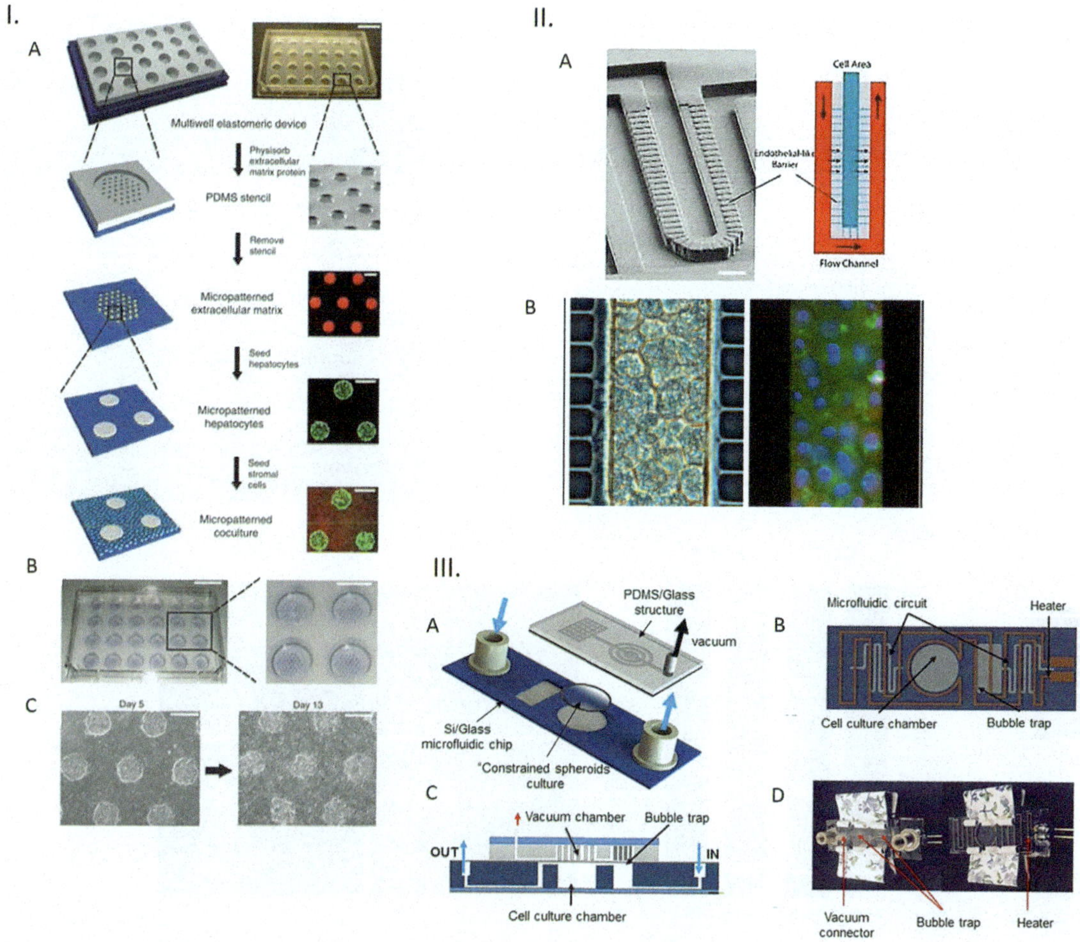

Fig. 5.5 Examples of microfluidics liver models. (I) Soft lithographic process for the fabrication of microscale liver hepatocytes cultures (multi-well format) A. Schematic of the process flow aside photomicrographs taken at each step. B. Photograph of a 24-well device with repeating hepatic microstructures. C. Phase-contrast micrographs of micropatterned co-cultures (Reprinted with permission (Khetani and Bhatia 2007) © 2007, Springer Nature). (II) An artificial liver sinusoid with a microfluidic endothelial-like barrier for primary hepatocyte culture. A. SEM micrograph illustrating the microfluidic sinusoid unit. B. Viability of rat primary hepatocytes when loaded at high density after 7 days (Reprinted with permission (Lee et al. 2007) © 2007 Wiley Periodicals, Inc.). (III) Schematic of the perfusion-incubator-liver-chip (PIC). (Reprinted with permission (Yu et al. 2017) © The Author(s) 2017 http://creativecommons.org/licenses/by/4.0/)

approach, a liver injury-on-a-chip device (Zhou et al. 2015b). The authors demonstrated that alcohol induced secretion of transforming growth factor beta (TGF-β) by the injured hepatocytes resulting in a paracrine activation of the HSCs. Non-alcoholic Fatty Liver Disease (NAFLD) was also assessed using the microfluidic approach. NAFLD, a chronic liver disease characterized by the accumulation of lipids in the hepatocytes and related to insulin resistance (Leclercq et al. 2007), can lead from simple steatosis to hepatocellular carcinoma. Gori et al. fabricated the first *in vitro* on-chip model of NAFLD established within a microfluidic device in a sinusoid-like custom (Gori et al. 2016) (Fig. 5.6). This model will allow to understand the cellular, molecular and epigenetic mechanisms that coordinate NAFLD development.

Blood vessels are vital for organ survival since they manage the delivery of nutrients and oxygen

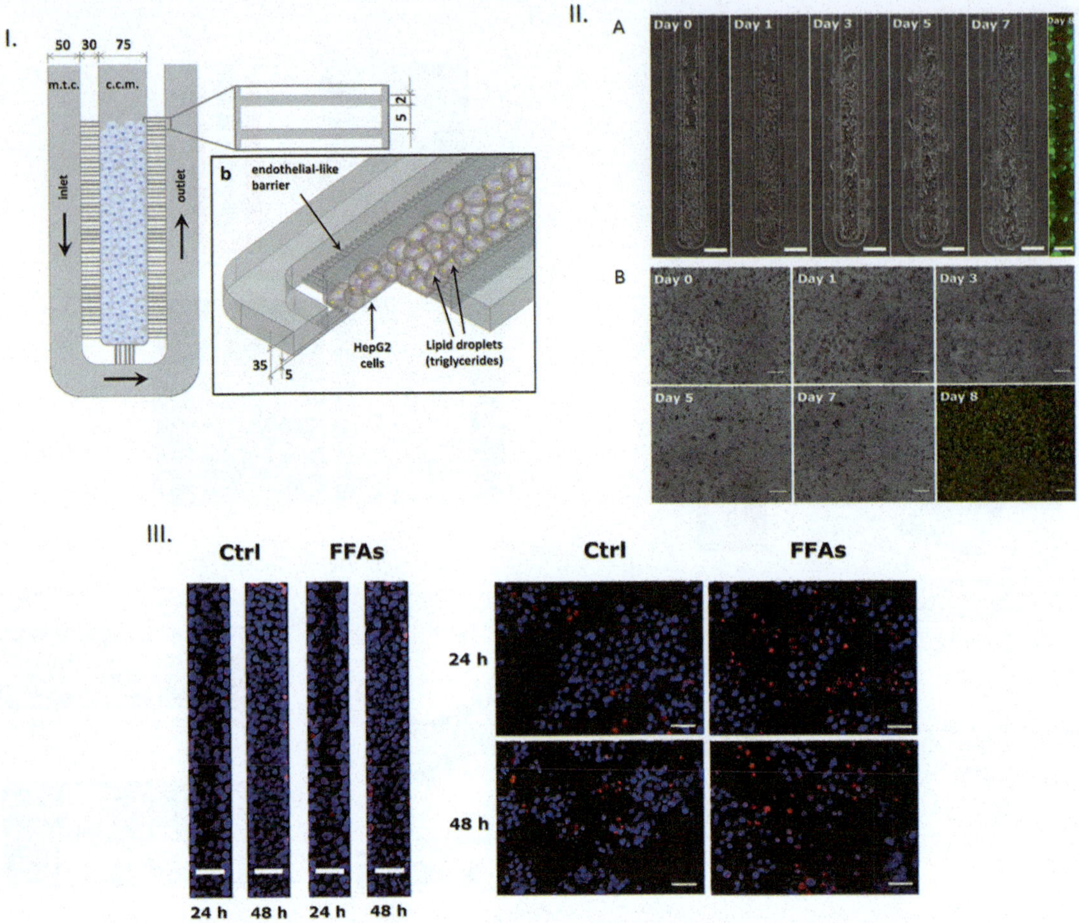

Fig. 5.6 Liver-on-a-Chip microfluidic model to investigate NAFLD. (I) Microarchitecture and geometric configuration of the NAFLD-on-a-chip device. (II) Evaluation of cell viability/cytotoxicity (A) and cell morphology (B) over a week in culture. (III) Evaluation of cell viability/cytotoxicity after treatment of HepG2 cells with free fatty acids (FFAs). (Reprinted with permission (Gori et al. 2016) © 2016 Gori et al. http://creativecommons.org/licenses/by/4.0/)

through the body (Ding et al. 2010). So, studies have been focused in developing organ-on-chip devices integrating the vasculature for further application in liver tissue engineering (Zhang et al. 2016). Moreover, recent investigations have focused on the development of vasculature-on-a-chip devices for drug delivery applications *in vivo* (Goudie et al. 2016; Uguz et al. 2017). However, until now no implantable microfluidic models have been established for liver regeneration application.

5.4 Conclusions

Liver, the largest organ of the body, has a central role in the control of metabolic homeostasis and the xenobiotic metabolism from the body, associated with endocrine and exocrine activities. However, despite the high regenerative potential, liver injuries resulted in significant impact on functionality and regeneration capability leading to liver failure. Nowadays, the only clinically accepted therapeutic option for patients with advanced liver disease remains the orthotopic liver transplantation, which has the disadvantage to be limited due to low organ availability. Therefore, currently, there is a continuous need for alternative strategies.

Over the last decade, several approaches have been developed to improve or replace, even only temporarily, the crucial liver functions in order to increase patient's life or to regenerate the hepatic tissue until find a compatible organ. These approaches include hepatocytes transplantation, extracorporeal liver support and stem cells therapy. Besides those, liver tissue engineering and regenerative medicine (TERM) strategies attempt to forward prominently. Even if the choice of an adequate source of cells represents an important step for liver TERM, the use of appropriate supports and the development of adequate culture conditions were essential for future clinical applications. Now, the use of natural and synthetic-based biomaterials were broadly applied for the development of liver TERM strategies. But, despite the potential and advantages presented by both types, new biomaterials have been investigated. Then, new approaches have been exploiting the use of decellularized liver matrices and nanomaterials showing both huge potential to widely support the development of advanced devices for clinical therapeutic applications and TERM.

Micro-technology approaches propose important advantages in the creation of liver models, essentially in terms of increment of hepatic microscale systems. Microfluidic platforms are highly attractive and efficient technologies to mimic liver microenvironment and microstructure. Therefore, several companies have as key goal the development and commercialization of chip devices, including organ-on-chip models, mainly for drug screening application. However, until now, only few models have been designed for *in vivo* liver regeneration assays. In the future, it is projected that construction and transplantation of such kind of micro-structures may sustain or surpass many liver functions in order to be considered as a liver regenerative medicine approach.

References

Adam R, Karam V et al (2012) Evolution of indications and results of liver transplantation in Europe. A report from the European Liver Transplant Registry (ELTR). J Hepatol 57(3):675–688

Agrawal P, Pramanik K (2016) Chitosan-poly(vinyl alcohol) nanofibers by free surface electrospinning for tissue engineering applications. Tissue Eng Regen Med 13(5):485–497

Ahadian S, Civitarese R et al (2017) Organ-on-a-Chip platforms: A convergence of advanced materials, cells, and microscale technologies. Adv Healthc Mater 7(2):1700506

Ahmed TAE, Dare EV et al (2008) Fibrin: a versatile scaffold for tissue engineering applications. Tissue Eng Part B Rev 14(2):199–215

Alan F-J, Catherine F et al (2015) Bioprinting of human pluripotent stem cells and their directed differentiation into hepatocyte-like cells for the generation of mini-livers in 3D. Biofabrication 7(4):044102

Ananthanarayanan A, Narmada BC et al (2011) Purpose-driven biomaterials research in liver-tissue engineering. Trends Biotechnol 29(3):110–118

Arisaka Y, Kobayashi J et al (2016) A heparin-modified thermoresponsive surface with heparin-binding epidermal growth factor-like growth factor for maintaining hepatic functions in vitro and harvesting hepatocyte sheets. Regen Ther 3:97–106

Aycock RS, Seyer JM (1989) Collagens of normal and cirrhotic human liver. Connect Tissue Res 23(1):19–31

Bajaj P, Schweller RM et al (2014) 3D biofabrication strategies for tissue engineering and regenerative medicine. Annu Rev Biomed Eng 16:247–276

Banerjee A, Weaver I et al (2017) Bioelectronic measurement and feedback control of molecules in living cells. Sci Rep 7(1):12511

Banihashemi M, Mohkam M et al (2015) Optimization of three dimensional culturing of the HepG2 cell line in fibrin scaffold. Hepat Mon 15(3):e22731

Barata D, van Blitterswijk C et al (2016) High-throughput screening approaches and combinatorial development of biomaterials using microfluidics. Acta Biomater 34:1–20

Bhattacharjee N, Urrios A et al (2016) The upcoming 3D-printing revolution in microfluidics. Lab Chip 16(10):1720–1742

Bidault L, Deneufchatel M et al (2013) Self-supported fibrin-polyvinyl alcohol interpenetrating polymer networks: an easily handled and rehydratable biomaterial. Biomacromolecules 14(11):3870–3879

Bierwolf J, Lutgehetmann M et al (2010) Primary rat hepatocyte culture on 3D nanofibrous polymer scaffolds for toxicology and pharmaceutical research. Biotechnol Bioeng 108(1):141–150

Bishi DK, Mathapati S et al (2013) Trans-differentiation of human mesenchymal stem cells generates functional hepatospheres on poly(l-lactic acid)-co-poly(ε-

caprolactone)/collagen nanofibrous scaffolds. J Mater Chem B 1(32):3972
Bissoyi A, Bit A et al (2016) Enhanced cryopreservation of MSCs in microfluidic bioreactor by regulated shear flow. Sci Rep 6:35416
Bruinsma BG, Kim Y et al (2015) Layer-by-layer heparinization of decellularized liver matrices to reduce thrombogenicity of tissue engineered grafts. J Clin Transl Res 1(1):04
Bruns H, Kneser U et al (2005) Injectable liver: a novel approach using fibrin gel as a matrix for culture and intrahepatic transplantation of hepatocytes. Tissue Eng 11(11–12):1718–1726
Burdick JA, Stevens MM (2005) Chapter 11: Biomedical hydrogels. In: Hench LL, Jones JR (eds) Biomaterials, artificial organs and tissue engineering. Woodhead Publishing, Cambridge, pp 107–115
Burra P, Freeman R (2012) Trends in liver transplantation 2011. J Hepatol 56:S101–S111
Caddeo S, Boffito M et al (2017) Tissue engineering approaches in the design of healthy and pathological in vitro tissue models. Front Bioeng Biotechnol 5:40
Caliari SR, Perepelyuk M et al (2016) Stiffening hydrogels for investigating the dynamics of hepatic stellate cell mechanotransduction during myofibroblast activation. Sci Rep 6:21387
Cao H, Kuboyama N (2010) A biodegradable porous composite scaffold of PGA/β-TCP for bone tissue engineering. Bone 46(2):386–395
Catapano G, De Bartolo L et al (2001) Morphology and metabolism of hepatocytes cultured in Petri dishes on films and in non-woven fabrics of hyaluronic acid esters. Biomaterials 22(7):659–665
Cen L, Liu W et al (2008) Collagen tissue engineering: development of novel biomaterials and applications. Pediatr Res 63(5):492–496
Chan Choi Y, Choi JS et al (2014) Human gelatin tissue-adhesive hydrogels prepared by enzyme-mediated biosynthesis of DOPA and Fe3+ ion crosslinking. J Mater Chem B 2(2):201–209
Chan EC, Kuo S-M et al (2016) Three dimensional collagen scaffold promotes intrinsic vascularisation for tissue engineering applications. PLoS One 11(2):e0149799
Chandrasekaran A, Abduljawad M et al (2016) Have microfluidics delivered for drug discovery? Expert Opin Drug Discovery 11(8):745–748
Chen Q, Utech S et al (2016) Controlled assembly of heterotypic cells in a core-shell scaffold: organ in a droplet. Lab Chip 16(8):1346–1349
Cheng Z, Runzhou Z et al (2016) Hepatocyte growth factor-loaded collagen-chitosan scaffold containing differentiated bone marrow-derived mesenchymal stem cells as a model for hepatic tissue engineering. J Biomater Tissue Eng 6(8):621–628
Choi YJ, Yi HG et al (2017) 3D cell printed tissue analogues: a new platform for Theranostics. Theranostics 7(12):3118–3137
Collaborators, G. M. a. C. o. D (2015) Global, regional, and national age–sex specific all-cause and cause-specific mortality for 240 causes of death, 1990–2013: a systematic analysis for the global burden of disease study 2013. Lancet 385(9963):117–171
Collins MN, Birkinshaw C (2013) Hyaluronic acid based scaffolds for tissue engineering—a review. Carbohydr Polym 92(2):1262–1279
Cozzolino AM, Noce V et al (2016) Modulating the substrate stiffness to manipulate differentiation of resident liver stem cells and to improve the differentiation state of hepatocytes. Stem Cells Int 2016:5481493
Croisier F, Jérôme C (2013) Chitosan-based biomaterials for tissue engineering. Eur Polym J 49(4):780–792
Damiati S, Kompella U et al (2018) Microfluidic devices for drug delivery systems and drug screening. Genes 9(2):pii: E103
Dash M, Samal SK et al (2017) Enzymatically biomineralized chitosan scaffolds for tissue-engineering applications. J Tissue Eng Regen Med 11(5):1500–1513
de la Puente P, Ludena D (2014) Cell culture in autologous fibrin scaffolds for applications in tissue engineering. Exp Cell Res 322(1):1–11
Deng X, Cao Y et al (2015) Enhanced liver functions of HepG2 cells in the alginate/xyloglucan scaffold. Biotechnol Lett 37(1):235–240
Deponti D, Giancamillo AD et al (2013) Collagen scaffold for cartilage tissue engineering: the benefit of fibrin glue and the proper culture time in an infant cartilage model. Tissue Eng A 20(5–6):1113–1126
Ding B-S, Nolan DJ et al (2010) Inductive angiocrine signals from sinusoidal endothelium are required for liver regeneration. Nature 468:310
Discher DE, Janmey P et al (2005) Tissue cells feel and respond to the stiffness of their substrate. Science 310(5751):1139
Dittrich PS, Manz A (2006) Lab-on-a-chip: microfluidics in drug discovery. Nat Rev Drug Discov 5:210
Drosos I, Kolios G (2013) Stem cells in liver regeneration and their potential clinical applications. Stem Cell Rev 9(5):668–684
Dutkowski P, Linecker M et al (2015) Challenges to liver transplantation and strategies to improve outcomes. Gastroenterology 148(2):307–323
Dvir-Ginzberg M, Gamlieli-Bonshtein I et al (2003) Liver tissue engineering within alginate scaffolds: effects of cell-seeding density on hepatocyte viability, morphology, and function. Tissue Eng 9(4):757–766
Ebrahimkhani MR, Neiman JAS et al (2014) Bioreactor technologies to support liver function in vitro. Adv Drug Deliv Rev 0:132–157
Elieh-Ali-Komi D, Hamblin MR (2016) Chitin and chitosan: production and application of versatile biomedical nanomaterials. Int J Adv Res 4(3):411–427
Fan J, Yang J (2017) Preparation and characterization of a chitosan/galactosylated hyaluronic acid/heparin scaffold for hepatic tissue engineering. J Biomater Sci Polym Ed 28(6):569–581

Feng Z-Q, Chu X et al (2009) The effect of nanofibrous galactosylated chitosan scaffolds on the formation of rat primary hepatocyte aggregates and the maintenance of liver function. Biomaterials 30(14):2753–2763
Gao B, Bataller R (2011) Alcoholic liver disease: pathogenesis and new therapeutic targets. Gastroenterology 141(5):1572–1585
Garg K (2017) Laminin enriched scaffolds for tissue engineering applications. Adv Tissue Eng Regen Med Open Access 2(3):194–200
Gentile P, Chiono V et al (2014) An overview of poly(lactic-co-glycolic) acid (PLGA)-based biomaterials for bone tissue engineering. Int J Mol Sci 15(3):3640–3659
Gong H, Agustin J et al (2014) Biomimetic design and fabrication of porous chitosan–gelatin liver scaffolds with hierarchical channel network. J Mater Sci Mater Med 25(1):113–120
Goral VN, Hsieh Y-C et al (2010) Perfusion-based microfluidic device for three-dimensional dynamic primary human hepatocyte cell culture in the absence of biological or synthetic matrices or coagulants. Lab Chip 10(24):3380–3386
Gori M, Simonelli MC et al (2016) Investigating nonalcoholic fatty liver disease in a liver-on-a-Chip microfluidic device. PLoS One 11(7):e0159729
Goudie MJ, Ghuman AP et al (2016) Investigation of diffusion characteristics through microfluidic channels for passive drug delivery applications. J Drug Deliv 2016:7913616
Guvendiren M, Perepelyuk M et al (2014) Hydrogels with differential and patterned mechanics to study stiffness-mediated myofibroblastic differentiation of hepatic stellate cells. J Mech Behav Biomed Mater 38:198–208
Hammond JS, Beckingham IJ et al (2006) Scaffolds for liver tissue engineering. Expert Rev Med Devices 3(1):21–27
Hammond JS, Gilbert TW et al (2011) Scaffolds containing growth factors and extracellular matrix induce hepatocyte proliferation and cell migration in normal and regenerating rat liver. J Hepatol 54(2):279–287
Harink B, Le Gac S et al (2013) Regeneration-on-a-chip? The perspectives on use of microfluidics in regenerative medicine. Lab Chip 13(18):3512–3528
Hasirci V, Berthiaume F et al (2001) Expression of liver-specific functions by rat hepatocytes seeded in treated poly(lactic-co-glycolic) acid biodegradable foams. Tissue Eng 7(4):385–394
Hemshekhar M, Thushara RM et al (2016) Emerging roles of hyaluronic acid bioscaffolds in tissue engineering and regenerative medicine. Int J Biol Macromol 86:917–928
Higashi T, Friedman SL et al (2017) Hepatic stellate cells as key target in liver fibrosis. Adv Drug Deliv Rev 121:27–42
Ho C-T, Lin R-Z et al (2013) Liver-cell patterning Lab Chip: mimicking the morphology of liver lobule tissue. Lab Chip 13(18):3578–3587
Hughes RD, Mitry RR et al (2012) Current status of hepatocyte transplantation. Transplantation 93(4):342–347
Ivanova EP, Bazaka K et al (2014) Chapter 3: Advanced synthetic polymer biomaterials derived from organic sources. In: Ivanova EP, Bazaka K, Crawford RJ (eds) New functional biomaterials for medicine and healthcare. Woodhead Publishing, Oxford, pp 71–99
Jain E, Damania A et al (2014) Biomaterials for liver tissue engineering. Hepatol Int 8(2):185–197
Jain E, Damania A et al (2015) Fabrication of macroporous cryogels as potential hepatocyte carriers for bioartificial liver support. Colloids Surf B: Biointerfaces 136:761–771
Jiang H-L, Kim Y-K et al (2010) Roles of spheroid formation of hepatocytes in liver tissue engineering. Int J Stem Cells 3(2):69–73
Kane BJ, Zinner MJ et al (2006) Liver-specific functional studies in a microfluidic Array of primary mammalian hepatocytes. Anal Chem 78(13):4291–4298
Kanninen LK, Harjumäki R et al (2016) Laminin-511 and laminin-521-based matrices for efficient hepatic specification of human pluripotent stem cells. Biomaterials 103:86–100
Kanta J (2016) Elastin in the liver. Front Physiol 7:491. https://doi.org/10.3389/fphys.2016.00491
Kazemnejad S (2009) Hepatic tissue engineering using scaffolds: state of the art. Avicenna J Med Biotechnol 1(3):135–145
Khademhosseini A, Langer R et al (2006) Microscale technologies for tissue engineering and biology. Proc Natl Acad Sci U S A 103(8):2480–2487
Khalaj Z, Lotfi AS et al (2016) Laminin matrix promotes hepatogenic terminal differentiation of human bone marrow mesenchymal stem cells. Iran J Basic Med Sci 19(1):34–42
Khetani SR, Bhatia SN (2007) Microscale culture of human liver cells for drug development. Nat Biotechnol 26:120
Kim TY, Kim DJ (2013) Acute-on-chronic liver failure. Clin Mol Hepatol 19(4):349–359
Kim L, Vahey MD et al (2006) Microfluidic arrays for logarithmically perfused embryonic stem cell culture. Lab Chip 6(3):394–406
Kim SM, Lee SH et al (2008a) Cell research with physically modified microfluidic channels: a review. Lab Chip 8(7):1015–1023
Kim I-Y, Seo S-J et al (2008b) Chitosan and its derivatives for tissue engineering applications. Biotechnol Adv 26(1):1–21
Kim Y, Kang K et al (2017) Three-dimensional (3D) printing of mouse primary hepatocytes to generate 3D hepatic structure. Ann Surg Treat Res 92(2):67–72
Kimura H, Sakai Y et al (2018) Organ/body-on-a-chip based on microfluidic technology for drug discovery. Drug Metab Pharmacokinet 33(1):43–48

Knight TA, Payne RG (2013) Characterization of a PGA-based scaffold for use in a tissue-engineered neo-urinary conduit. In: Basu J, Ludlow JW (eds) Organ regeneration: methods and protocols. Humana Press, Totowa, pp 179–188

Knight E, Przyborski S (2015) Advances in 3D cell culture technologies enabling tissue-like structures to be created in vitro. J Anat 227(6):746–756

Kumar A, Han SS (2017) PVA-based hydrogels for tissue engineering: a review. Int J Polym Mater Polym Biomater 66(4):159–182

Kundu B, Rajkhowa R et al (2013) Silk fibroin biomaterials for tissue regenerations. Adv Drug Deliv Rev 65(4):457–470

Kwapiszewska K, Michalczuk A et al (2014) A microfluidic-based platform for tumour spheroid culture, monitoring and drug screening. Lab Chip 14(12):2096–2104

Leclercq IA, Da Silva Morais A et al (2007) Insulin resistance in hepatocytes and sinusoidal liver cells: mechanisms and consequences. J Hepatol 47(1):142–156

Lee JS, Cho S-W (2012) Liver tissue engineering: recent advances in the development of a bio-artificial liver. Biotechnol Bioprocess Eng 17(3):427–438

Lee Hyeong J, Kim Yong B et al (2015) A new approach for fabricating collagen/ECM-based bioinks using Preosteoblasts and human adipose stem cells. Adv Healthc Mater 4(9):1359–1368

Lee KY, Mooney DJ (2012) Alginate: properties and biomedical applications. Prog Polym Sci 37(1):106–126

Lee PJ, Hung PJ et al (2006) Nanoliter scale microbioreactor array for quantitative cell biology. Biotechnol Bioeng 94(1):5–14

Lee PJ, Hung PJ et al (2007) An artificial liver sinusoid with a microfluidic endothelial-like barrier for primary hepatocyte culture. Biotechnol Bioeng 97(5):1340–1346

Lee Y-B, Polio S et al (2010a) Bio-printing of collagen and VEGF-releasing fibrin gel scaffolds for neural stem cell culture. Exp Neurol 223(2):645–652

Lee KH, Shin SJ et al (2010b) Microfluidic synthesis of pure chitosan microfibers for bio-artificial liver chip. Lab Chip 10(10):1328–1334

Lee Y, Kwon J et al (2012) Reduction of inflammatory responses and enhancement of extracellular matrix formation by vanillin-incorporated poly(lactic-co-glycolic acid) scaffolds. Tissue Eng Part A 18(19–20):1967–1978

Lee J, Kim SH et al (2013) Fabrication and characterization of microfluidic liver-on-a-chip using microsomal enzymes. Enzym Microb Technol 53(3):159–164

Lee BH, Shirahama H et al (2017) Colloidal templating of highly ordered gelatin methacryloyl-based hydrogel platforms for three-dimensional tissue analogues. NPG Asia Mater 9:e412

Lin CH, Chang HC, Hsu CHA (2016) Microfluidic platform for high-throughput single-cell isolation and culture. J Vis Exp 112:e54105. https://doi.org/10.3791/54105

Liu WF, Chen CS (2005) Engineering biomaterials to control cell function. Mater Today 8(12):28–35

Liu M, Yang J et al (2016) Superior performance of co-cultured mesenchymal stem cells and hepatocytes in poly(lactic acid-glycolic acid) scaffolds for the treatment of acute liver failure. Biomed Mater (Bristol, UK) 11(1):015008

Lou R, Yu W et al (2017) Fabrication of stable galactosylated alginate microcapsules via covalent coupling onto hydroxyl groups for hepatocytes applications. Carbohydr Polym 155:456–465

Lyu S, Untereker D (2009) Degradability of polymers for implantable biomedical devices. Int J Mol Sci 10(9):4033–4065

Macdonald NP, Cabot JM et al (2017) Comparing microfluidic performance of three-dimensional (3D) printing platforms. Anal Chem 89(7):3858–3866

Magiera A, Markowski J et al (2017) PLA-based hybrid and composite electrospun fibrous scaffolds as potential materials for tissue engineering. J Nanomater 2017:11

Makadia HK, Siegel SJ (2011) Poly lactic-co-glycolic acid (PLGA) as biodegradable controlled drug delivery carrier. Polymers 3(3):1377–1397

Malikmammadov E, Tanir TE et al (2018) PCL and PCL-based materials in biomedical applications. J Biomater Sci Polym Ed 29(7–9):863–893

Mari CE, Laura Saenz del B et al (2017) Gelatin as biomaterial for tissue engineering. Curr Pharm Des 23(24):3567–3584

Mariod AA, Fadul H (2013) Review: gelatin, source, extraction and industrial applications. Acta Sci Pol Technol Aliment 12(2):135–147

Martinez-Hernandez A, Amenta PS (1993) The hepatic extracellular matrix. Virchows Arch A 423(2):77–84

Miranda JP, Rodrigues A et al (2010) Extending hepatocyte functionality for drug-testing applications using high-viscosity alginate–encapsulated three-dimensional cultures in bioreactors. Tissue Eng Part C Methods 16(6):1223–1232

Miranda-Nieves D, Chaikof EL (2017) Collagen and elastin biomaterials for the fabrication of engineered living tissues. ACS Biomater Sci Eng 3(5):694–711

Mironov Anton V, Grigoryev Aleksey M et al (2016) 3D printing of PLGA scaffolds for tissue engineering. J Biomed Mater Res A 105(1):104–109

Nakamura S, Ijima H (2013) Solubilized matrix derived from decellularized liver as a growth factor-immobilizable scaffold for hepatocyte culture. J Biosci Bioeng 116(6):746–753

Nakamura S, Kubo T et al (2013) Heparin-conjugated gelatin as a growth factor immobilization scaffold. J Biosci Bioeng 115(5):562–567

Nallagangula KS, Nagaraj SK et al (2018) Liver fibrosis: a compilation on the biomarkers status and their significance during disease progression. Future Sci OA 4(1):FSO250

Naresh K, Utpal B (2012) Silk fibroin based biomimetic artificial extracellular matrix for hepatic tissue engineering applications. Biomed Mater 7(4):045004
Nikkhah M, Edalat F et al (2012) Engineering microscale topographies to control the cell–substrate interface. Biomaterials 33(21):5230–5246
Olsen AL, Bloomer SA et al (2011) Hepatic stellate cells require a stiff environment for myofibroblastic differentiation. Am J Physiol Gastrointest Liver Physiol 301(1):G110–G118
Ostrovidov S, Jiang J et al (2004) Membrane-based PDMS microbioreactor for perfused 3D primary rat hepatocyte cultures. Biomed Microdevices 6(4):279–287
Pan Z, Ding J (2012) Poly(lactide-co-glycolide) porous scaffolds for tissue engineering and regenerative medicine. Interface Focus 2(3):366–377
Patel D, Haque A et al (2015) Using reconfigurable microfluidics to study the role of HGF in autocrine and paracrine signaling of hepatocytes. Integr Biol (Camb) 7(7):815–824
Pensalfini M, Ehret AE et al (2017) Factors affecting the mechanical behavior of collagen hydrogels for skin tissue engineering. J Mech Behav Biomed Mater 69:85–97
Perez RA, Jung CR et al (2017) Biomaterials and culture technologies for regenerative therapy of liver tissue. Adv Healthc Mater 6(2):1600791
Poh Z, Goh BB et al (2015) Rates of cirrhosis and hepatocellular carcinoma in chronic hepatitis B and the role of surveillance: a 10-year follow-up of 673 patients. Eur J Gastroenterol Hepatol 27(6):638–643
Powers MJ, Domansky K et al (2002) A microfabricated array bioreactor for perfused 3D liver culture. Biotechnol Bioeng 78(3):257–269
Protzer U, Maini MK et al (2012) Living in the liver: hepatic infections. Nat Rev Immunol 12(3):201–213
Rinaudo M (2006) Chitin and chitosan: properties and applications. Prog Polym Sci 31(7):603–632
Rothbauer M, Zirath H et al (2018) Recent advances in microfluidic technologies for cell-to-cell interaction studies. Lab Chip 18(2):249–270
Schuerlein S, Schwarz T et al (2017) A versatile modular bioreactor platform for tissue engineering. Biotechnol J 12(2):1600326
Schütte J, Hagmeyer B et al (2011) "Artificial micro organs"—a microfluidic device for dielectrophoretic assembly of liver sinusoids. Biomed Microdevices 13(3):493–501
Semnani D, Naghashzargar E et al (2016) Evaluation of PCL/chitosan electrospun nanofibers for liver tissue engineering. Int J Polym Mater 66(3):149–157
Seo S-J, Choi Y-J et al (2006) Alginate/Galactosylated chitosan/heparin scaffold as a new synthetic extracellular matrix for hepatocytes. Tissue Eng 12(1):33–44
Shah Mohammadi M, Bureau MN et al (2014) Chapter 11: Polylactic acid (PLA) biomedical foams for tissue engineering. In: Netti PA (ed) Biomedical foams for tissue engineering applications. Woodhead Publishing, Cambridge, pp 313–334
Shang Y, Tamai M et al (2014) Hybrid sponge comprised of galactosylated chitosan and hyaluronic acid mediates the co-culture of hepatocytes and endothelial cells. J Biosci Bioeng 117(1):99–106
Shim J-H, Kim AJ et al (2013) Effect of solid freeform fabrication-based polycaprolactone/poly (lactic-co-glycolic acid)/collagen scaffolds on cellular activities of human adipose-derived stem cells and rat primary hepatocytes. J Mater Sci Mater Med 24(4):1053–1065
Shirakigawa N, Takei T et al (2013) Base structure consisting of an endothelialized vascular-tree network and hepatocytes for whole liver engineering. J Biosci Bioeng 116(6):740–745
Sia SK, Whitesides GM (2003) Microfluidic devices fabricated in poly(dimethylsiloxane) for biological studies. Electrophoresis 24(21):3563–3576
Siegel R, Naishadham D et al (2012) Cancer statistics, 2012. CA Cancer J Clin 62(1):10–29
Silva JM, Rodrigues LC et al (2017) Engineered tubular structures based on chitosan for tissue engineering applications. J Biomater Appl 32(7):841–852
Singal AK, Anand BS (2013) Recent trends in the epidemiology of alcoholic liver disease. Clin Liver Dis 2(2):53–56
Sivaraman A, Leach JK et al (2005) A microscale in vitro physiological model of the liver: predictive screens for drug metabolism and enzyme induction. Curr Drug Metab 6(6):569–591
Skardal A, Smith L et al (2012) Tissue specific synthetic ECM hydrogels for 3-D in vitro maintenance of hepatocyte function. Biomaterials 33(18):4565–4575
Stange J (2011) Extracorporeal liver support. Organogenesis 7(1):64–73
Stevens KR, Miller JS et al (2015) Degradable hydrogels derived from PEG-diacrylamide for hepatic tissue engineering. J Biomed Mater Res A 103(10):3331–3338
Stevens KR, Scull MA et al (2017) In situ expansion of engineered human liver tissue in a mouse model of chronic liver disease. Sci Transl Med 9(399):eaah5505
Strohmeier O, Keller M et al (2015) Centrifugal microfluidic platforms: advanced unit operations and applications. Chem Soc Rev 44(17):6187–6229
Stutchfield BM, Simpson K et al (2011) Systematic review and meta-analysis of survival following extracorporeal liver support. Br J Surg 98(5):623–631
Su N, Gao P-L et al (2017) Fibrous scaffolds potentiate the paracrine function of mesenchymal stem cells: a new dimension in cell-material interaction. Biomaterials 141(Supplement C):74–85
Sultana N, Mokhtar M et al (2015) Chitosan-based nanocomposite scaffolds for tissue engineering applications. Mater Manuf Process 30(3):273–278
Sun W, Chen Y-Q et al (2016) Organs-on-chips and its applications. Chin J Anal Chem 44(4):533–541
Tang M, Wang G et al (2016) A review of biomedical centrifugal microfluidic platforms. Micromachines 7(2):pii: E26

Teotia RS, Kalita D et al (2015) Bifunctional Polysulfone-chitosan composite hollow fiber membrane for bioartificial liver. ACS Biomater Sci Eng 1(6):372–381

Toosi S, Naderi-Meshkin H et al (2016) PGA-incorporated collagen: toward a biodegradable composite scaffold for bone-tissue engineering. J Biomed Mater Res A 104(8):2020–2028

Török E, Lutgehetmann M et al (2010) Primary human hepatocytes on biodegradable poly(l-lactic acid) matrices: a promising model for improving transplantation efficiency with tissue engineering. Liver Transpl 17(2):104–114

Tourovskaia A, Figueroa-Masot X et al (2005) Differentiation-on-a-chip: a microfluidic platform for long-term cell culture studies. Lab Chip 5(1):14–19

Tripathi A, Melo JS (2015) Preparation of a sponge-like biocomposite agarose-chitosan scaffold with primary hepatocytes for establishing an in vitro 3D liver tissue model. RSC Adv 5(39):30701–30710

Uguz I, Proctor CM et al (2017) A microfluidic ion pump for in vivo drug delivery. Adv Mater 29(27):1701217

Vasanthan KS, Subramanian A et al (2012) Role of biomaterials, therapeutic molecules and cells for hepatic tissue engineering. Biotechnol Adv 30(3):742–752

Venkatesan J, Bhatnagar I et al (2015) Alginate composites for bone tissue engineering: a review. Int J Biol Macromol 72:269–281

Walker GM, Zeringue HC et al (2004) Microenvironment design considerations for cellular scale studies. Lab Chip 4(2):91–97

Wang T, Feng Z-Q et al (2013) Nanoporous fibers of type-I collagen coated poly(l-lactic acid) for enhancing primary hepatocyte growth and function. J Mater Chem B 1(3):339–346

Wang B, Hu Q et al (2016) Porous lactose-modified chitosan scaffold for liver tissue engineering: influence of galactose moieties on cell attachment and mechanical stability. Int J Polym Sci 2016:8

Watanabe M, Zemack H et al (2016) Maintenance of hepatic functions in primary human hepatocytes cultured on Xeno-free and chemical defined human recombinant laminins. PLoS One 11(9):e0161383

Willerth SM, Sakiyama-Elbert SE (2008) Combining stem cells and biomaterial scaffolds for constructing tissues and cell delivery. StemBook, Cambridge, MA

Xia L, Arooz T et al (2012) Hepatocyte function within a stacked double sandwich culture plate cylindrical bioreactor for bioartificial liver system. Biomaterials 33(32):7925–7932

Xu T, Yang H et al (2017) Polylactic acid nanofiber scaffold decorated with chitosan Islandlike topography for bone tissue engineering. ACS Appl Mater Interfaces 9(25):21094–21104

Yang J, Goto M et al (2002) Galactosylated alginate as a scaffold for hepatocytes entrapment. Biomaterials 23(2):471–479

Yu Y-Q, Jiang X-S et al (2014) Local delivery of vascular endothelial growth factor via nanofiber matrix improves liver regeneration after extensive hepatectomy in rats. J Biomed Nanotechnol 10(11):3407–3415

Yu F, Deng R et al (2017) A perfusion incubator liver chip for 3D cell culture with application on chronic hepatotoxicity testing. Sci Rep 7(1):14528

Zavan B, Brun P et al (2005) Extracellular matrix-enriched polymeric scaffolds as a substrate for hepatocyte cultures: in vitro and in vivo studies. Biomaterials 26(34):7038–7045

Zeigerer A, Wuttke A et al (2017) Functional properties of hepatocytes in vitro are correlated with cell polarity maintenance. Exp Cell Res 350(1):242–252

Zhang Y, Zhang Y et al (2014) Galactosylated poly(epsilon-caprolactone) membrane promoted liver-specific functions of HepG2 cells in vitro. Mater Sci Eng C Mater Biol Appl 41:52–58

Zhang B, Montgomery M et al (2016) Biodegradable scaffold with built-in vasculature for organ-on-a-chip engineering and direct surgical anastomosis. Nat Mater 15(6):669–678

Zhang J, Wei X et al (2017) Stem cell culture and differentiation in microfluidic devices toward organ-on-a-chip. Future Sci OA 3(2):FSO187

Zhang D, Wu X et al (2018) The development of collagen based composite scaffolds for bone regeneration. Bioact Mater 3(1):129–138

Zhao Y, Sha XUL et al (2011) In vitro and in vivo characterization of silk fibroin/gelatin composite scaffolds for liver tissue engineering. J Dig Dis 13(3):168–178

Zhou P, Huang Y et al (2015a) Decellularization and recellularization of rat livers with hepatocytes and endothelial progenitor cells. Artif Organs 40(3):E25–E38

Zhou Q, Patel D et al (2015b) Liver injury-on-a-chip: microfluidic co-cultures with integrated biosensors for monitoring liver cell signaling during injury. Lab Chip 15(23):4467–4478

Zhu Xin H, Gan Seng K et al (2007) Proteins combination on PHBV microsphere scaffold to regulate Hep3B cells activity and functionality: a model of liver tissue engineering system. J Biomed Mater Res A 83A(3):606–616

Microfluidic Systems in CNS Studies

6

Anna Andrzejewska and Miroslaw Janowski

Abstract

Current technological progress facilitates the introduction of micro-devices into biotechnology research including studies on central nervous system. Wide range of micro-chambers with diversity of channel systems and multiple compartments enable users to create models which closely mimic nervous tissue structure which nowadays is often called as brain-on-a-chip technology. Heretofore experiments showing the influence of substance gradients, cell interactions, spatial conditions, neuroinflammation, stem cells migration, drug delivery, mechanisms controlling progression of diseases like Alzheimer, Parkinson, multiple sclerosis or nerve injury were performed in microfluidic devices. Moreover, the integration of bio-sensors and development of dedicated software for microfluidic studies can enable performing high throughput and good quality automated experiments investigating regeneration and degeneration processes in models well emulating central nervous system structures.

Keywords

Microfluidic chambers · Microfluidic channels · PDMS · CNS

Highlights

- Microfluidic devices consist of channel systems which enable to work with a great accuracy simultaneously reducing time and cost of experiments performance
- Microfluidic systems better mimic natural brain tissue homeostasis than conventional cell culture due to assuring of constant flow of culture medium and discharge of unnecessary products, providing natural cells spatial organization and inducing shear stress
- Nowadays microfluidic systems are employed in research investigating pathomechanisms of Alzheimer disease, Parkinson disease, multiple sclerosis, nerve degeneration, neuro-inflammation and brain metastasis as well as in attempts of development of new therapeutic strategies

A. Andrzejewska
NeuroRepair Department, Mossakowski Medical Research Centre, Polish Academy of Sciences, Warsaw, Poland
e-mail: aandrzejewska@imdik.pan.pl

M. Janowski (✉)
Center for Advanced Imaging Research, Department of Diagnostic Radiology and Nuclear Medicine, University of Maryland, Baltimore, MD, USA
e-mail: miroslaw.janowski@som.maryland.edu

J. M. Oliveira, R. L. Reis (eds.), *Biomaterials- and Microfluidics-Based Tissue Engineered 3D Models*, Advances in Experimental Medicine and Biology 1230,
https://doi.org/10.1007/978-3-030-36588-2_6

6.1 Introduction

Microfluidic devices (micro-devices) consist of channel systems with a cross-section in range of a few micrometres, and historically have been developed as a tool for chemical analysis hence they enable to work with a great accuracy using even very small volumes of reagents (Harrison et al. 1993). Micro-devices save time required to run experiments, as well as significantly reduce their costs mainly due to low consumption of often very expensive reagents. The same advantages have led to adaptation of microfluidic systems in a wide variety of disciplines including biotechnology. Technological progress provided a new set of materials for micro-device production which fit biological applications. The most commonly used micro-chambers are fabricated by photolithography method from polydimethylsiloxane (PDMS). Photolithography is the process used to obtain a variety of channel shape projection in micro-chamber. Briefly, the UV light exposure causes the etching of PDMS, therefore in order to obtain the desired micro-pattern, the PDMS is placed on the photo-stable surface (like glass or silicone), and then irradiated by UV light through the layer of photoresistant mask of particular shape. This leads to the mapping of the mask shape in the PDMS (Das and Srivastava 2016) (Fig. 6.1). Low cost and highly reproducible fabrication of PDMS chambers fulfils mass production requirements. PDMS is a synthetic polymer with low auto-fluorescence and good transparency facilitating its use in fluorescent microscope-based imaging. Moreover, PDMS is gas-permeable and biocompatible therefore it is well suited for cell culture experiments, however due to hydrophobicity it is not favourable for cell adhesion. Heretofore the spectrum of surface modification techniques has been employed in order to enable cell growth inside PDMS chambers including plasma treatment or coating with functional groups and/or extracellular matrix. Alternatively in some studies we can found micro-devices produced from natural substrates like collagen, fibrin or agarose which better mimic the natural conditions (Zhou et al. 2012; Berthier et al. 2012). Currently available technology allowed creating very complex micro-channel systems including one and multi compartment chambers connected with each other by channels. These systems may be horizontally or vertically oriented, and possess singular or many inputs and outputs. All these combinations were developed in an effort to create models that best reflects the *in vivo* situation what nowadays is often called as "organ-on-a-chip" technology (Prabhakarpandian et al. 2013; Brown et al. 2015; Wang et al. 2017a). This task is even more demanding considering the challenge of recreation of the most complicated and inscrutable system of human body – central nervous system (CNS).

6.2 Microfludics in CNS

The microfluidic systems possess a few advantages over standard two- or even three dimensional culture system what make them more appropriate to mimic CNS-like microenvironment. Constant flow of culture medium containing nutritional compounds together with simultaneous discharge of unnecessary products of cell metabolism allow obtaining proper conditions for nervous tissue homeostasis which is not provided in conventional cell culture methods. The constant flow provides also stability of soluble factors gradient in studies aiming to investigate their influence on cellular response. Taking advantage of this property of microfluidic system, the influence of retinoic acid and sonic hedgehog (SHH) gradient on differentiation of neural stem cells was studied in analogy to phenomenon of neural tube formation during foetal development (Uzel et al. 2016). Similar analysis was made for concentration gradient of SHH, BMP4 (bone morphogenetic protein 4) and FGF8 (fibroblast growth factor 8) regarding to their role in neural progenitor differentiation into neurons and generation of neural networks (Park et al. 2009). The ability of factors like netrin-1 and slit-2 in axonal guidance was investigated analogically (Kothapalli et al. 2011) (Fig. 6.2).

Another interesting improvement in analysis of CNS cells physiology through utilization of

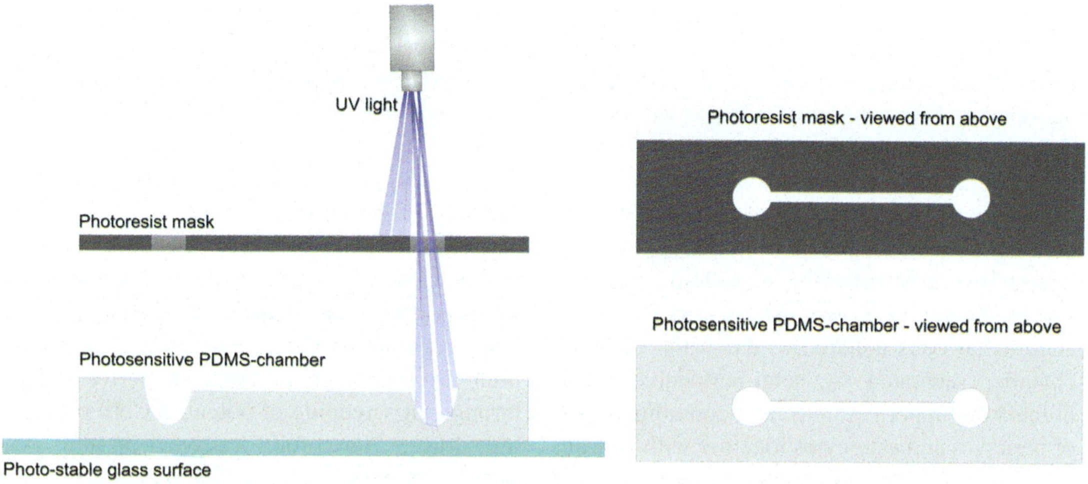

Fig. 6.1 The schematic overview of photolithography technique

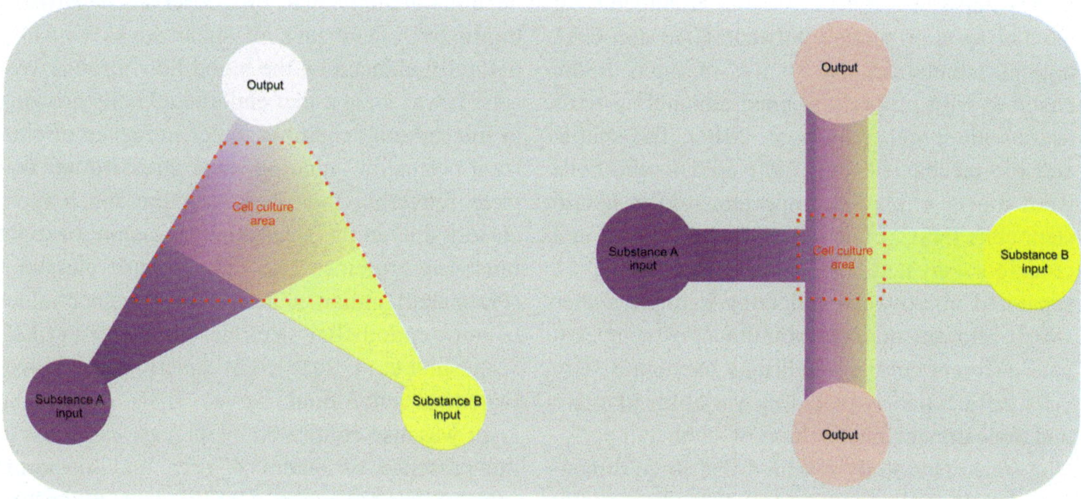

Fig. 6.2 The example of systems of channels providing cells exposition on gradient of factors in microfluidic chamber

micro-devices is the possibility to isolate in distinct compartments soma from axons and dendrites of singular neurons by micro-grooves and trace the signalling between them. This system was used to assess the gene expression changes in neuronal nuclei after BDNF-treatment of their dendrites (Cohen et al. 2011). The separation of axons from soma has also showed that neurothrophin-3 and BDNF provide strong stimulation signal for axon elongation independently from exposition of cell body (Taylor et al. 2005). Similar strategies have been used for synapse formation study, myelination process analysis and investigation on transmission of axonal signals (Neto et al. 2016). Interactions between different types of CNS cells were observed in microfluidic systems, too. The complex net of mutual physical and functional relations existing among neurons, astrocytes, oligodendrocytes, microglia cells, pericytes and endothelial cells have led to creation of neurovascular unit (NVU) concept. It is the basic entity providing proper function of CNS and damage of even one of its element induces dysfunction of the whole system. Thus all studies investigating separated parts of NVU are biased by the lack of crucial factors provided by each cells of this entity (Muoio et al. 2014). The majority of NVU models are based on cell co-culture method, however this type of systems does not provide natural spatial organization of cells. Much closer mapping of natural CNS cell configuration was gained by the use of vertically situated two-compartments microfluidic chamber system created by Bown and co-workers, where endothelial cells adhered to the walls of lower chamber separated by semi-permeable membrane from upper compartment containing layer of pericytes and astrocytes together with neurons embedded in three-dimensional collagen gel (Brown et al. 2015). Noteworthy work has also been made by Kim et al., who created a model of hippocampal neural network with architecture of parallel axon projection between CA3 and CA1 regions maintained due to use of microfluidic chamber with properly aligned channel systems in monolithic gel (Kim et al. 2017). The spatial and bio-mechanical cues delivered to the cells were shown to play an important role in neural stem cells (NSCs) fate determination. Wang and co-authors created micro-device containing 16 separated micro-wells reflecting cellular niches which allowed them to simultaneously investigate different culture conditions for neural stem cells differentiation including variables like two and three dimensional culture or static and perfusion state. This study revealed that three dimensional culture in the presence of extracellular matrix proteins stimulates self-renewal of NSCs, whereas two-dimensional and spheroid cultures induce their differentiation under continuous perfusion conditions (Wang et al. 2017a). Among other of the crucial mechanical factors which are better mimicked by the microfluidic system than by the conventional culture are more natural spatial ratio between nutrition availability and tissue area as well as presence of shear stress. The exchange of oxygen and nutrients delivered by fluid in micro-channels take place by diffusion mechanism analogical to *in vivo* process what is in opposite in other culture systems where cells are immersed in medium (Harink et al. 2013). The presence of blood-brain-barrier (BBB) is one of the most unique features of CNS. In physiological state its existence is truly beneficial since it protects the CNS from harmful substances diluted in blood, however at the same time during pathological process it creates an obstacle in drug delivery following systemic administration. Many theoretically effective substances failed in treatment of CNS diseases due to their inability to cross BBB. This makes the challenge of creation of proper BBB model *in vitro* to become the high priority issue in order to perform high-throughput screening of potential CNS penetrating drugs. Heretofore transwell system was mainly adopted to obtain this goal however the layer of brain endothelial cells placed inside of transwell insert do not create intracellular connections analogical to this observed in BBB. It might be related lack of shear stress which is naturally induced by the blood flow through vessels. It was shown that endothelial cells growing in microfluidic chamber under influence of constant perfusion inducing shear stress reveal features reflecting the BBB structure much more closely including stronger expression of tight junction proteins like ZO-1 and claudin-1 (Prabhakarpandian et al. 2013) and higher values of trans-endothelial electrical resistance (TEER) (Takeshita et al. 2014). The functionality of recreated in microfluid device BBB endothelial layer was also confirmed by its responsiveness to the presence of mannitol, urea and histamine solutions causing drop of TEER value (Takeshita et al. 2014; Wang et al. 2016). Microfluidic BBB model has already been proven to be useful for screening BBB permeability of drugs such as caffeine, cimetidine or doxorubicin (Wang et al. 2017b). Alternatively, the same model of BBB can be used to study neuro-inflammation by addition of pro-inflammatory cytokine to the chamber to induce activation of endothelial cells. The flow of leukocytes through microfluidic channel covered by activated endothelial cells perfectly mimics the process of leukocyte extravasation (Cho et al. 2015) (Fig. 6.3). Next possible application of BBB microfluidic model is to emulate the process of metastasis (Xu et al. 2016).

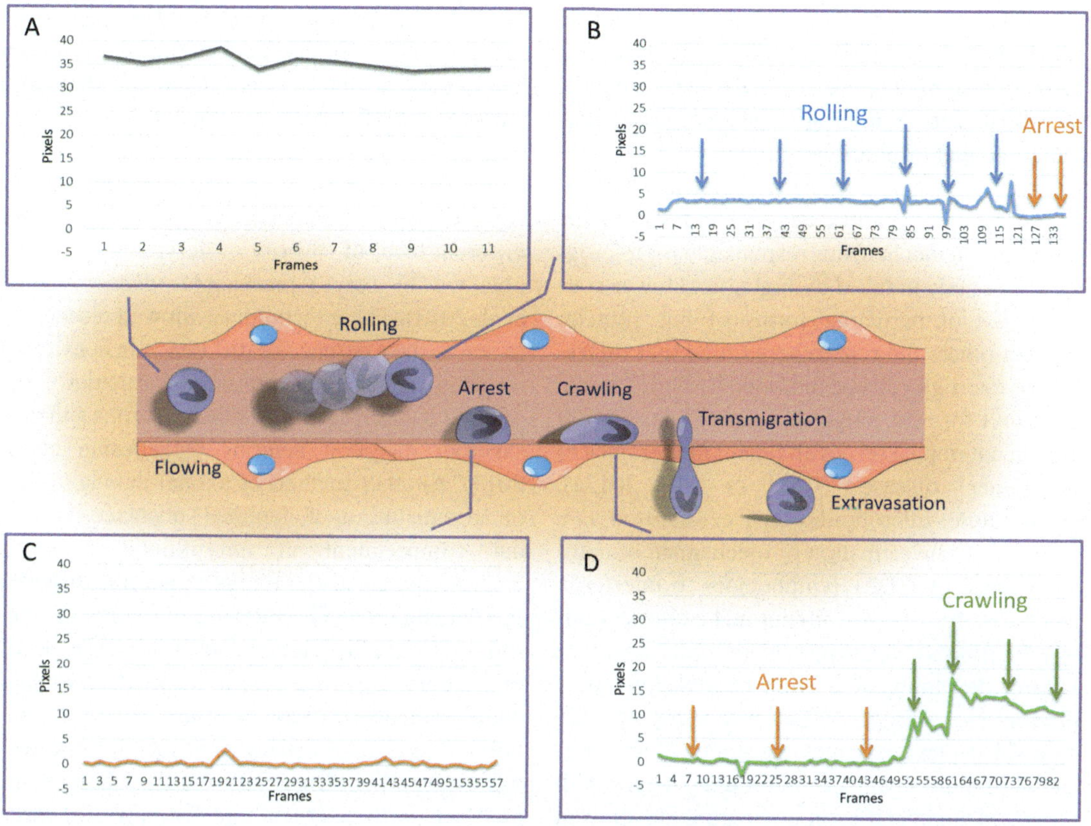

Fig. 6.3 Movement patterns of Mesenchymal Stem Cells flowing through microfluidic chamber in analogy to steps of leucocytes diapedesis process observed in a VCAM-1 protein covered channel of microfluidic chamber. Line charts were automatically generated by dedicated software and show examples of velocity variation typical for unstopped (**a**), rolling (**b**), arrested (**c**), and crawling (**d**) cells. (Reprinted from Andrzejewska et al. 2019)

Moreover this system can be utilized to predict the efficacy of delivery and homing of stem cells like mesenchymal stem cells or glia progenitors to CNS structures after their systemic administration as a therapeutic approach in neurological disorders (Gorelik et al. 2012; Jablonska et al. 2018; Andrzejewska et al. 2019). Very important area of CNS research where microfluid device can be employed is investigation of pathophysiology of neurodegenerative diseases. Beyond the blood flow imitation, the stream of fluid in microchamber can reflect a much slower brain interstitial fluid perfusion. This system was resourcefully used in study by Choi and co-workers to analyse the neurotoxicity of amyloid beta oligomeric assemblies gradient on neurons as an equivalent of Alzheimer disease progression model (Choi et al. 2013). Similar technology was used in combination with culture of neurospheroids, which showed that amyloid β can be more toxic for cells causing their death and interrupting neural circuits, if it is delivered via interstitial perfusion in comparison to static conditions (Park et al. 2015). This experiment highlights the huge additional cognitive value of research performed in environment better mimicking the natural brain tissue homeostasis like microfluidic systems than conventional culture. Through displaying neurons culture in microfluid device containing separated compartments for soma and axons the neuron-neuron and axon to soma transport of amyloid β was identified (Song et al. 2014). Movement of Parkinson disease pathognomonic protein – α-synuclein between neurons and along axons was also shown using microfluidic device (Freundt et al. 2012). Except of amyloid plaques formation another feature of Alzheimer disease is the impairment of microglia function. The accumulation and migration of these cells toward soluble or bound amyloid β fraction was also assessed in microfluidic device and revealed the difference in mechanisms responsible for microglia chemotaxis induced by higher and lower concentration of particular forms of this protein. Among others the researchers found that migration of microglia can be inhibited by high amount of bound form of amyloid β, which can have a significant impact on development of therapies of Alzheimer disease (Cho et al. 2013a). Furthermore, microfluidic device was used as a platform for investigation of mechanism of damage caused by CD8+ lymphocytes to interferon gamma activated cortical neuronal axons in analogy to process occurring in multiple sclerosis. During development of this neurodegenerative disease axonal injury is induced primarily only with subsequent cell body disruption. The application of multi-chamber microfluidic device with separated soma and axon compartments allowed researchers to perform study providing more accurate system for proper event sequence, where axons could be attacked first being isolated from cells' body (Sauer et al. 2013). Taking advantage of possibility to culture even singular neuron in chambers with soma and axon division this system was widely adapted to study nerve injury and regeneration process. Currently a few methods of induction of nerve degeneration in microfluidic devices are available which reflect the differences in injury mechanisms *in vivo*. The physical injury of axons like those created by stretching or straining is mirrored by application of pneumatic-stress or direct compression applied within micro-chambers. These types of systems can be obtained by addition of bottom chamber passing under axons projections. This chamber is covered by flexible PDMS membrane which upward deformation is induced by increase of internal pressure in bottom chamber. This elongate the interval distance between upper chamber compartments and cause the tension of axons passing through this area (Dollé et al. 2013) (Fig. 6.4). Alternatively bottom chamber may be replaced by smaller sectors placed inside of the main chamber, which deformation induced by pressure increase causes crushing of adjacent axons (Hosmane et al. 2011). By manipulation of applied pressure intensity and its duration the different level of severity of nerve injury can be obtained in very reproducible manner (Dollé et al. 2013). These techniques allowed researches to reveal the existence of difference in sensitivity on compression-injury between particular types of neurons such as hippocampal neurons and dorsal root ganglion neurons (Magdesian et al. 2012). Another method of axonal damage relies on administration of chemical substances to axonal compartment in microfluidic chamber. Exposure to particular toxins can mimic different *in vivo* situation. The infusion of excess of neurotransmitters like glutamate was used as a model of excitotoxicity which is a cause of secondary damage in stroke (Samson et al. 2016), in addition to blood emulated haemorrhage-related axonal injury (Charier et al. 2015), whereas the treatment with detergent like Triton X-100 was a model for chemically-induced axotomy (Deleglise et al. 2013). Axon dissection can be also very efficiently obtained inside of microfluidic chamber by laser based injury which is an immensely precise and high throughput method however it demands the availability of more sophisticated equipment (Yanik et al. 2004). All these platforms gave us a unique opportunity to study the degeneration and regeneration process of axons which were exposed to toxins or treatment independently from cellular body. Moreover, cell culture in micro-devices can be easily combined with assessment of fractions (like somas or axons) isolated from particular chamber compartments by analytical techniques. The material acquired from microfluidic devices after cell culture was proven to be suitable for mRNA analysis, microarrays, biochemical assays and morphology assessment by fluorescent and Atomic Force Microscopy (Park et al. 2006; Taylor et al. 2009; Magdesian et al. 2012).

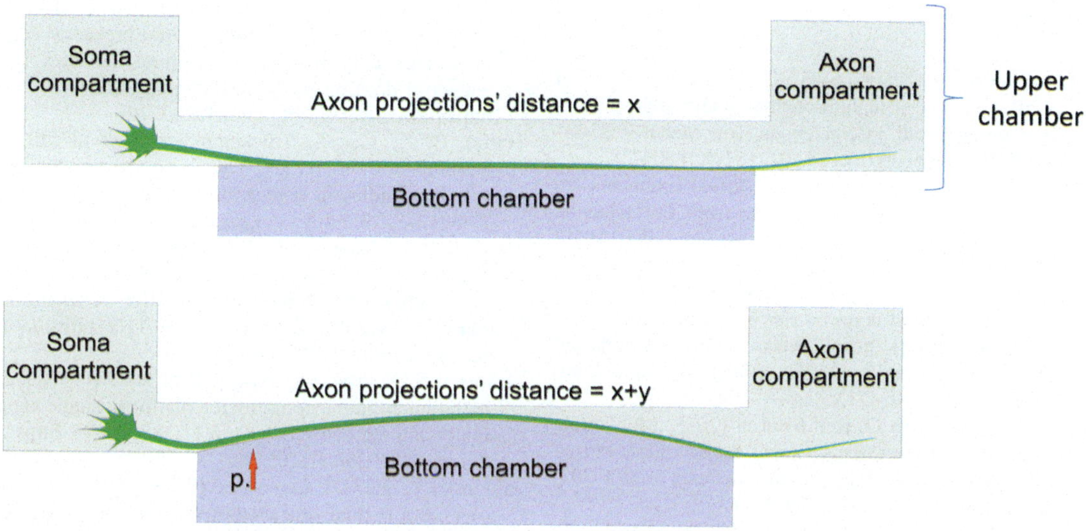

Fig. 6.4 The mechanism of induction of model of nerve injury caused by stretching or straining created in microfluidic chamber

The fusion of technologies is critical for further progress in immersing microfluidic devices within CNS studies. Implementation of sensors like electrodes inside the microfluidic chambers can enable users to extend the range of gained information with neurons functionality experiments as these cells exert electrical and electrochemical activity (Cho et al. 2013b; Kafi et al. 2016) and to provide the possibility of characterisation of compounds on which they react on (MacKerron et al. 2017). Another possible application of additional biosensors is connected with very detailed environment control including parameters like temperature, metabolic activity, oxygen level (Harink et al. 2013), or sensation of TEER value which was already mentioned above (Wang et al. 2017b). All these methods require the computer control of the experimental conditions. Therefore the development of high quality dedicated software systems is necessary to obtain quantitative results from high throughput studies, which currently are not broadly available on market (Andrzejewska et al. 2019). The future directions of microfluidic device development should be linked with multi-electrode assays and three-dimensional printing techniques in order to obtain optimal platforms to mimic such complicated systems like CNS in spatial as well as functional aspects. In turn, this approach requires very close collaboration of specialists from different disciplines such physicists, material scientists, biotechnologists and software engineers.

6.3 Conclusions

Implementation of new technologies in CNS studies is indispensable since the broadening of our knowledge introduces more and more variables which are shown to have significant impact on biological processes. Recent discoveries indicate on spatial arrangement, fluid pressure and chemical – physical surface properties as factors determining the fate of cells and course of the regeneration process. Microfluidic systems allow us to control most of these parameters thus meeting current challenges in recreation of CNS structures and functionality.

Acknowledgments This work was supported by NCR&D STRATEGMED1/235773/19/NCBR/2016 "EXPLORE ME" project and NIH R01NS091100.

References

Andrzejewska A, Nowakowski A, Grygorowicz T, et al (2019) Single-cell, high-throughput analysis of cell docking to vessel wall. J Cereb Blood Flow Metab 39:2308–2320. https://doi.org/10.1177/0271678X18805238

Berthier E, Young EWK, Beebe D (2012) Engineers are from PDMS-land, biologists are from Polystyrenia. Lab Chip 12:1224–1237. https://doi.org/10.1039/C2LC20982A

Brown JA, Pensabene V, Markov DA et al (2015) Recreating blood-brain barrier physiology and structure on chip: a novel neurovascular microfluidic bioreactor. Biomicrofluidics 9:054124. https://doi.org/10.1063/1.4934713

Charier D, Beauchet O, Bell M et al (2015) Memantine plus vitamin D prevents axonal degeneration caused by lysed blood. ACS Chem Neurosci 6:393–397. https://doi.org/10.1021/cn500303k

Cho H, Hashimoto T, Wong E et al (2013a) Microfluidic chemotaxis platform for differentiating the roles of soluble and bound amyloid-β on microglial accumulation. Sci Rep 3:1823. https://doi.org/10.1038/srep01823

Cho H-Y, Eun-Bi-Ko, Kim T-H, Choi J-W (2013b) Fabrication of carbon nanotubes/RGD peptide composites to enhance electrochemical performance of cell chip. J Biomed Nanotechnol 9:1398–1402

Cho H, Seo JH, Wong KHK et al (2015) Three-dimensional blood-brain barrier model for in vitro studies of neurovascular pathology. Sci Rep 5:15222. https://doi.org/10.1038/srep15222

Choi YJ, Chae S, Kim JH et al (2013) Neurotoxic amyloid beta oligomeric assemblies recreated in microfluidic platform with interstitial level of slow flow. Sci Rep 3:1921. https://doi.org/10.1038/srep01921

Cohen MS, Bas Orth C, Kim HJ et al (2011) Neurotrophin-mediated dendrite-to-nucleus signaling revealed by microfluidic compartmentalization of dendrites. Proc Natl Acad Sci U S A 108:11246–11251. https://doi.org/10.1073/pnas.1012401108

Das S, Srivastava VC (2016) Microfluidic-based photocatalytic microreactor for environmental application: a review of fabrication substrates and techniques, and operating parameters. Photochem Photobiol Sci 15:714–730. https://doi.org/10.1039/c5pp00469a

Deleglise B, Lassus B, Soubeyre V et al (2013) Synapto-protective drugs evaluation in reconstructed neuronal network. PLoS One 8:e71103. https://doi.org/10.1371/journal.pone.0071103

Dollé J-P, Morrison B, Schloss RS, Yarmush ML (2013) An organotypic uniaxial strain model using microfluidics. Lab Chip 13:432–442. https://doi.org/10.1039/c2lc41063j

Freundt EC, Maynard N, Clancy EK et al (2012) Neuron-to-neuron transmission of α-synuclein fibrils through axonal transport. Ann Neurol 72:517–524. https://doi.org/10.1002/ana.23747

Gorelik M, Orukari I, Wang J et al (2012) Use of MR cell tracking to evaluate targeting of glial precursor cells to inflammatory tissue by exploiting the very late antigen-4 docking receptor. Radiology 265:175–185. https://doi.org/10.1148/radiol.12112212

Harink B, Le Gac S, Truckenmüller R et al (2013) Regeneration-on-a-chip? The perspectives on use of microfluidics in regenerative medicine. Lab Chip 13:3512–3528. https://doi.org/10.1039/c3lc50293g

Harrison DJ, Glavina PG, Manz A (1993) Towards miniaturized electrophoresis and chemical analysis systems on silicon: an alternative to chemical sensors. Sens Actuators B 10:107–116. https://doi.org/10.1016/0925-4005(93)80033-8

Hosmane S, Fournier A, Wright R et al (2011) Valve-based microfluidic compression platform: single axon injury and regrowth. Lab Chip 11:3888–3895. https://doi.org/10.1039/c1lc20549h

Jablonska A, Shea DJ, Cao S et al (2018) Overexpression of VLA-4 in glial-restricted precursors enhances their endothelial docking and induces diapedesis in a mouse stroke model. J Cereb Blood Flow Metab 38:835–846. https://doi.org/10.1177/0271678X17703888

Kafi MA, Cho H-Y, Choi J-W (2016) Engineered peptide-based nanobiomaterials for electrochemical cell chip. Nano Converg 3:17. https://doi.org/10.1186/s40580-016-0077-7

Kim SH, Im S-K, Oh S-J et al (2017) Anisotropically organized three-dimensional culture platform for reconstruction of a hippocampal neural network. Nat Commun 8:14346. https://doi.org/10.1038/ncomms14346

Kothapalli CR, van Veen E, de Valence S et al (2011) A high-throughput microfluidic assay to study neurite response to growth factor gradients. Lab Chip 11:497–507. https://doi.org/10.1039/c0lc00240b

MacKerron C, Robertson G, Zagnoni M, Bushell TJ (2017) A microfluidic platform for the characterisation of CNS active compounds. Sci Rep 7:15692. https://doi.org/10.1038/s41598-017-15950-0

Magdesian MH, Sanchez FS, Lopez M et al (2012) Atomic force microscopy reveals important differences in axonal resistance to injury. Biophys J 103:405–414. https://doi.org/10.1016/j.bpj.2012.07.003

Muoio V, Persson PB, Sendeski MM (2014) The neurovascular unit – concept review. Acta Physiol Oxf Engl 210:790–798. https://doi.org/10.1111/apha.12250

Neto E, Leitão L, Sousa DM et al (2016) Compartmentalized microfluidic platforms: the unrivaled breakthrough of in vitro tools for neurobiological research. J Neurosci 36:11573–11584. https://doi.org/10.1523/JNEUROSCI.1748-16.2016

Park JW, Vahidi B, Taylor AM et al (2006) Microfluidic culture platform for neuroscience research. Nat Protoc 1:2128–2136. https://doi.org/10.1038/nprot.2006.316

Park JY, Kim S-K, Woo D-H et al (2009) Differentiation of neural progenitor cells in a microfluidic chip-generated cytokine gradient. Stem Cells 27:2646–2654. https://doi.org/10.1002/stem.202

Park J, Lee BK, Jeong GS et al (2015) Three-dimensional brain-on-a-chip with an interstitial level of flow and its application as an in vitro model of Alzheimer's disease. Lab Chip 15:141–150. https://doi.org/10.1039/c4lc00962b

Prabhakarpandian B, Shen M-C, Nichols JB et al (2013) SyM-BBB: a microfluidic blood brain barrier model. Lab Chip 13:1093–1101. https://doi.org/10.1039/c2lc41208j

Samson AJ, Robertson G, Zagnoni M, Connolly CN (2016) Neuronal networks provide rapid neuroprotection against spreading toxicity. Sci Rep 6:33746. https://doi.org/10.1038/srep33746

Sauer BM, Schmalstieg WF, Howe CL (2013) Axons are injured by antigen-specific CD8(+) T cells through a MHC class I- and granzyme B-dependent mechanism. Neurobiol Dis 59:194–205. https://doi.org/10.1016/j.nbd.2013.07.010

Song H-L, Shim S, Kim D-H et al (2014) β-Amyloid is transmitted via neuronal connections along axonal membranes. Ann Neurol 75:88–97. https://doi.org/10.1002/ana.24029

Takeshita Y, Obermeier B, Cotleur A et al (2014) An in vitro blood-brain barrier model combining shear stress and endothelial cell/astrocyte co-culture. J Neurosci Methods 232:165–172. https://doi.org/10.1016/j.jneumeth.2014.05.013

Taylor AM, Blurton-Jones M, Rhee SW et al (2005) A microfluidic culture platform for CNS axonal injury, regeneration and transport. Nat Methods 2:599–605. https://doi.org/10.1038/nmeth777

Taylor AM, Berchtold NC, Perreau VM et al (2009) Axonal mRNA in uninjured and regenerating cortical mammalian axons. J Neurosci 29:4697–4707. https://doi.org/10.1523/JNEUROSCI.6130-08.2009

Uzel SGM, Amadi OC, Pearl TM et al (2016) Simultaneous or sequential orthogonal gradient formation in a 3D cell culture microfluidic platform. Small 12:612–622. https://doi.org/10.1002/smll.201501905

Wang JD, Khafagy E-S, Khanafer K et al (2016) Organization of endothelial cells, pericytes, and astrocytes into a 3D microfluidic in vitro model of the blood-brain barrier. Mol Pharm 13:895–906. https://doi.org/10.1021/acs.molpharmaceut.5b00805

Wang Y, Ma J, Li N et al (2017a) Microfluidic engineering of neural stem cell niches for fate determination. Biomicrofluidics 11:014106. https://doi.org/10.1063/1.4974902

Wang YI, Abaci HE, Shuler ML (2017b) Microfluidic blood-brain barrier model provides in vivo-like barrier properties for drug permeability screening. Biotechnol Bioeng 114:184–194. https://doi.org/10.1002/bit.26045

Xu H, Li Z, Yu Y et al (2016) A dynamic in vivo-like organotypic blood-brain barrier model to probe metastatic brain tumors. Sci Rep 6:36670. https://doi.org/10.1038/srep36670

Yanik MF, Cinar H, Cinar HN et al (2004) Neurosurgery: functional regeneration after laser axotomy. Nature 432:822. https://doi.org/10.1038/432822a

Zhou J, Khodakov DA, Ellis AV, Voelcker NH (2012) Surface modification for PDMS-based microfluidic devices. Electrophoresis 33:89–104. https://doi.org/10.1002/elps.201100482

Microfluidics for Angiogenesis Research

7

Lígia Costa, Rui Luís Reis, Joana Silva-Correia, and Joaquim Miguel Oliveira

Abstract

Angiogenesis is a natural and vital phenomenon of neovascularization that occurs from pre-existing vasculature, being present in many physiological processes, namely in development, reproduction and regeneration. Being a highly dynamic and tightly regulated process, its abnormal expression can be on the basis of several pathologies. For that reason, angiogenesis has been a subject of major interest among the scientific community, being transverse to different areas and founding particular attention in tissue engineering and cancer research fields. Microfluidics has emerged as a powerful tool for modelling this phenomenon, thereby surpassing the limitations associated to conventional angiogenic models. Holding a tremendous flexibility in terms of experimental design towards a specific goal, microfluidic systems can offer an unlimited number of opportunities for investigating angiogenesis in many relevant scenarios, namely from its fundamental comprehension in normal physiological processes to the identification and testing of new therapeutic targets involved on pathological angiogenesis. Additionally, microvascular 3D *in vitro* models are now opening up new prospects in different fields, being used for investigating and establishing guidelines for the development of next generation of 3D functional vascularized grafts. The promising applications of this emerging technology in angiogenesis studies are herein overviewed, encompassing fundamental and applied research.

Keywords

Angiogenesis · Sprouting · Endothelization · Microfluidics · Microvascular models · Tumour angiogenesis

L. Costa · J. Silva-Correia (✉)
3B's Research Group, I3Bs – Research Institute on Biomaterials, Biodegradables and Biomimetics, University of Minho, Headquarters of the European Institute of Excellence on Tissue Engineering and Regenerative Medicine, Barco, Guimarães, Portugal

ICVS/3B's – PT Government Associate Laboratory, Braga, Guimarães, Portugal
e-mail: joana.correia@i3bs.uminho.pt

R. L. Reis · J. M. Oliveira
3B's Research Group, I3Bs – Research Institute on Biomaterials, Biodegradables and Biomimetics, University of Minho, Headquarters of the European Institute of Excellence on Tissue Engineering and Regenerative Medicine, Guimarães, Portugal

ICVS/3B's PT Government Associate Lab, Braga, Guimarães, Portugal

The Discoveries Centre for Regenerative and Precision Medicine, Headquarters at University of Minho, Guimarães, Portugal

J. M. Oliveira, R. L. Reis (eds.), *Biomaterials- and Microfluidics-Based Tissue Engineered 3D Models*, Advances in Experimental Medicine and Biology 1230,
https://doi.org/10.1007/978-3-030-36588-2_7

Highlights

- Microfluidic technology for modelling the angiogenesis process can unveil key players of angiogenesis regulation in physiological and pathological conditions.
- Complex microvascular systems on a chip can possibly investigating the effect of hemodynamic forces, cellular interactions, biochemical and biophysical stimuli involved on angiogenesis.
- Microfluidic platforms have gained prominence in angiogenesis models but the great majority of these platforms are confined to proof-of-concept models.

7.1 Introduction

Vascularization has been one of the major barriers on the clinical translation of scalable substitutes fabricated using tissue engineering (TE) strategies (Novosel et al. 2011). Although great efforts have been made over the past decade to find a gold standard approach for engineering 3D vascularized tissues, with a significant contribution of materials science and microfabrication tools, this goal still remains a virtual reality (Kim et al. 2016a). Different vascularization approaches have been introduced in order to overcome this hurdle, which can be generically divided in two main groups: (i) cell-based, and (ii) scaffold-based strategies (Novosel et al. 2011). The first assents on the intrinsic potential of vessel-forming cells for neovascularization, which underlies on two basic phenomena: vasculogenesis and angiogenesis. The first, meaning the assembly of undifferentiated endothelial cells (ECs) into a primordial microvascular system, whereas the second corresponds to the sprouting of new blood vessels from pre-existing ones (Novosel et al. 2011). Cell-based strategies consist on the incorporation of ECs on TE constructs, being the concept of *in vitro* pre-vascularization introduced as initial approach that can possibly generating a microvasculature network on the scaffolds before implantation, thus can help reducing the *in vivo* period of hypoxia and simultaneously favouring the angiogenesis by cell releasing of angiogenic factors (Laschke and Menger 2016). The functionalization of biomaterials with biological cues aiming to stimulate angiogenesis, such as the vascular endothelial growth factors (VEGFs) or the adhesive extracellular matrix (ECM)-derived peptides, especially those that specifically binds to endothelial cells (ECs), has been one of the most attempted strategies to direct *in vivo* angiogenesis on engineered tissue substitutes (Rouwkema and Khademhosseini 2016; Akintewe et al. 2017). However, it has been seen that the simple availability of angiogenic factors on engineered constructs is not sufficient to drive the formation of a mature vascular network (Rouwkema and Khademhosseini 2016). Thus, the creation of different angiogenic factors' patterns onto scaffolds and the control over their spatiotemporal releasing is nowadays seen as a potential strategy to direct/control neovascularization (Kant and Coulombe 2018). As a complex coordinated process, physical cues have also a critical role on angiogenesis modulation, being some strategies focused on the smart design of TE constructs, called scaffold-based approaches. In fact, an adequate microarchitecture of the TE substitutes, comprising a suitable porosity/interconnectivity for being easily perfusable by the host vascular system, is essential to promote the organization of the ECs into a new vascular network (Datta et al. 2017). Advanced high precision technologies, such as 3D bioprinting and photolithography, have arisen as potential tools to create highly defined microvascular network' templates on TE substitutes to guide the *in vivo* formation/organization of new vasculature (Griffith et al. 2005; Chan et al. 2012; Laschke and Menger 2016; Sudo et al. 2016).

Vascularization control has a pivotal role not only in the context of TE, but also in pathological conditions, such as cancer, in which hindering vascular network formation is a potential treatment strategy (Weis and Cheresh 2011).

Integrated approaches for understanding the biological and physical mechanisms underlying angiogenesis constitute a fundamental step to the progress on these different fields.

The progress that have been made on microfabrication techniques, such as soft lithography, 3D printing, and micropatterning techniques have turned microfluidics a potential tool to the *in vitro* miniaturization of the complex biochemical and biophysical phenomena involved on microvasculature morphogenesis (Khademhosseini et al. 2006). By means of applying microfluidic technologies to the design and fabrication of complex microvascular systems, with a tight control of fluids perfusion, a plenty of opportunities can arise for the study of hemodynamic forces, cellular interactions, biochemical and biophysical stimuli involved on angiogenesis (Weis and Cheresh 2011; Kim et al. 2017).

In vitro microvascular models have evolved from simple 2D microvascular systems, widely used to conduct simple assays such as the influence of different fluid shear stress on cellular viability and the chemotactic effect of growth factors (GFs) on cell behaviour (Lee et al. 2018), to more physiologically relevant 3D microvascular models (Bersini and Moretti 2015).

In fact, it is known the vital role of the ECM in the context of TE, so the merge of microfluidics and hydrogels, particularly extracellular matrix derived-hydrogels, have naturally arisen to better approach the 3D *in vivo* niches of vascular formation (Lewis and Gerecht 2016). Through a precise control of cell patterning, biochemical gradients and dynamic physical stimuli, it has been possible to recreate the angiogenesis process in a more realistic and physiological relevant conditions (Bersini and Moretti 2015; Lewis and Gerecht 2016).

Herein, an overview of the latest advances on microfluidics platforms contributing for the comprehension of the angiogenesis process is provided. The microvascular models unveiling key factors involved on this complex process in normal physiological and pathological scenarios are highlighted. Additionally, the development of microvascular platforms for the screening of potential drugs targeting pathological angiogenesis are also briefly reviewed.

7.2 Microfluidic Platforms for Angiogenesis Research

Angiogenesis is a complex orchestrated phenomenon of new blood vessels' formation involving the sprouting or splitting of pre-existing ones, being a vital and natural process during development and wound repair (Otrock et al. 2007). For this reason, angiogenesis research has been one of the main focus on several scientific areas, from fundamental studies, for the comprehension of the biological and biophysical mechanisms, to applied areas, such as, TE and cancer research, where its modulation is imperative. Some signalling pathways involved on angiogenesis induction have been clarified in the past decades (Carmeliet and Jain 2011). The signalling pathways associated with recognized angiogenesis' molecular inducers, such as the vascular endothelial growth factor (VEGF) (Koch and Claesson-Welsh 2012), angiopoietins (ANG) (Pafumi et al. 2015), fibroblast growth factors (FGF) and VEGF crosstalk (Song and Finley 2018), platelet derived growth factor (Dubrac et al. 2018), have been elucidated. Intercellular communication represents also a key factor during vascular morphogenesis. It has been stated that under pro-angiogenic conditions, the phenotypic specialization of the ECs into tip or stalk cells is regulated via Notch-signalling (Kofler et al. 2011). The role of several adhesion-membrane proteins (e.g., $\alpha v \beta 3$-integrin, ephrin-2B and VE-cadherin which are involved in cell-cell and cell-ECM communication) in ECs migration, sprouting and new blood vessels' stabilization, has also been described (Carmeliet and Jain 2011). The ECM remodelling comprises a natural step of vascular morphogenesis (Mongiat et al. 2016). The role of some proteolytic systems, such as plasmin or metalloproteinases (MMPs), on angiogenesis regulation, as well as, its influence by delivering ECM-degradation derived GFs and cytokines have also been reported (Mongiat et al. 2016). Besides all these regulatory' systems, hypoxia has been identified as a major player on angiogenesis, by the synergistic effect of different molecular pathways involved on angiogenesis activation, via HIF-1α complex (Krock et al. 2011). Also, the shear stress resulting from hemodynamic forces is a regulator factor of angiogenesis, by means of altering the gene expression of ECs through mechanotransduction signals, via CD31(PECAM) cell-cell junction complex, and further activation of downstream pathways (Wragg et al. 2014). Although several pathways coordinating angiogenesis have been uncovered in the past decades, angiogenesis is a highly complex and dynamic regulated process, being its underlying mechanisms in some pathological conditions and even in normal circumstances not fully understood and therefore unpredictable.

Microfluidics have emerged as a potential tool to bridge the gap between the poor representative conventional 2D culture models and the *in vivo* models with an intricate and uncontrollable variability (Bhatia and Ingber 2014; Bogorad et al. 2017). By providing micrometer-sized platforms that can be designed in different microchannels' geometries with distinct compartmentalization and using an unlimited combination of cell types/biomaterials continuously perfused with a selected fluid, it have been possible to recreate target microenvironments, such as the vascular morphogenesis in different contexts, broaden application from basic to applied research (van der Meer et al. 2009; Bogorad et al. 2017; Mannino et al. 2018b).

In fact, the flexible design of these microfluidic platforms can allow to set up different chemical and physical cues with control of their spatiotemporal deliver, have unveiled the potential of this technology, not only to the fundamental study of the subjacent mechanisms of vascular formation and regulation, but also for the construction of high-throughput models for drugs screening targeting angiogenesis (Young 2013). Moreover, since this type of technology commonly makes use of optically transparent materials, it is compatible with several imaging microscopy techniques, enabling real time monitoring and a full set of more reliable functional readouts (Young 2013).

Photolithography, a preferentially used technique to create patterns in photosensitive materials, which function as templates for replica moulding of biologically inert and optically transparent materials (e.g., polydimethylsiloxane, PDMS), has been a recurrent microfabrication method to create microvascular models (Mannino et al. 2018b). Novel microfabrication techniques have been developed to emulate the 3D microenvironment of vascular niches and overcome some constraints associated with lithography techniques, such as the requirement of specialized training with labour intensive practice and costs. The 3D printing techniques involving the use of sacrificial materials to generate microchannels directly inside of an hydrogel/scaffold, together with microfiber fabrication, comprises some of these non-traditional microfabrication methods (Mannino et al. 2018b). Although several microfabrication methods can be employed depending on the experimental design and envisioned objectives, the endothelization of these microchannels, consisting of culturing/patterning them with ECs, is an essential step towards recreating the *in vivo* vascular morphogenesis (Mannino et al. 2018a).

Some authors have categorized the *in vitro* microvascular models based on the approaches used to produce them, namely: (1) templating, and (2) self-organizing methods (Hasan et al. 2014). The templating approach refers to the methods in which hollow lumen templates are generated inside of ECM-like matrices through subtractive moulding techniques (Fig. 7.1a, b) (Sadr et al. 2011; Yoshida et al. 2013; Buchanan et al. 2014), sacrificial materials (Wu et al. 2011; Miller et al. 2012) or bioprinting (Norotte et al. 2009). The second approach is based on the innate morphogenic potential of ECs to spontaneously generate microvascular networks, under GFs gradients or co-culture of supporting cells, mainly supported by photolithography techniques. These self-assembly approach can offer a closer proximity to the *in vivo* physiological conditions (Lee et al. 2018).

Despite this approach-based categorization, microvascular models can go from the 2D endothelialized microvascular models, comprehending those in which ECs are simple lined on the microchannels without a supportive cellular matrix (Fig. 7.1c, d) (Theberge et al. 2015), usually used to conduct simple shear stress studies on ECs response (Raasch et al. 2015) or endothelium permeability studies (Sato et al. 2015), to more complex and realistic 3D models. Using a 2D approach, Theberge et al. (Theberge et al. 2015) have investigated the chemical crosstalk between ECs and macrophages during angiogenesis on a co-culture microfluidics platform (Fig. 7.1c, d) (Theberge et al. 2015). The role of macrophage derived-MMP12 as an anti-angiogenic factor was confirmed by observing tubule formation impairment in co-culture conditions (ECs cultured with fibroblasts under macrophages stimulation), which can be reversed by applying an MMP12 inhibitor (Theberge et al. 2015).

However, the importance of having the ECs in their natural *in vivo* organization, *i.e.* in lumenized structures on their phenotype and cell behaviour (Bischel et al. 2014), have prompted the use of 3D microvascular models. Supportive cellular matrices have been used for the development of more physiologically relevant 3D microvascular models. Interestingly, the 3D lumenized structures involving microfluidic devices with ECM-matrices (Bischel et al. 2013) can allow studying the vascular morphogenesis phenomenon, through vasculogenesis and angiogenesis (Fig. 7.1e–g) (Chung et al. 2009; Jeong et al. 2011a; Dai et al. 2011; Yeon et al. 2012; Kim et al. 2013a; Jeon et al. 2014; Cochrane et al. 2018; Mannino et al. 2018b). These are typically composed by interleaved controlled and conditioned flow channels (where different stimuli can be applied) connected to one or more compartments, typically micropatterned with different 3D matrices, such as fibrin, collagen or Matrigel, to emulate the ECM, where ECs are attached and allowed to sprout, with the final objective of investigate microvascular network formation in direction of an applied stimuli (Cochrane et al. 2018). For example, Zanotelli MR et al. (Zanotelli et al. 2016) have developed a 3D morphogenic-based microvascular model by using induced pluripotent stem cell-derived endothelial cells (iPSC-ECs) encapsulated in a functionalized poly(ethylene glycol) (PEG) hydrogel containing cell-adhesive and MMP-degradable peptides in a tri-channel microfluidic system (Fig. 7.1h–j). In that study, the ability of this tuned matrix towards generating *in vivo*-like lumenized vascular networks was demonstrated, thus evidencing the utility of combining supportive matrices with standard microfluidic platforms for the *in vitro* study and recapitulation of vascular morphogenesis in a more realistic and 3D context (Zanotelli et al. 2016).

The design of such types of microvascular models has been adapted according to the final objective of the study. Since these devices have been applied to the systematic study of different factors affecting angiogenesis, such as the role of GFs (Chung et al. 2009; Shin et al. 2011; Jeong et al. 2011a; Dai et al. 2011), flow conditions (Song and Munn 2011; Buchanan et al. 2014), intercellular communication (Kim et al. 2015b), tumour angiogenesis (Buchanan et al. 2014; Chung et al. 2017; Miller et al. 2018), and drugs screening targeting angiogenesis (Kim et al. 2015a; Theberge et al. 2015; Amann et al. 2017), a panoply of increasingly complex microfluidic' designs have been generated. Hence, instead of focusing on the different complexity of these microvascular models, further attention will be given to their final aim. In this context, some studies using different microvascular models for assessing the effect of several factors on angiogenesis will be discussed in the next sub-sections. Moreover, the application of these microfluidic platforms for modelling angiogenesis-related pathologic conditions and its potential use for discovery of new therapeutic targets is also reviewed.

7.2.1 Regulatory Factors Controlling Angiogenesis

Although several molecular-signalling pathways involved in angiogenesis induction have been uncovered in past years, their intricate crosstalk towards new vasculature formation, has not been fully understood. By its turn, the 3D microfluidic models of angiogenesis have both enabled the

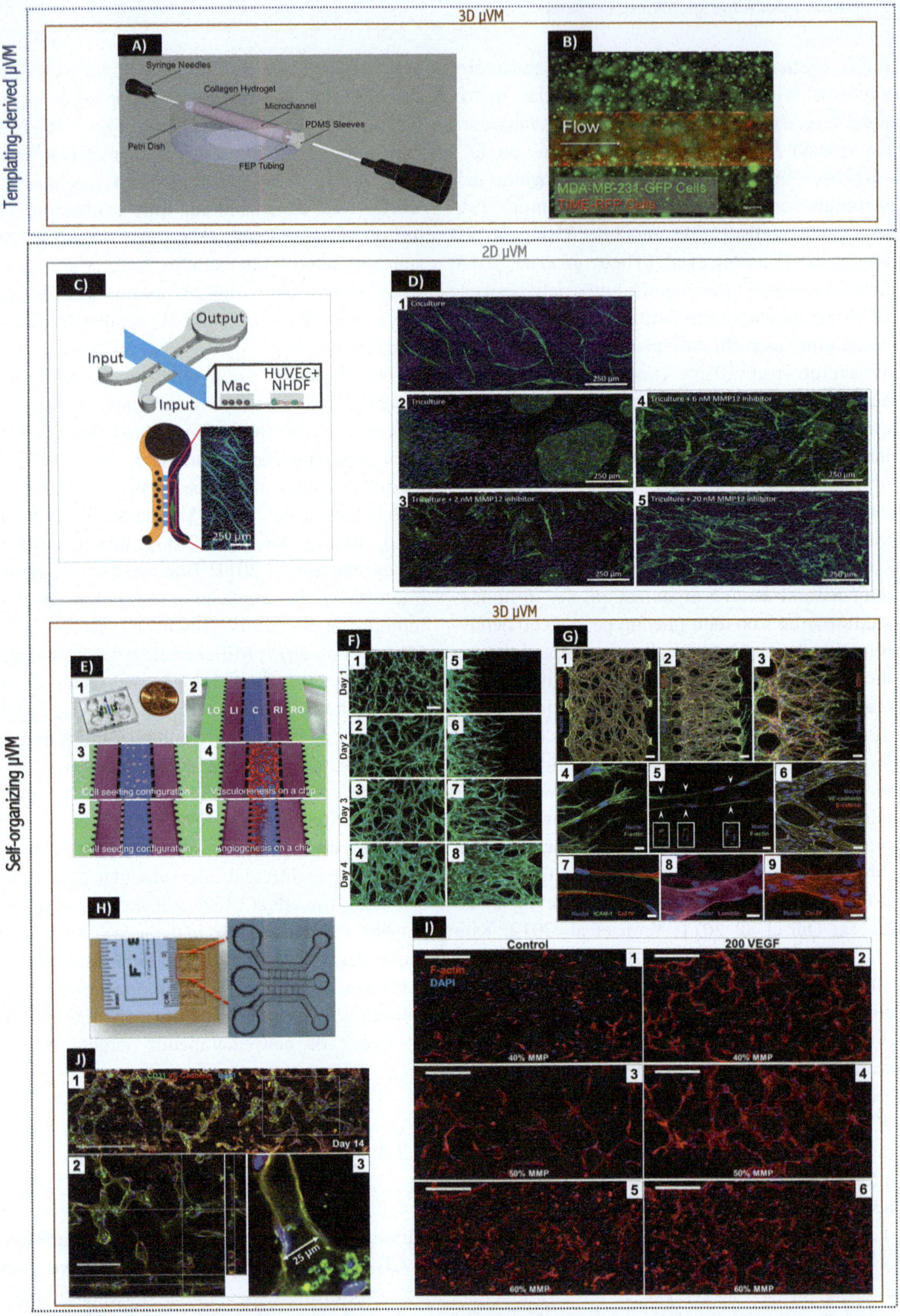

Fig. 7.1 Approach-based categorization of the microvascular models (μVM) into templating [**a**-**b**] and self-organizing methods, from the simpler 2D μVM [**c**-**d**] to the more realistic 3D μVM [**e**-**g**; **h**-**j**]. [**a**] Scheme of the 3D μVM tumour model to study the effect of wall shear stress (WSS) on tumour-endothelium paracrine signalling of angiogenesis. [**b**] Lumen formation after 72h of pre-condition flow rate in co-culture system of breast cancer

precise application of a set of chemical and physical stimulus, and their reliable and quantitative monitoring in terms of the synchronized cellular responses involved on vascular morphogenesis.

The VEGF and ANG-1 crosstalk guiding the coordinated cellular responses during angiogenesis was assessed by Shin et al. (Shin et al. 2011). The authors have investigated EC sprout into a 3D collagen-type I matrix, encased on a microfluidic device and linked to three channels, *i.e.* one serving as control and the others perfused with culture media supplemented with the biochemical factors, being the ECs cultured on the ANG-1 supplemented channel (Fig. 7.2a). The role of ANG-1 on the collective and coordinated migration of tip and stalk ECs during vascular morphogenesis was confirmed on this study. Differences on the morphogenetic behaviour of ECs under VEGF alone and VEGF combined with ANG-1 were clearly observed, thus evidencing the role of ANG-1 on enhancing and stabilizing the tip-stalk EC connections (Fig. 7.2b, c). Since the morphogenesis and life cycle of stalk ECs are regulated by direct tip-stalk cell connections, ANG-1 was identified as a major player on its regulation (Shin et al. 2011). By it turn, Nguyen et al. (Nguyen et al. 2013) have demonstrated the effect of different combined angiogenic factors on the multicellular responses involved on neovascularization and the opposite effect of angiogenic inhibitors. This biomimetic model of angiogenesis was based on two cylindrical channels (one lined with ECs and perfused with medium and the other used to apply the angiogenic stimuli) encased and linked by a 3D collagen matrix. Several proangiogenic factors (*i.e.* basic fibroblast growth factor (bFGF), hepatocyte growth factor (HGF), VEGF, monocyte chemotactic protein-1 (MCP-1), sphingosine-1-phosphate (S1P) and phorbol 12-myristate 13-acetate (PMA)), were tested individually and in different combinations. Results showed that only S1P and PMA gradients have resulted in significant EC migration, when tested independently. Although, PMA promoted a more cohesive and coordinated EC migration when compared to the EC response under S1P. Two distinct cocktails were tested and the combina-

Fig. 7.1 (continued) cells (MDA-MB-231 cells) and endothelial cells expressing a red fluorescent protein (RFP) (scale bar: 200 μm). Reprinted with permission from Buchanan et al. (Buchanan et al. 2014). Copyright © 2014 Taylor & Francis Group, LLC. [**c**] 2D self-organizing μVM configuration for the study of different cell derived-soluble factor signalling on endothelial tubule formation. [**d**] Fluorescence micrographs of the right channel containing the co-culture system (ECs and fibroblasts) immunocytochemistry stained for CD31 (green) and nuclei (DAPI-blue) in different conditions (co-culture/tri-culture systems and in the presence of MMP12 inhibitor). Reprinted with permission from Theberge et al. (Theberge et al. 2015). Copyright © 2015 American Chemical Society. [**e**] Schematic representation of a 3D self-morphogenic μVM for studying vasculogenesis and angiogenesis. The microfluidic platform comprises the outermost channels (LO-left outside and RO-right outside) for human normal lung fibroblasts (LFs) seeding, flanked with two fluidic channels and one central channel. For vasculogenesis experiment [**e**.3-4] HUVECs-loaded fibrin matrix were patterned in the central channel, while for angiogenesis study [**e**.5-6], HUVECs were attached on the side of central channel patterned with acellular 3D fibrin matrix, while the fibroblasts were only cultured in the same matrix on the RO channel. [**f**] Micrographs of endothelial network formation by vasculogenesis [**f**.1-4] and angiogenesis [**f**.5-8]. [**g**] Immunofluorescence images of 3D endothelial networks stained for CD31 (red), F-actin (green) and nuclei (blue) formed by vasculogenesis [**g**.1] and angiogenesis [**g**.2-3] (scale bar: 50 μm). F-actin-rich filopodia extension of angiogenic sprouts [**g**.4] (scale bar: 20 μm) and hollow lumen formation [**g**.5] (scale bar: 10 μm). [**g**.6] Immunofluorescence images of the microvascular networks stained for junction proteins (scale bar: 50 μm) and for vessel maturation markers, ICAM-1 [**g**.7] (scale bar: 10 μm), laminin [**g**.8] and collagen IV [**g**.9] (scale bar: 20 μm). Reprinted with permission from Kim et al. (Kim et al. 2013a). Copyright © 2013 Royal Society of Chemistry. [**h**] Scheme of the 3D μVM comprising a tri-channel interconnected system: one central channel and two fluidic channels. Confocal images of microvascular network formation from iPSC-ECs, encapsulated in different MMP degradable peptide-crosslinked PEG matrices (polymerized into the central channel) and cultured in basal medium [**i**.1, 3, 5] or in VEGF-supplemented medium [**i**.2, 4,6] (scale bar: 250 μm). Capillary network formation in 50% crosslinked PEG hydrogel stained for CD31 (green), VE-Cadherin (red) and nuclei (blue-DAPI) at different magnifications [**j**.1, 2, 3]. Scale bars: 250 μm [**j**.1] and 100 μm [**j**.2]. Reprinted with permission from Zanotelli et al. (Zanotelli et al. 2016). Copyright © 2016 Acta Materialia Inc. Published by Elsevier Ltd.

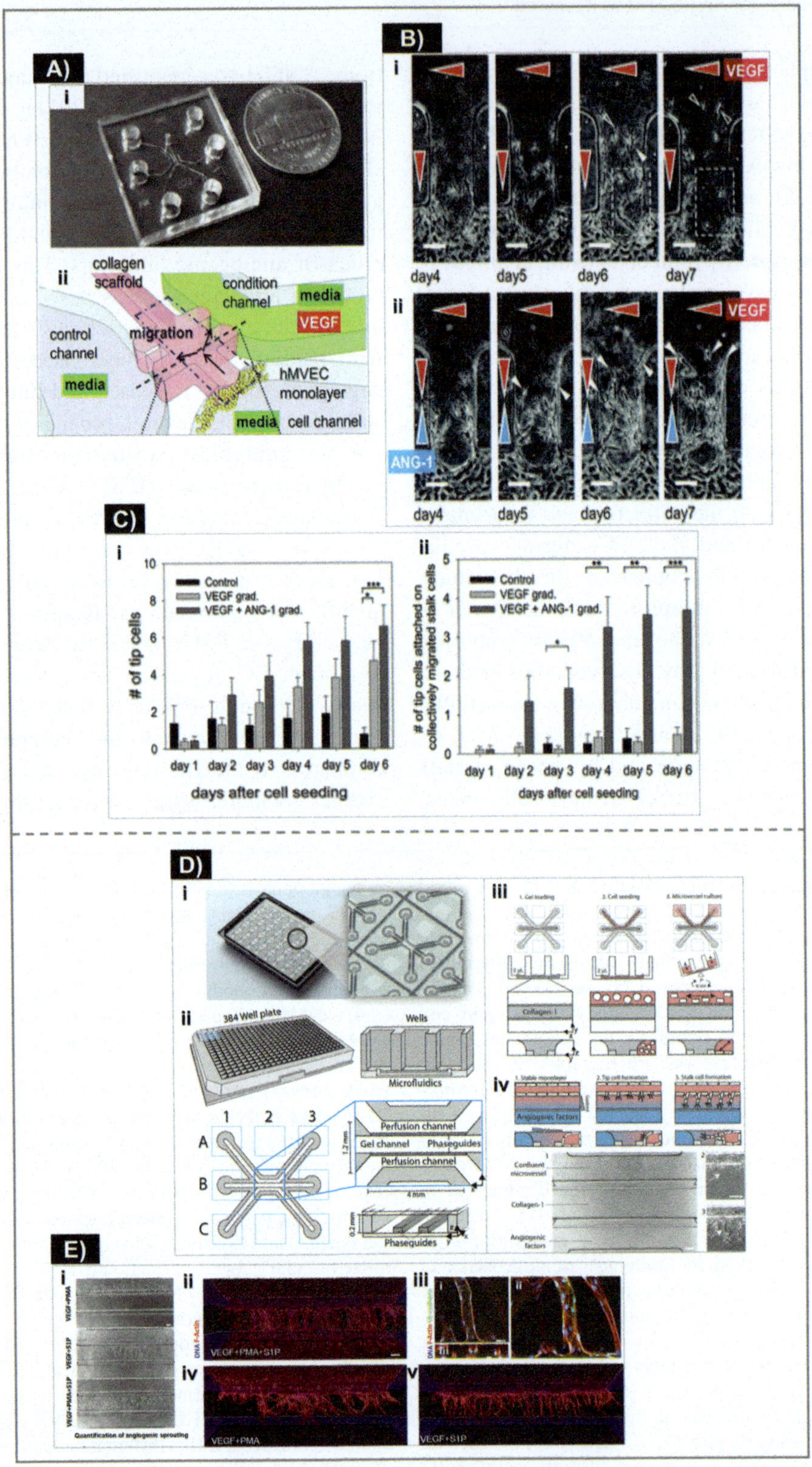

Fig. 7.2 Microvascular platforms for the study of the biochemical cues driving angiogenesis. [**a**.i-ii] Microfluidic system and its schematic representation to investigate the VEGF-ANG-1 crosstalk on the 3D cooperative angiogenic sprout response. [**b**] Angiogenic sprouts formation into the collagen scaffold, under the effect of horizontal and vertical VEGF gradients [**b**.i] and under the same VEGF applied gradients supplemented

tion containing MCP-1, VEGF, PMA, and S1P has shown a superior effect on multicellular sprouting response, resembling better the *in vivo* vascular morphogenesis. The effects of a VEGF receptor inhibitor on the angiogenic sprouting of the developed systems were also evaluated. The results unveiled a VEGF-induced filopodia formation and sprout elongation in a context-dependent manner (Nguyen et al. 2013). The proximity of this *in vitro* model to the *in vivo* vascular morphogenesis process, which recapitulated all the coordinated steps involved on angiogenesis, such as sprouting, elongation and lumen formation/stabilization, could be potentially applied to the study of the molecular mechanisms driving neovascularization.

In a similar study, Farahat et al. (Farahat et al. 2012) have developed a 3D high-throughput microfluidic platform for the study of two proangiogenic factors combination, VEGF and S1P, on the EC response. The authors have demonstrated the synergistic effect of both factors on angiogenesis sprouting induction, particularly on influencing the growth and direction of the sprouts (Farahat et al. 2012). Inspired on this microfluidic architecture, Del Amo et al. (2016), have investigated the effect of some GFs, involved on regeneration processes, namely VEGF, PDGF-BB, TGFβ and BMP-2 on the initial stage of endothelial sprout. The results suggested that VEGF is among the main angiogenesis regulators, being its concentration and spatiotemporal location decisive for sprouting evolution (Del Amo et al. 2016).

Recently, van Duinen et al. (2019) have developed a robust and standardized angiogenesis model on a microfluidic platform for high-throughput assessment of gradients driving this process. The scalability of this microvascular platform was achieved by combining an array of 40 microfluidic devices, integrated underneath a 384-well plate. Maintaining the platform under passive levelling perfusion on a rocker platform, reproductible gradients were generated and different experiments were allowed. Each device was composed by three PhaseGuides-interconnected channels: one central channel patterned with collagen type I, serving as sprouting channel, and two adjacent perfusion channels (one for culturing ECs and the other to generate different angiogenic gradients after the establishment of a confluent endothelial monolayer) (Fig. 7.2d). Results have proven the utility of the platform to conduct *in vitro* relevant studies of factors driving induction and regression of microvasculature, due to its physiological resemblance with the *in vivo* angiogenesis EC multicellular responses. Moreover, an optimal cocktail for the generation of a robust angiogenic sprouting response, was identified, being composed by VEGF-165, PMA and S1P, with the last having a critical role on the directional guidance and maintenance of angiogenesis sprouting (Fig. 7.2e). Under prolonged exposure to gradients, anastomosis have occurred, forming perfusable stable lumens with proper barrier function and sprouts regression, emulating the *in vivo* neovessel stabilization and maturation (van Duinen et al. 2019). As it can be noticed from

Fig. 7.2 (continued) with ANG-1 [**b**.ii]. The white arrowheads point to tip-stalk cell sprouting structures while empty arrowheads indicate isolated tip cells (scale bar: 100 mm). [**c**] Quantification of the number of tip cells [**c**.i] and the number of tip cells attached to stalk cells sprouting structures [**c**.ii] under VEGF gradients or ANG-1 supplemented VEGF gradients (*$P < 0.05$; **$P < 0.01$; ***$P < 0.001$). Reprinted with permission from Shin et al. (Shin et al. 2011). Copyright © 2011 Royal Society of Chemistry. [**d**] Schematic representation of the standardized microfluidic platform for the high-throughput screening' factors governing angiogenesis [**d**.i-ii]. Microvessel culturing 'steps [**d**.iii] on the collagen' adjacent perfusion channel, for the investigation of ECs sprouting response under gradients imposed on the other perfusion channel [**d**.iv.1]. Tip cell filopodia formation observed, at day 1, with lumen formation visible at day 2 [**d**.iv.2,3], after GFs gradient application. Angiogenic sprouts formation under different angiogenic factor' combinations (VEGF+PMA; VEGF+S1P and VEGF+PMA + S1P), after 4 days of stimulation [E.i] and the respective sprouts, after 6 days of stimulation, stained for F-actin (red) and nucleus (blue) [**e**.ii,iv,v]. Close up view of angiogenic sprouting under VEGF+PMA + S1P, stained for F-actin (red), nucleus (blue) and VE-cadherin (green) (Scale bars: 100 μm) [**e**.iii]. Tip-stalk cell structures with filopodia and lumen formation [**e**.iii], traversing the entire gel channel [**e**.ii] under the combination VEGF+PMA + S1P. Reprinted with permission from van Duinen et al. (2019). Copyright © 2018, Springer Nature

Fig. 7.2, microfluidic angiogenic models have been accepted and broadly used to study the effect of several biochemical-angiogenesis modulators and their crosstalk on ECs behaviour during the different vascular morphogenic steps.

Mechanical forces, imposed by both blood flow and interstitial plasma are also key regulators of vascular morphogenesis and remodelling, having a profound impact on EC gene expression and behaviour. It has been reported that variations of the normal shear stress, experienced by a quiescent endothelium, can induce intracellular signalling pathways mediated by EC-mechanosensing molecules, such as the CD31 platelet endothelial cell adhesion complex (Wragg et al. 2014; Lertkiatmongkol et al. 2016). Nitric oxide (NO) has also been identified as mediator of shear stress-induced angiogenesis (Wragg et al. 2014). Although it is known that angiogenesis is a tightly and dynamically regulated process by biochemical and mechanical signalling, the interplay between these factors have been poorly explored. Song and Munn (2011) have investigated the crosstalk between mechanical forces, imposed by tangential shear stress and transverse interstitial flow, and VEGF, in the vascular morphogenesis regulation. That study was conducted on a microfluidic device model for angiogenesis sprouting. The microfluidic platform comprised two EC lined microchannels, where controllable fluid forces were imposed, as well as, established VEGF gradients, interconnected by a collagen compartment for EC sprouting. The study demonstrated that physiological shear stress reduces VEGF-induced angiogenesis in a NO-signalling dependent manner. By its turn, interstitial flow and VEGF interplay have shown a synergistic effect, promoting EC invasion and sprouting with tip cell filopodia formation against the transverse applied flow and directed to VEGF increasing gradients (Song and Munn 2011).

Using a microfluidic device Galie et al. (2014) have reported a common shear stress threshold value (~10 dyn/cm^2) for both luminal and transmural flow, in which angiogenesis sprouting is induced through up-regulation of the downstream effectors, such as the MMP-1, critical on 3D cell migration and sprouting (Galie et al. 2014).

In another study, the individual and combined effect of interstitial flow (IF) and proangiogenic factors, such as VEGF an S1P, were investigated (Kim et al. 2016b). It was reported a microfluidic platform comprehending six parallel microchannels interconnected by arrays of microposts. The two outermost channels were cultured with stromal fibroblasts (known to secret pro-angiogenic factors), the immediately adjacent two intermediate microchannels served as fluidic channels, and the two central microchannels were patterned/cultured using two different configurations. One was cultured with ECs embedding on a fibrin matrix (for EC assembly on a vascular network through vasculogenesis) and the other patterned with the acellular fibrin matrix (serving as a sprouting channel) (Fig. 7.3a). Different directional IF pressures were applied, as well as, spatially defined pro-angiogenic gradients were established by varying the fibroblast densities' ratio between the outermost channels. The results

Fig. 7.3 (continued) (scale bar: 100 μm). Representative fluorescence images of the angiogenic sprouting response (stained for nuclei-blue, F-actin-green and CD31-red) to different directional applied flow pressures (indicated by white arrows) [**b**.i] (scale bar: 200 μm). Higher magnification images of the angiogenic sprouts depicted on the dotted boxes [**b**.ii], with an increase of actin-rich filopodia formation (yellow arrows) in response to opposite sprout-directional applied IF (scale bar: 40 μm). Quantification of total sprout area [**b**.iii] and total microvascular network area [**b**.iv]. Reprinted with permission from Kim et al. (2016b). Copyright © 2016 Royal Society of Chemistry. [**c**] Schematic configuration of the microfluidic system to investigate the effect of biomechanical matrix properties, namely density, on endothelial sprouting response, under appropriate established VEGF gradients. Representative phase contrast images of endothelial sprout morphology at different matrix densities [**d**]. Favoured endothelial multicellular response with the formation of stable elongated sprouts at intermediate collagen densities [**b**.iii-iv]. Sprouting features of ECs-coated microbeads at different collagen matrix densities [**e**], depicting the total number of sprouts per bead [**e**.i], the average sprout thickness [**e**.ii], the average aspect ratio (sprout length to width) [**e**.iii] and the number of new sprouts at different time points normalized to the total number of observed sprouts [**e**.iv] ($*p < 0.05$; $**p < 0.01$). Reprinted with permission from Shamloo et al. (Shamloo and Heilshorn 2010). Copyright © 2010 Royal Society of Chemistry

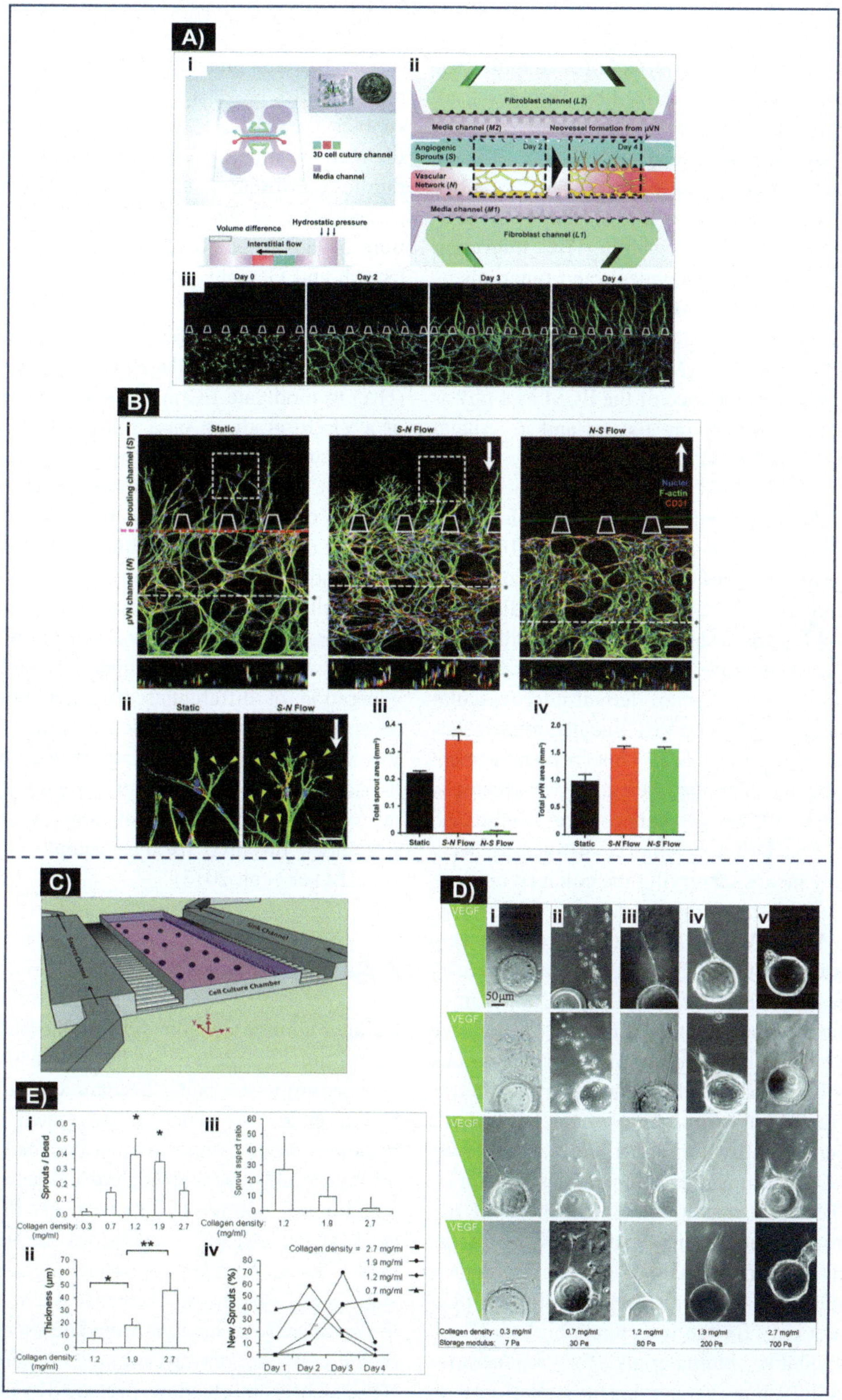

Fig. 7.3 Microvascular platforms for the study of the biophysical cues driving angiogenesis. [**a**] Microfluidic device to investigate the influence of interstitial flow (IF) on microvascular morphogenesis, under pro-angiogenic factors-fibroblast stimulation, with its schematic configuration [**a**.i-ii]. Sequential assembly of a microvascular network, through vasculogenesis, on channel N and angiogenic sprouting into channel S, after 4 days [**a**.iii]

have shown a synergistic effect of IF and the pro-angiogenic factors synthetized by fibroblasts, on the directional guidance and stimulation of endothelial sprouting (Kim et al. 2016b) (Fig. 7.3b).

The ECM role on angiogenesis regulation has also been described. It is known that ECM provides adhesive motifs enabling EC migration during vascular morphogenesis and functions as a pool of ECM-remodelling derived biomolecules having a critical and intricate role on angiogenesis regulation (Mongiat et al. 2016). Although the importance of the ECM as a physical support of the living tissues and its crucial role on cell behaviour is consensual, the biomechanical properties of ECM on angiogenesis regulation have been neglected. Shamloo and Heilshorn (Shamloo and Heilshorn 2010) have investigated the influence of ECM properties, such as matrix density/stiffness and VEGF cross-talk on ECs sprouting and lumen stabilization. The researchers have used a microfluidic device for culturing adult human dermal microvascular endothelial cells (HDMVEC)-coated microbeads on a 3D collagen/fibronectin matrix compartment connected by microcapillaries to two channels, where the VEGF gradients were established (Fig. 7.3c). Different matrix stiffnesses were tested, by means of varying the collagen concentration on the matrix blend. Several VEGF (e.g., VEGF-A and VEGF-165) concentrations were also tested. Results suggested that matrix density mediates VEGF-induced sprouting and lumen formation by having critical role on the balance between endothelial tip cell migration and stalk cell proliferation. While higher matrices' densities promoted the formation of thicker sprouts with better ability for lumen formation, lower stiffnesses impaired lumen stabilization (Fig. 7.3d–e) (Shamloo and Heilshorn 2010). That study stated the importance of ECM biomechanical properties on angiogenesis regulation, evidencing the utility of this microfluidic platform, not only for the investigation of mechanically suitable biomaterials for regenerative applications, but also serving as a platform for the discovery of new potential biochemical therapies targeting ECM-related pathological angiogenesis.

Given the high adaptability of microfluidic platforms to be designed in such a way that enables the *in vitro* recreation of a target 3D microenvironment, it is not surprising that microfluidics constitute an usefulness platform for the investigation of the guidelines for the construction of advanced vascularized TE substitutes (Smith and Gerecht 2014). Jeong et al. (2011b) have reported the ability of an ECM-like hydrogel system composed by a semi-interpenetrating network (SIPNs) of collagen and hyaluronic acid (HA) to modulate EC migration in the presence of a VEGF gradient, on a study conducted on a three channel microfluidic device. In other study, Baker et al. (2013) have investigated the influence of different molecular inducers' diffusion patterns on vascular morphogenesis, namely on the location and morphology of EC sprouting. Using different lithography-defined sacrificial templates, different microchannel outlines were drawn within a collagen matrix, enabling the generation of differential diffusion patterns of several GFs. Results have demonstrated that, EC sprouting is defined on the highest gradient locations, being the concerted EC migration/sprouting with the typical formation of stalk-like structures, favoured by intermediate GF gradients (Baker et al. 2013).

7.2.2 Cellular Interactions

Cellular communication and cellular-ECM interaction are key players on homeostasis maintenance in any tissue, being this intercommunication fundamental to generate integrated cellular responses when homeostasis is perturbed, such as on regeneration/wound healing processes or on pathological conditions (e.g. cancer) (Sakthivel et al. 2019). All these conditions are accompanied by vascular remodelling through angiogenesis-based processes. In such scenario, the organized and interconnected quiescent endothelial cells monolayer (via junction molecules-VE-cadherin and claudins) protected by pericytes, initiates a cascade of cellular responses. Briefly, in response to an angiogenic signal, pericytes detach from the basement membrane of blood vessels by proteolytic degradation mediated by MMPs. Subsequently, ECs lose their junctions and the new forming vessel dilates. The permeability of the growing endothelium increases (via VEGF signalling) leading to a parallel increase of plasma proteins extravasation creating a temporary ECM. ECs migrate in direction of this provisional ECM, through integrin signalling and the simultaneous remodelling of ECM release pro-angiogenic factors providing an angiogenic-propitious microenvironment. The new vessel is then formed by the concerted action of the selected tip EC (DLL4 and JAGGED1), which migrates, through integrins mediated-adhesion, in the direction of signalling molecules (semaphorins and ephrins), and the stalk EC, which divides and elongates the new capillary. The stabilization/maturation of the newly formed blood vessel is crucial for its functionality, consisting in the covering of the ECs layer by pericytes, the deposition of a basement membrane and ECs-junctional re-establishment (Fig. 7.4) (Carmeliet and Jain 2011).

The role of pericytes during the angiogenesis process was studied on a microfluidic platform by Kim et al. (2015b). In that study, the authors recreated *in vitro*, the sequential process of angiogenesis: sprouting, elongation and maturation, by means of using a microfluidic device. This device was composed of a central acellular channel, patterned with a 3D fibrin matrix, connected to two adjacent media channels, and an outermost fibroblast cultured channel. A co-culture of endothelial cells (HUVECS) and pericytes were attached to one side of the fibrin central channel and the sprouting was directionally observed in response of the biochemical factors generated by dermal and lung-derived fibroblasts. *In vivo*-like sprouting, elongation and stabilization of the newly formed blood vessels, with pericytes covering, was successfully recapitulated in the presence of the biochemical factors secreted by dermal fibroblasts. Thus, the results evidenced that pericytes play an important role on both vessel diameter and branching regulation. Moreover, the effects of VEGF and inflammatory cytokines on pericytes-ECs interaction were assessed. The results were in agreement with previous studies, showing morphological alterations of the ECs and pericytes in response to this biochemical stimulus. Together, the results demonstrated the utility of this platform as a dynamic model of pericytes and EC communication during angiogenesis, to study pathological situations, as well as, for drug screening of potential therapeutics targeting abnormal ECs-pericytes cell signalling observed, for instance, in tumour-induced angiogenesis (Kim et al. 2015b).

Angiogenesis is a dynamically regulated process dependent on the activation/suppression of integrated molecular and cellular signalling pathways. Comprehending the cell-interaction and cell-ECM involved on vascular morphogenesis is of major interest for several research areas, in which, the control of this process is crucial. Microfluidics have been applied to the study of cellular interactions involved on angiogenesis and their abnormal communication in pathological conditions, by providing suitable co-culture systems for the *in vitro* recreation of vascular morphogenesis under targeted microenvironments (Khademhosseini et al. 2006; Sakthivel et al. 2019).

The importance of EC source for the *in vitro* reconstruction of organ-specific microvascular models was asserted by H. Uwamori et al. (2019), who conducted a study in a trichannel microfluidic device using different ECs sources: brain microvascular endothelial cells (BMECs) and HUVEC in co-culture with mesenchymal cells, for modelling the blood brain barrier (BBB). The different co-culture systems (MSC with HUVEC and MSC with BMECs) were attached to one side of the fibrin gel prepatterned on the central compartment, where microvascular networks were allowed to grow. Results from permeability tests of the developed microvasculatures and junction proteins' expression profiles showed significant differences in terms of ECs morphology and barrier function. It was demonstrated that BMECs formed more robust endothelial barrier, being less permeable than the microvascular networks formed with HUVECS, due to higher expression of tight junction proteins. The barrier function of the BMEC-microvasculature, under induced inflammatory conditions by addition of

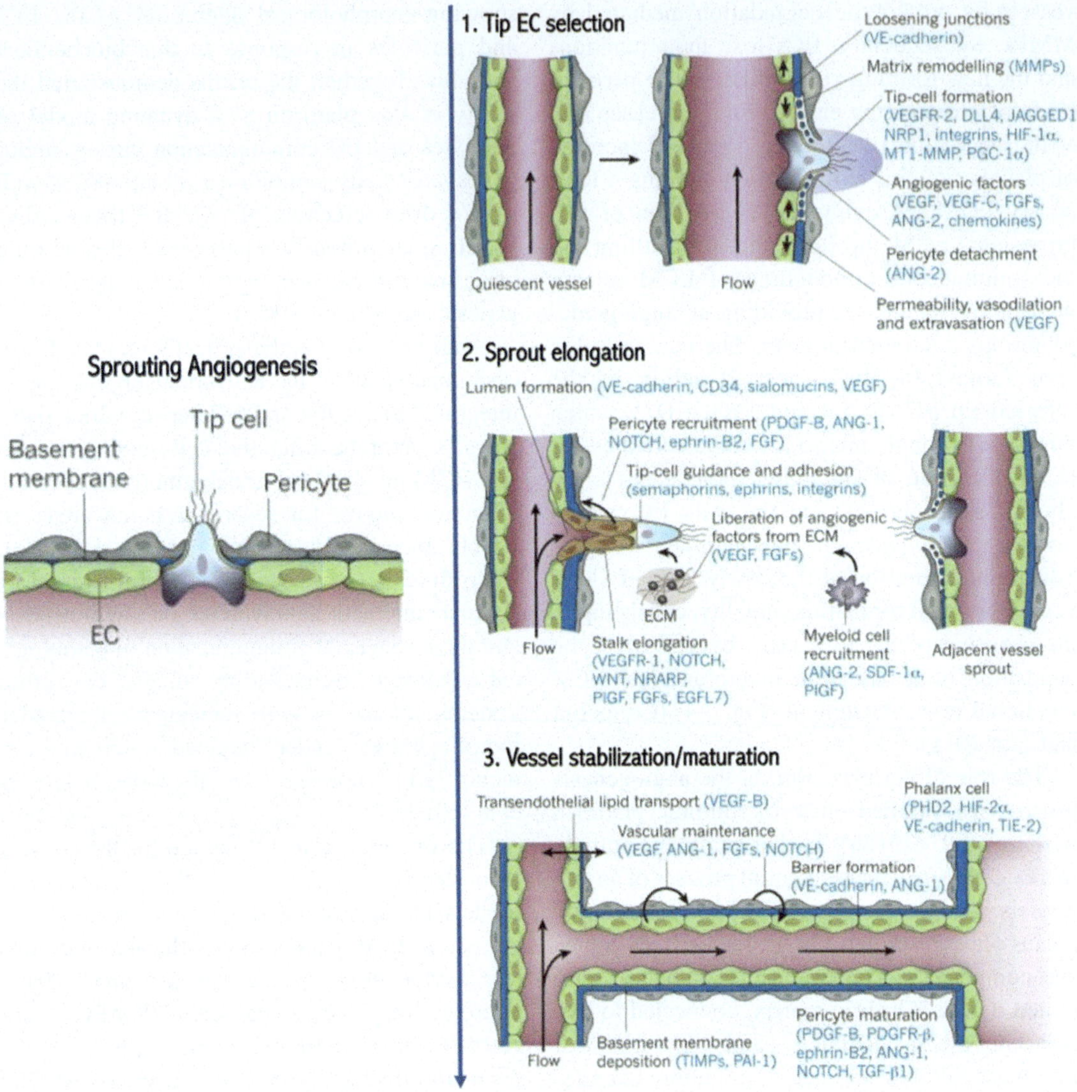

Fig. 7.4 Molecular signalling involved on neovascularization trough angiogenesis sprouting. Under a pro-angiogenic microenvironment, the quiescent endothelium dilates, and the tip EC is selected to initiate vessel branching. The process is accompanied by degradation of basement membrane, pericyte detachment and EC junctions loosening. The permeability of the branching vessel increases and the plasm proteins-extravasion provides a temporary ECM, while its degradation liberates angiogenic stimulators. Tip EC migrates in the direction of signalling molecules accompanied by stalk cells proliferation allowing sprout elongation and lumen formation. After sprouts fusion (anastomosis), stalk cells recruit pericytes for neovessel stabilization. EC junctions are re-established, the basement membrane deposited, and the vessel maturation promoted by maintenance signals. Reprinted with permission from Carmeliet and Jain (2011). Copyright © 2011 Springer Nature

thrombin in cell culture media, has shown an increased permeability with leaky characteristics. This organ specific-endothelium model mimicking BBB functions can be useful to conduct studies of BBB-associated pathologies (Uwamori et al. 2019).

7.2.3 Modelling Pathological Conditions

Beyond the critical importance of angiogenesis in normal physiological processes, such as, in the context of tissue regeneration, its aberrant regulation is in the origin of many pathological conditions (Chung and Ferrara 2011), such as in inflammatory scenarios (Costa et al. 2007; Kim et al. 2013b; Varricchi et al. 2015), arthritis (Marrelli et al. 2011; Elshabrawy et al. 2015), autoimmune diseases (Liakouli et al. 2011; Liu et al. 2015), and cancer progression and metastasis (Gavalas et al. 2013; Multhoff et al. 2014; Bielenberg and Zetter 2015), among others. Molecular and cellular mechanisms involved on pathological angiogenesis in different circumstances, have been investigated on *in vitro* microvascular platforms (Abaci et al. 2013). Choroidal pathological-angiogenesis associated to macular degeneration was investigated by Chen et al. (2017). In that study, the ocular fundus was recreated through the set-up of a co-culture system of human retinal pigmental epithelial (ARPE-19) cells and HUVECs, separated by a porous membrane in a microfluidic device. The abnormal secretion of VEGFs, appointed as the trigger of pathological choroidal vascularization, was imposed on the co-culture microfluidic system through a combination of low glucose and hypoxic conditions. Results demonstrated that the changes on the co-culture microenvironment can increase the VEGF levels secreted by ARPE-19 cells, leading to an induced directional growth of HUVECs, which seems to cause ARPE-19 monolayer detachment (Chen et al. 2017).

The potential of these microvascular systems for modelling complex vascular microenvironments, in physiological and pathological conditions, and predict ECs response under drug effect, have been demonstrated by Abaci et al. (Abaci et al. 2015). The authors have developed a microvascular model to study EC phenotype changes in pathological settings, such as on ischemic conditions associated to atherosclerosis. The physiological relevance of the microvascular model was assessed by treating the healthy microvascular system with a commonly used vasotoxic anticancer drug, 5-Fluorouracil (5-FU), in combination with a vasoprotective drug (Resveratrol), being the outcomes compared with an *in vivo* mouse model. Similarly to what have been reported on *in vivo* studies (Beck et al. 2000; Hayashi et al. 2003; Skuli et al. 2012), ischemic conditions trigger the transition of a quiescent endothelium to an activated angiogenic state, with higher proliferation rates, up-regulation of angiogenic genes and compromised endothelial barrier function. In agreement with previous data, the treatment with 5-FU revealed that the drug does not promote EC apoptosis. Nevertheless, it impairs EC barrier function by means of interfering with adhesive cell junctions, which can be attenuated by co-administration with Resveratrol. This microvascular model has, thus, proven its potential for being used as a physiologically realistic platform to investigate EC responses/phenotypes in healthy and diseased conditions and to screen potential drug therapies (Abaci et al. 2015).

The great majority of these microvascular platforms have been devoted to recapitulate the complex tumour microenvironment (TME) (Michna et al. 2018), for studying the interplay between its components during tumour development and progression. The intricate interaction between endothelial cells, tumour (Lin et al. 2017) and stromal (Chung et al. 2017) cells has been investigated using microfluidic platforms.

Chung et al. (Chung et al. 2017) have studied tumour cell-stroma interaction in a paracrine fashion and angiogenesis driven by tumour-stroma complex on a microfluidic platform. In that study, it was evidenced the differential influence of stroma on morphological changes of different tumour cell types. Moreover, proangiogenic factors resulting from different tumour cells-fibroblasts coculture systems (simulating tumour stroma), have stimulated the angiogenic sprouting. Although, differences between the ability of different tumour cell types on EC sprouting induction, phenotype and on the permeability of new formed blood vessels were noticed (Chung et al. 2017).

By it turn, it has been proposed an integrated microfluidic platform which allows the simultaneous investigation of tumour-EC interactions (through a co-culture system of HUVECs and cervical carcinoma cells (CaSki), identification of cell secreted angiogenic proteins, and evaluation of paclitaxel drug resistance (Lin et al. 2017). It was noticed that HUVECs can induce paclitaxel-drug resistance in co-culture conditions. The potential of this integrated platform for studying/monitoring tumour-EC interaction, for identification of tumour biomarkers and, simultaneously, for functioning as an anti-cancer drug screening platform was evidenced on that study.

The promising use of such technology for engineering vascularized tumour models that can be useful for investigating the complex molecular and cellular mechanisms involved on cancer progression and metastasis (Wang et al. 2018), namely, by modelling tumour angiogenesis will be highlighted in the following sections.

7.2.3.1 Tumour Angiogenesis Study

Tumour angiogenesis is a natural process of tumour growth and development, by establishing a vascular blood supply to support tumour cells' viability and proliferation. Tumour microenvironment comprises a pool of signalling molecules and pathways driving the uncontrolled activation of the quiescent endothelium by creating pro-angiogenic conditions (Weis and Cheresh 2011). The formation of tortuous and dysregulated tumour vascular networks, also characterized by the lack of proper barrier functions (due to a deficient pericyte coverage), has also been closely associated to tumour metastasis (Bielenberg and Zetter 2015). Considering that a vast majority of cancer deaths results from metastatic cases, the understanding of the complex phenomenon behind tumour angiogenesis development is imperative for the identification of new targets for therapeutic treatment.

Microfluidic technologies hold a tremendous potential to contribute to the progress on this field, by providing co-culture cell platforms, where specific tumour microenvironments can be modelled, and the intricate tumour-EC interactions studied in order to possibly identifying key molecular and cellular factors involved in tumour angiogenesis, while new therapeutic drugs can be tested (Caballero et al. 2017; Sontheimer-Phelps et al. 2019). Several cancer models have been developed on microfluidic platforms (Sontheimer-Phelps et al. 2019), aiming to elucidate the complex multistep involved on cancer metastasis, being the existing model mainly focused on cancer intravasion (Zervantonakis et al. 2012) and extravasion processes (Chen et al. 2013; Jeon et al. 2013).

Tumour angiogenesis, a putative target for cancer treatment (Zhao and Adjei 2015), have been, yet, underexplored using microfluidic approaches, considering the potential of this technology, not only, to the fundamental comprehension of this process, but also to the high-throughput screening of potential drugs targeting tumour angiogenesis.

Lee et al. (2014) have developed a microfluidic platform modelling tumour angiogenesis and metastasis. The reported microfluidic chip was organized in seven microchannels, interconnected by a set of microposts, spatially organized to culture different cells and to allow the analysis of different processes. Briefly, microvessel was generated by vasculogenesis of HUEVCs, which were able to self-assemble on a perfusable microvessel under lung fibroblasts-angiogenic factors' stimuli. The immediately adjacent microchannel was divided into two smaller microchannels: the upper and bridge channels, serving for cancer culture, in different configurations, according to the objective of study (cancer angiogenesis or metastasis). For cancer angiogenesis study, human glioblastoma multiform cells (U87MG), embedded in a fibrinogen solution, were patterned into the upper channel while acellular fibrinogen was injected into the adjacent bridge channel. The angiogenic stimulation of the surrounding microvasculature by cancer cells-secreted factors was successfully modelled on this *in vitro* platform. A significant formation of sprouting structures compared with the control group, with no angiogenic-cancer stimulation, was clearly noticed. Moreover, under bevacizumab treatment, a common anti-VEGF drug, the angiogenic sprouting was blocked (Lee et al. 2014). Although such type of *in vitro* model was able to recapitulate the tumour angiogenesis and intravasion of cancer cells, more realistic platforms which can better resemble the tumour microenvironment and EC-tumour communication during these processes are in great need.

In a different study, Zheng et al. (2016) have studied leukaemia-induced bone marrow angiogenesis on a microfluidic device comprising three parallel and interconnected microchannels. The microfluidic platform was composed by two cell culture channels, one for endothelial cell culture (HUVECs) and the other for leukemic cell monocultures (e.g., U937, HL60, and K562 leukemic tested cell lines) or co-cultures with bone marrow stromal cell line (e.g., HS5). These channels were interfaced by a collagen-patterned compartment, and both were perfused with endothelial growth medium (EGM). Results have shown the different vascular morphogenic ability, assigned by different leukemic cell lines. Moreover, in co-culture systems, it was evident a more disconcerted EC response, with formation of endothelial sprouts without the typical multicellular stalk structures and a higher number of isolated tip EC, when compared to the EC response obtained under stromal cell monoculture conditions. Although the results were in agreement with leukaemia patient-derived bone marrow samples, the proposed microfluidic model of leukemic-induced bone marrow-angiogenesis has to be improved in order to better mimic the complex bone marrow microenvironment (Zheng et al. 2016).

By it turn, Buchanan et al. (2014) have investigated how shear stress influence the paracrine signalling involved in angiogenesis activation, on a microvascular tumour model. Using a microfluidic system composed by one perfusable microchannel, lined with ECs and concentrically co-cultured with breast cancer cells, encased onto a collagen type I matrix, different flow shear stress rates were experimented. Results have shown that flow shear stress can modulate the angiogenic phenotype of tumour cells in a paracrine fashion, since higher values of shear stress have promoted a decrease on tumour-expression of angiogenic factors (Buchanan et al. 2014).

Miller et al. (2018) has reported the first 3D human renal carcinoma model on a microfluidic chip. In that study, primary human renal carcinoma cells from patient donors were used. The primary-donor derived tumour cell clusters were embedded in a collagen matrix and patterned around an engineered human microvessel subjected to continuous flow. Results showed that the human renal carcinoma cells stimulate ECs sprouting, being their phenotype, namely the expression of angiogenic factors, donor-dependent. The endothelial sprouting was then successfully blocked by means of adding a vascular endothelial growth factor receptor 2 (VEGFR2) blocking chimera. This patient-specific vascularized tumour model has, thus, proven its potential to the discovery of new targets for blocking tumour angiogenesis in a personalized therapeutic approach (Miller et al. 2018).

7.2.4 Drugs Screening

A deep understanding of the biochemical, biophysical and cellular interplay during this process on specific conditions would opening new opportunities for the discover of new potential targets for therapeutic treatment (Yoo and Kwon 2013).

Surpassing the natural constrains associated to the conventional cell culture systems, microfluidics has arisen as powerful technology for *in vitro* modelling specific conditions, in a more relevant physiological context, allowing the study of angiogenesis in different scenarios. Although the potential and versatility of microfluidic platforms for fundamental research and also their promising application as preclinical models for drug screening and testing is recognized, their routine application in biological research and acceptance in pharmaceutical industry is, yet, limited (Lee et al. 2018). Therefore, efforts are now directed to turn these platforms as primary experimental models for clinical research trials, focusing more in their end-applicability design instead of model's development and validation.

For instance, Kim et al. (2015a) have developed a high-throughput microfluidic platform for the quantitative assessment effect of an anti-angiogenic drug. This microfluidic device comprised three main components: i) the ECM-like chamber patterned with collagen and linked to a tumour-mimic channel, where the VEGF gradients were generated, and ii) an endothelial cell culture channel, allowing the simultaneous application of different concentrations of the anti-angiogenic drug bortezomib. This quantitative platform has demonstrated its potential by means of enabling the determination of a specific concentration of bortezomib inhibiting VEGF-induced angiogenesis, with minimal cytotoxic effect on ECs (Kim et al. 2015a).

Targeting a different pathological-angiogenesis condition, such as the vascular remodelling associated with idiopathic pulmonary fibrosis, Zeinali et al. (2018) have created a human microvascular model to study the effects of nintedanib, a potent inhibitor of tyrosine-kinases-angiogenic receptors, on vascular remodelling. The microfluidic chip was composed by one central chamber, for co-culture of healthy-patient derived normal lung fibroblasts (NL-FBs) and HUVECs in a fibrin matrix, interconnected with two perfusable microchannels, which were connected to two outermost channels, cultured with NL-FBs in the same matrix. After the assembly of a perfusable microvascular network, the system was treated with three different concentrations of nintedanib. The results have shown that nintedanib strongly decreases microvascular density by inducing the weakening of the ECs intercellular junctions, by inhibiting the supportive and stabilizing role of fibroblasts on the existing microvasculature and by, simultaneously, suppressing endothelial sprouting. Moreover, it was observed that the drug increases the permeability of the existing microvasculature (Zeinali et al. 2018). Although the microvascular model has demonstrated its potential for the study of cell behaviour under therapeutic drugs, a more realistic model of the pathological cellular niche would prompt this type of microvascular models to patient-specific models, where new therapeutics drugs can be screened in a personalized approach.

Carvalho et al. (2019) have developed a colorectal tumour-on-a-chip model comprising a microvascular part to access the effect and efficacy of anti-cancer drugs delivered by nanoparticles. The chip was composed by a core chamber, where the tumoral niche was created, by culturing colorectal cancer cells (HCT-116 cells) into a 3D Matrigel supplemented with VEGF, laterally connected to two microfluidic channels, cultured with human colonic microvascular endothelial cells (HCoMECs) and perfused with anticancer drug-loaded nanoparticles. Although the main objective of that study was to recreate a physiologically relevant model of colorectal cancer to function as a nanomedicines screening platform, the natural ability of the tumour niche, in the presence of VEGF, to induce endothelial sprouting of the HCoMECs, suggested the ability of this platform for further investigation of key targets involved on tumour-angiogenesis induction and the simultaneous screening of drugs targeting this process (Carvalho et al. 2019).

By it turn, Phan et al. (2017) have developed a promising custom-fit 96 well plate array of 12 standardized and highly reproductible vascularized micro-organs, for high-throughput screening of potential drugs. As a proof-of-concept, vascularized microtumours (VMT), consisting of colorectal cancer cell line HCT-116 embedded in ECs and stromal cells, were established in the platform and used to blindly test a set of FDA-approved anti-cancer drugs. Anti-angiogenic and anticancer drugs were successfully identified, unveiling the potential of this platform for screening drug effect on multiple tissues (Phan et al. 2017).

7.3 Concluding Remarks

In the past decade, the advance of microfabrication techniques has enabled a great progress in microfluidic technologies, by providing new tools for the designing of more complex and sophisticated platforms. In fact, microfluidics has an unparallel potential for *in vitro* mimicking 3D complex cellular microenvironments in a closer proximity to physiological conditions. For this reason, microfluidics holds the potential to fill the gap between the poor representative conventional culture systems and the complex *in vivo* models with their intrinsic variability and unpredictable associated' factors.

By means of offering an unlimited number of channel geometry designs, cellular/biomaterials combination and a controlled spatiotemporal delivery of biochemical and biophysical cues, by a continuous and controllable fluid perfusion systems, this technology have been providing singular assessment of cellular behaviour, under targeted conditions. All these advantages have positioned microfluidic platforms as promising primary experimental *in vitro* models.

Although several angiogenesis regulatory molecules and signalling pathways have been identified/clarified in the past years, the understanding of its regulation in different contexts, *i.e.* signalling mechanisms in a specific surrounding microenvironment, such as, on pathological scenarios, is imperative.

By incorporating vasculature on these microfluidic platforms, it has been possible to increase the complexity and physiological relevance of organ-on-a-chip models and, by the other side, to study angiogenesis in a variety of normal physiological and pathological contexts. Beyond the tremendous potential of this microvascular models to elucidate angiogenesis mechanisms in pathological scenarios, they hold promising application as direct drug screening platforms for simultaneous identification and testing of new potential targets/drugs for pathological angiogenesis. Additionally, microfluidic platforms incorporating vasculature founds also potential application for the screening of biomaterials driving angiogenesis overcoming the major hurdle that TE faces, the lack of vascularization of developed grafts.

Although these microfluidic platforms have gained prominence in disease models and for pharmaceutical research, the great majority of these platforms are confined to proof-of-concept models. Therefore, now the main challenge is to move forward this technology from its theoretical potential to its practical application on fundamental and applied research. The process would pass for the simplification and standardization of these platforms, turning this technology more accessible by the end-users, thus, encouraging its acceptance as integral part of a pharmaceutic research line.

Acknowledgments The authors would like to acknowledge the financial support provided by the Portuguese Foundation for Science and Technology (FCT) through the project B-FABULUS (PTDC/BBB-ECT/2690/2014) and Fun4TE (PTDC/EMD-EMD/31367/2017). The FCT distinctions attributed to J. Silva-Correia (IF/00115/2015) and J. Miguel Oliveira (IF/01285/2015) under the Investigator FCT program are also greatly acknowledged.

References

Abaci HE, Drazer G, Gerecht S (2013) Recapitulating the vascular microenviroment in microfluidic platforms. Nano Life 03:1340001. https://doi.org/10.1142/S1793984413400011

Abaci HE, Shen Y, Tan S, Gerecht S (2015) Recapitulating physiological and pathological shear stress and oxygen to model vasculature in health and disease. Sci Rep 4:4951. https://doi.org/10.1038/srep04951

Akintewe OO, Roberts EG, Rim N-G et al (2017) Design approaches to myocardial and vascular tissue engineering. Annu Rev Biomed Eng 19:389–414. https://doi.org/10.1146/annurev-bioeng-071516-044641

Amann A, Zwierzina M, Koeck S et al (2017) Development of a 3D angiogenesis model to study tumour – endothelial cell interactions and the effects of anti-angiogenic drugs. Sci Rep 7:2963. https://doi.org/10.1038/s41598-017-03010-6

Baker BM, Trappmann B, Stapleton SC et al (2013) Microfluidics embedded within extracellular matrix to define vascular architectures and pattern diffusive gradients. Lab Chip 13:3246. https://doi.org/10.1039/c3lc50493j

Beck H, Acker T, Wiessner C et al (2000) Expression of Angiopoietin-1, Angiopoietin-2, and tie receptors after middle cerebral artery occlusion in the rat. Am J Pathol 157:1473–1483. https://doi.org/10.1016/S0002-9440(10)64786-4

Bersini S, Moretti M (2015) 3D functional and perfusable microvascular networks for organotypic microfluidic models. J Mater Sci Mater Med 26:180. https://doi.org/10.1007/s10856-015-5520-5

Bhatia SN, Ingber DE (2014) Microfluidic organs-on-chips. Nat Biotechnol 32:760–772. https://doi.org/10.1038/nbt.2989

Bielenberg DR, Zetter BR (2015) The contribution of angiogenesis to the process of metastasis. Cancer J 21:267–273. https://doi.org/10.1097/PPO.0000000000000138

Bischel LL, Young EWK, Mader BR, Beebe DJ (2013) Tubeless microfluidic angiogenesis assay with three-dimensional endothelial-lined microvessels. Biomaterials 34:1471–1477. https://doi.org/10.1016/j.biomaterials.2012.11.005

Bischel LL, Sung KE, Jiménez-Torres JA et al (2014) The importance of being a lumen. FASEB J 28:4583–4590. https://doi.org/10.1096/fj.13-243733

Bogorad MI, DeStefano J, Wong AD, Searson PC (2017) Tissue-engineered 3D microvessel and capillary network models for the study of vascular phenomena. Microcirculation 24:e12360. https://doi.org/10.1111/micc.12360

Buchanan CF, Verbridge SS, Vlachos PP, Rylander MN (2014) Flow shear stress regulates endothelial barrier function and expression of angiogenic factors in a 3D microfluidic tumor vascular model. Cell Adhes Migr 8:517–524. https://doi.org/10.4161/19336918.2014.970001

Caballero D, Blackburn SM, de Pablo M et al (2017) Tumour-vessel-on-a-chip models for drug delivery. Lab Chip 17:3760–3771. https://doi.org/10.1039/C7LC00574A

Carmeliet P, Jain RK (2011) Molecular mechanisms and clinical applications of angiogenesis. Nature 473:298–307. https://doi.org/10.1038/nature10144

Carvalho MR, Barata D, Teixeira LM et al (2019) Colorectal tumor-on-a-chip system: A 3D tool for precision onco-nanomedicine. Sci Adv 5:eaaw1317. https://doi.org/10.1126/sciadv.aaw1317

Chan JM, Zervantonakis IK, Rimchala T et al (2012) Engineering of in vitro 3D capillary beds by self-directed angiogenic sprouting. PLoS One 7:e50582. https://doi.org/10.1371/journal.pone.0050582

Chen MB, Whisler JA, Jeon JS, Kamm RD (2013) Mechanisms of tumor cell extravasation in an in vitro microvascular network platform. Integr Biol 5:1262. https://doi.org/10.1039/c3ib40149a

Chen L, Ito S, Kai H et al (2017) Microfluidic co-cultures of retinal pigment epithelial cells and vascular endothelial cells to investigate choroidal angiogenesis. Sci Rep 7:3538. https://doi.org/10.1038/s41598-017-03788-5

Chung AS, Ferrara N (2011) Developmental and pathological angiogenesis. Annu Rev Cell Dev Biol 27:563–584. https://doi.org/10.1146/annurev-cellbio-092910-154002

Chung S, Sudo R, Zervantonakis IK et al (2009) Surface-treatment-induced three-dimensional capillary morphogenesis in a microfluidic platform. Adv Mater 21:4863–4867. https://doi.org/10.1002/adma.200901727

Chung M, Ahn J, Son K et al (2017) Biomimetic model of tumor microenvironment on microfluidic platform. Adv Healthc Mater 6:1700196. https://doi.org/10.1002/adhm.201700196

Cochrane A, Albers HJ, Passier R et al (2018) Advanced in vitro models of vascular biology: human induced pluripotent stem cells and organ-on-chip technology. Adv Drug Deliv Rev. https://doi.org/10.1016/j.addr.2018.06.007

Costa C, Incio J, Soares R (2007) Angiogenesis and chronic inflammation: cause or consequence? Angiogenesis 10:149–166. https://doi.org/10.1007/s10456-007-9074-0

Dai X, Cai S, Ye Q et al (2011) A novel in vitro angiogenesis model based on a microfluidic device. Chin Sci Bull 56:3301. https://doi.org/10.1007/s11434-011-4717-3

Datta P, Ayan B, Ozbolat IT (2017) Bioprinting for vascular and vascularized tissue biofabrication. Acta Biomater 51:1–20. https://doi.org/10.1016/j.actbio.2017.01.035

Del Amo C, Borau C, Gutiérrez R et al (2016) Quantification of angiogenic sprouting under different growth factors in a microfluidic platform. J Biomech 49:1340–1346. https://doi.org/10.1016/j.jbiomech.2015.10.026

Dubrac A, Künzel SE, Künzel SH et al (2018) NCK-dependent pericyte migration promotes pathological neovascularization in ischemic retinopathy. Nat Commun 9:3463. https://doi.org/10.1038/s41467-018-05926-7

Elshabrawy HA, Chen Z, Volin MV et al (2015) The pathogenic role of angiogenesis in rheumatoid arthritis. Angiogenesis 18:433–448. https://doi.org/10.1007/s10456-015-9477-2

Farahat WA, Wood LB, Zervantonakis IK et al (2012) Ensemble analysis of Angiogenic growth in three-dimensional microfluidic cell cultures. PLoS One 7:e37333. https://doi.org/10.1371/journal.pone.0037333

Galie PA, Nguyen D-HT, Choi CK et al (2014) Fluid shear stress threshold regulates angiogenic sprouting. Proc Natl Acad Sci 111:7968–7973. https://doi.org/10.1073/pnas.1310842111

Gavalas N, Liontos M, Trachana S-P et al (2013) Angiogenesis-related pathways in the pathogenesis of ovarian Cancer. Int J Mol Sci 14:15885–15909. https://doi.org/10.3390/ijms140815885

Griffith CK, Miller C, Sainson RCA et al (2005) Diffusion limits of an in vitro thick prevascularized tissue.

Tissue Eng 11:257–266. https://doi.org/10.1089/ten.2005.11.257

Hasan A, Paul A, Vrana NE et al (2014) Microfluidic techniques for development of 3D vascularized tissue. Biomaterials 35:7308–7325. https://doi.org/10.1016/j.biomaterials.2014.04.091

Hayashi T, Noshita N, Sugawara T, Chan PH (2003) Temporal profile of angiogenesis and expression of related genes in the brain after ischemia. J Cereb Blood Flow Metab 23:166–180. https://doi.org/10.1097/01.WCB.0000041283.53351.CB

Jeon JS, Zervantonakis IK, Chung S et al (2013) In vitro model of tumor cell extravasation. PLoS One 8:e56910. https://doi.org/10.1371/journal.pone.0056910

Jeon JS, Bersini S, Whisler JA et al (2014) Generation of 3D functional microvascular networks with human mesenchymal stem cells in microfluidic systems. Integr Biol 6:555–563. https://doi.org/10.1039/C3IB40267C

Jeong GS, Han S, Shin Y et al (2011a) Sprouting angiogenesis under a chemical gradient regulated by interactions with an endothelial monolayer in a microfluidic platform. Anal Chem 83:8454–8459. https://doi.org/10.1021/ac202170e

Jeong GS, Kwon GH, Kang AR et al (2011b) Microfluidic assay of endothelial cell migration in 3D interpenetrating polymer semi-network HA-Collagen hydrogel. Biomed Microdevices 13:717–723. https://doi.org/10.1007/s10544-011-9541-7

Kant RJ, Coulombe KLK (2018) Integrated approaches to spatiotemporally directing angiogenesis in host and engineered tissues. Acta Biomater 69:42–62. https://doi.org/10.1016/j.actbio.2018.01.017

Khademhosseini A, Langer R, Borenstein J, Vacanti JP (2006) Microscale technologies for tissue engineering and biology. Proc Natl Acad Sci 103:2480–2487. https://doi.org/10.1073/pnas.0507681102

Kim S, Lee H, Chung M, Jeon NL (2013a) Engineering of functional, perfusable 3D microvascular networks on a chip. Lab Chip 13:1489. https://doi.org/10.1039/c3lc41320a

Kim Y-W, West XZ, Byzova TV (2013b) Inflammation and oxidative stress in angiogenesis and vascular disease. J Mol Med 91:323–328. https://doi.org/10.1007/s00109-013-1007-3

Kim C, Kasuya J, Jeon J et al (2015a) A quantitative microfluidic angiogenesis screen for studying anti-angiogenic therapeutic drugs. Lab Chip 15:301–310. https://doi.org/10.1039/C4LC00866A

Kim J, Chung M, Kim S et al (2015b) Engineering of a biomimetic Pericyte-covered 3D microvascular network. PLoS One 10:1–15. https://doi.org/10.1371/journal.pone.0133880

Kim JJ, Hou L, Huang NF (2016a) Vascularization of three-dimensional engineered tissues for regenerative medicine applications. Acta Biomater 41:17–26. https://doi.org/10.1016/j.actbio.2016.06.001

Kim S, Chung M, Ahn J et al (2016b) Interstitial flow regulates the angiogenic response and phenotype of endothelial cells in a 3D culture model. Lab Chip 16:4189–4199. https://doi.org/10.1039/C6LC00910G

Kim S, Kim W, Lim S, Jeon J (2017) Vasculature-on-a-chip for in vitro disease models. Bioengineering 4:8. https://doi.org/10.3390/bioengineering4010008

Koch S, Claesson-Welsh L (2012) Signal transduction by vascular endothelial growth factor receptors. Cold Spring Harb Perspect Med 2:a006502–a006502. https://doi.org/10.1101/cshperspect.a006502

Kofler NM, Shawber CJ, Kangsamaksin T et al (2011) Notch signaling in developmental and tumor angiogenesis. Genes Cancer 2:1106–1116. https://doi.org/10.1177/1947601911423030

Krock BL, Skuli N, Simon MC (2011) Hypoxia-induced angiogenesis: good and evil. Genes Cancer 2:1117–1133. https://doi.org/10.1177/1947601911423654

Laschke MW, Menger MD (2016) Prevascularization in tissue engineering: current concepts and future directions. Biotechnol Adv 34:112–121. https://doi.org/10.1016/j.biotechadv.2015.12.004

Lee H, Park W, Ryu H, Jeon NL (2014) A microfluidic platform for quantitative analysis of cancer angiogenesis and intravasation. Biomicrofluidics 8:054102. https://doi.org/10.1063/1.4894595

Lee S, Ko J, Park D et al (2018) Microfluidic-based vascularized microphysiological systems. Lab Chip 18:2686–2709. https://doi.org/10.1039/C8LC00285A

Lertkiatmongkol P, Liao D, Mei H et al (2016) Endothelial functions of platelet/endothelial cell adhesion molecule-1 (CD31). Curr Opin Hematol 23:253–259. https://doi.org/10.1097/MOH.0000000000000239

Lewis DM, Gerecht S (2016) Microfluidics and biomaterials to study angiogenesis. Curr Opin Chem Eng 11:114–122. https://doi.org/10.1016/j.coche.2016.02.005

Liakouli V, Cipriani P, Marrelli A et al (2011) Angiogenic cytokines and growth factors in systemic sclerosis. Autoimmun Rev 10:590–594. https://doi.org/10.1016/j.autrev.2011.04.019

Lin L, Lin X, Lin L et al (2017) Integrated microfluidic platform with multiple functions to probe tumor–endothelial cell interaction. Anal Chem 89:10037–10044. https://doi.org/10.1021/acs.analchem.7b02593

Liu J, Wang X, Yang X et al (2015) Investigating the role of angiogenesis in systemic lupus erythematosus. Lupus 24:621–627. https://doi.org/10.1177/0961203314556293

Mannino RG, Pandian NKR, Jain A, Lam WA (2018a) Engineering "endothelialized" microfluidics for investigating vascular and hematologic processes using non-traditional fabrication techniques. Curr Opin Biomed Eng 5:13–20. https://doi.org/10.1016/j.cobme.2017.11.006

Mannino RG, Qiu Y, Lam WA (2018b) Endothelial cell culture in microfluidic devices for investigating

microvascular processes. Biomicrofluidics 12:042203. https://doi.org/10.1063/1.5024901
Marrelli A, Cipriani P, Liakouli V et al (2011) Angiogenesis in rheumatoid arthritis: a disease specific process or a common response to chronic inflammation? Autoimmun Rev 10:595–598. https://doi.org/10.1016/j.autrev.2011.04.020
Michna R, Gadde M, Ozkan A et al (2018) Vascularized microfluidic platforms to mimic the tumor microenvironment. Biotechnol Bioeng 115:2793–2806. https://doi.org/10.1002/bit.26778
Miller JS, Stevens KR, Yang MT et al (2012) Rapid casting of patterned vascular networks for perfusable engineered three-dimensional tissues. Nat Mater 11:768–774. https://doi.org/10.1038/nmat3357
Miller CP, Tsuchida C, Zheng Y et al (2018) A 3D human renal cell carcinoma-on-a-chip for the study of tumor angiogenesis. Neoplasia 20:610–620. https://doi.org/10.1016/j.neo.2018.02.011
Mongiat M, Andreuzzi E, Tarticchio G, Paulitti A (2016) Extracellular matrix, a hard player in angiogenesis. Int J Mol Sci 17:1822. https://doi.org/10.3390/ijms17111822
Multhoff G, Radons J, Vaupel P (2014) Critical role of aberrant angiogenesis in the development of tumor hypoxia and associated radioresistance. Cancers (Basel) 6:813–828. https://doi.org/10.3390/cancers6020813
Nguyen D-HT, Stapleton SC, Yang MT et al (2013) Biomimetic model to reconstitute angiogenic sprouting morphogenesis in vitro. Proc Natl Acad Sci 110:6712–6717. https://doi.org/10.1073/pnas.1221526110
Norotte C, Marga FS, Niklason LE, Forgacs G (2009) Scaffold-free vascular tissue engineering using bioprinting. Biomaterials 30:5910–5917. https://doi.org/10.1016/j.biomaterials.2009.06.034
Novosel EC, Kleinhans C, Kluger PJ (2011) Vascularization is the key challenge in tissue engineering. Adv Drug Deliv Rev 63:300–311. https://doi.org/10.1016/j.addr.2011.03.004
Otrock Z, Mahfouz R, Makarem J, Shamseddine A (2007) Understanding the biology of angiogenesis: review of the most important molecular mechanisms. Blood Cells Mol Dis 39:212–220. https://doi.org/10.1016/j.bcmd.2007.04.001
Pafumi I, Favia A, Gambara G et al (2015) Regulation of Angiogenic functions by angiopoietins through calcium-dependent signaling pathways. Biomed Res Int 2015:1–14. https://doi.org/10.1155/2015/965271
Phan DTT, Wang X, Craver BM et al (2017) A vascularized and perfused organ-on-a-chip platform for large-scale drug screening applications. Lab Chip 17:511–520. https://doi.org/10.1039/C6LC01422D
Raasch M, Rennert K, Jahn T et al (2015) Microfluidically supported biochip design for culture of endothelial cell layers with improved perfusion conditions. Biofabrication 7:015013. https://doi.org/10.1088/1758-5090/7/1/015013
Rouwkema J, Khademhosseini A (2016) Vascularization and angiogenesis in tissue engineering: beyond creating static networks. Trends Biotechnol 34:733–745. https://doi.org/10.1016/j.tibtech.2016.03.002
Sadr N, Zhu M, Osaki T et al (2011) SAM-based cell transfer to Photopatterned hydrogels for microengineering vascular-like structures. Biomaterials 32:7479–7490. https://doi.org/10.1016/j.biomaterials.2011.06.034
Sakthivel K, O'Brien A, Kim K, Hoorfar M (2019) Microfluidic analysis of heterotypic cellular interactions: a review of techniques and applications. TrAC Trends Anal Chem. https://doi.org/10.1016/j.trac.2019.03.026
Sato M, Sasaki N, Ato M et al (2015) Microcirculation-on-a-chip: A microfluidic platform for assaying blood- and lymphatic-vessel permeability. PLoS One 10:e0137301. https://doi.org/10.1371/journal.pone.0137301
Shamloo A, Heilshorn SC (2010) Matrix density mediates polarization and lumen formation of endothelial sprouts in VEGF gradients. Lab Chip 10:3061. https://doi.org/10.1039/c005069e
Shin Y, Jeon JS, Han S et al (2011) In vitro 3D collective sprouting angiogenesis under orchestrated ANG-1 and VEGF gradients. Lab Chip 11:2175. https://doi.org/10.1039/c1lc20039a
Skuli N, Majmundar AJ, Krock BL et al (2012) Endothelial HIF-2α regulates murine pathological angiogenesis and revascularization processes. J Clin Invest 122:1427–1443. https://doi.org/10.1172/JCI57322
Smith Q, Gerecht S (2014) Going with the flow: microfluidic platforms in vascular tissue engineering. Curr Opin Chem Eng 3:42–50. https://doi.org/10.1016/j.coche.2013.11.001
Song M, Finley SD (2018) Mechanistic insight into activation of MAPK signaling by pro-angiogenic factors. BMC Syst Biol 12:145. https://doi.org/10.1186/s12918-018-0668-5
Song JW, Munn LL (2011) Fluid forces control endothelial sprouting. Proc Natl Acad Sci 108:15342–15347. https://doi.org/10.1073/pnas.1105316108
Sontheimer-Phelps A, Hassell BA, Ingber DE (2019) Modelling cancer in microfluidic human organs-on-chips. Nat Rev Cancer 19:65–81. https://doi.org/10.1038/s41568-018-0104-6
Sudo R, Chung S, Shin Y, Tanishita K (2016) Integrated vascular engineering: vascularization of reconstructed tissue. In: Tanishita K, Yamamoto K (eds) Vascular engineering. Springer Japan, Tokyo, pp 297–332
Theberge AB, Yu J, Young EWK et al (2015) Microfluidic multiculture assay to analyze biomolecular signaling in angiogenesis. Anal Chem 87:3239–3246. https://doi.org/10.1021/ac503700f
Uwamori H, Ono Y, Yamashita T et al (2019) Comparison of organ-specific endothelial cells in terms of microvascular formation and endothelial barrier functions. Microvasc Res 122:60–70. https://doi.org/10.1016/j.mvr.2018.11.007
van der Meer AD, Poot AA, Duits MHG et al (2009) Microfluidic technology in vascular research. J Biomed Biotechnol 2009:1–10. https://doi.org/10.1155/2009/823148

van Duinen V, Zhu D, Ramakers C et al (2019) Perfused 3D angiogenic sprouting in a high-throughput in vitro platform. Angiogenesis 22:157–165. https://doi.org/10.1007/s10456-018-9647-0

Varricchi G, Granata F, Loffredo S et al (2015) Angiogenesis and lymphangiogenesis in inflammatory skin disorders. J Am Acad Dermatol 73:144–153. https://doi.org/10.1016/j.jaad.2015.03.041

Wang X, Sun Q, Pei J (2018) Microfluidic-based 3D engineered microvascular networks and their applications in vascularized microtumor models. Micromachines 9:493. https://doi.org/10.3390/mi9100493

Weis SM, Cheresh DA (2011) Tumor angiogenesis: molecular pathways and therapeutic targets. Nat Med 17:1359–1370. https://doi.org/10.1038/nm.2537

Wragg JW, Durant S, McGettrick HM et al (2014) Shear stress regulated gene expression and angiogenesis in vascular endothelium. Microcirculation 21:290–300. https://doi.org/10.1111/micc.12119

Wu W, DeConinck A, Lewis JA (2011) Omnidirectional printing of 3D microvascular networks. Adv Mater 23:H178–H183. https://doi.org/10.1002/adma.201004625

Yeon JH, Ryu HR, Chung M et al (2012) In vitro formation and characterization of a perfusable three-dimensional tubular capillary network in microfluidic devices. Lab Chip 12:2815. https://doi.org/10.1039/c2lc40131b

Yoo SY, Kwon SM (2013) Angiogenesis and its therapeutic opportunities. Mediat Inflamm 2013:1–11. https://doi.org/10.1155/2013/127170

Yoshida H, Matsusaki M, Akashi M (2013) Multilayered blood capillary analogs in biodegradable hydrogels for in vitro drug permeability assays. Adv Funct Mater 23:1736–1742. https://doi.org/10.1002/adfm.201201905

Young EWK (2013) Advances in microfluidic cell culture Systems for Studying Angiogenesis. J Lab Autom 18:427–436. https://doi.org/10.1177/2211068213495206

Zanotelli MR, Ardalani H, Zhang J et al (2016) Stable engineered vascular networks from human induced pluripotent stem cell-derived endothelial cells cultured in synthetic hydrogels. Acta Biomater 35:32–41. https://doi.org/10.1016/j.actbio.2016.03.001

Zeinali S, Bichsel CA, Hobi N et al (2018) Human microvasculature-on-a chip: anti-neovasculogenic effect of nintedanib in vitro. Angiogenesis 21:861–871. https://doi.org/10.1007/s10456-018-9631-8

Zervantonakis IK, Hughes-Alford SK, Charest JL et al (2012) Three-dimensional microfluidic model for tumor cell intravasation and endothelial barrier function. Proc Natl Acad Sci 109:13515–13520. https://doi.org/10.1073/pnas.1210182109

Zhao Y, Adjei AA (2015) Targeting angiogenesis in Cancer therapy: moving beyond vascular endothelial growth factor. Oncologist 20:660–673. https://doi.org/10.1634/theoncologist.2014-0465

Zheng Y, Sun Y, Yu X et al (2016) Angiogenesis in liquid tumors: an in vitro assay for leukemic-cell-induced bone marrow angiogenesis. Adv Healthc Mater 5:1014–1024. https://doi.org/10.1002/adhm.201501007

Biomaterials and Microfluidics for Drug Discovery and Development

8

Mariana R. Carvalho, Roman Truckenmuller, Rui Luís Reis, and Joaquim Miguel Oliveira

Abstract

Microfluidic devices are now one of the most promising tools to mimic *in vivo* like conditions, either in normal or disease scenarios, such as tumorigenesis or pathogenesis. Together with the potential of biomaterials, its combination with microfluidics represents the ability to more closely mimic cells' natural microenvironment concerning its three-dimensional (3D) nature and continuous perfusion with nutrients and cells' crosstalk. Due to miniaturization and increased experimental throughput, microfluidics have generated significant interest in the drug discovery and development domain. Herein, the most recent advances in the field of microfluidics for drug discovery are overviewed, and the role of biomaterials in 3D *in vitro* models and the contribution of organ-on-a-chip technologies highlighted.

Keywords

3D models · Biomaterials · Drug discovery · Microfluidics · Organ-on-a-chip · Cancer

M. R. Carvalho · R. L. Reis · J. M. Oliveira (✉)
3B's Research Group, I3Bs – Research Institute on Biomaterials, Biodegradables and Biomimetics, University of Minho, Headquarters of the European Institute of Excellence on Tissue Engineering and Regenerative Medicine, Guimarães, Portugal

ICVS/3B's PT Government Associate Lab, Braga/Guimarães, Portugal

The Discoveries Centre for Regenerative and Precision Medicine, Headquarters at University of Minho, Guimarães, Portugal
e-mail: miguel.oliveira@dep.uminho.pt

R. Truckenmuller
Department of Complex Tissue Regeneration, MERLN Institute for Technology-Inspired Regenerative Medicine, Maastricht University, Maastricht, The Netherlands

Highlights

- Microfluidic devices are now one of the most promising tools to mimic *in vivo* like conditions, either in normal or disease scenarios, such as tumorigenesis or pathogenesis.
- In the case of cancer drugs, studies report the rate of successful translation from animal models to clinical cancer trials is less than 8%, revealing the urgent need to develop better models.
- The advances in the fields of biomaterials engineering combined with microfluidics have advanced to impressive levels of

J. M. Oliveira, R. L. Reis (eds.), *Biomaterials- and Microfluidics-Based Tissue Engineered 3D Models*, Advances in Experimental Medicine and Biology 1230,
https://doi.org/10.1007/978-3-030-36588-2_8

sophistication, helping in the drug discovery and development.

8.1 Introduction

Many diseases, particularly acute disorders, are now treatable or manageable very effectively. The discovery of new medications for health problems that affect a great part of the world's population, such as cancer, cardiovascular diseases, metabolic disorders, arthritis, depression, anxiety, gastrointestinal disorders, pain, infectious diseases and many others, have led to improvement in overall health (Mullard 2011). As a result, the increased quality of life may lead to increased life expectancy (Mullard 2011).

The drug discovery development process traditionally starts in the laboratory where cells cultured on flat bottom flasks are observed in their response to a drug, and finally moves through the established chain of drug development: animal and human clinical trials. Only then the product can be approved and commercialized (Hughes et al. 2011). However, it is well described in the literature that the vast majority of "hits" fail between animal testing and clinical trials. The very one successful drug, from being a promising "hit" to be launched as a finished product takes between 12 and 15 years and can exceed $1 billion in costs (Paul et al. 2010; Mak et al. 2014). Among the primary causes of failure, nonclinical/clinical safety (>50%) and efficacy (>10%) stand out from all other factors (e.g., strategic, commercial or operational) (Hay et al. 2014).

More recently, scientific development lead us to cell-based high-throughput screening (HTS) techniques, which have enabled researchers to screen more than a million compounds in a couple of months (Hughes et al. 2011). However, still only about 1 out of 5000 promising drugs successfully makes it to the market. Even though HTS improves the screening speed and is now considered a standard platform for drug discovery, it leaves several unresolved issues, being the monetary costs and efficiency on top. Current pharmaceutical industries employ HTS primarily for chemical optimization at the early stage of the drug development (Zheng et al. 2013). Estimates predict HTS products only contribute 19–33% (or less) to successful marketed drugs (Khanna 2012).

Regarding the animal experimentation, although this step is indispensable for pre-clinical screening in the drug discovery process, various issues such as ethical considerations and species differences remain. In the case of cancer drugs, studies report that the average rate of successful translation from animal models to clinical cancer trials is less than 8% (Mak et al. 2014; Szymański et al. 2012).

Although three-dimensional (3D) culture techniques were introduced to the HTS techniques, to construct a more natural extracellular microenvironment, most of these conventional screening platforms are still limited in applicability due to their reliance on static culture maintenance. To tackle these issues, 3D cell-based *in vitro* assays have been actively pursued, most of the times using hydrogels. However, it remains difficult to accurately predict drug efficacy, toxicity, and organs interactions, because cultivated cells often do not retain their original organ functions and morphologies in conventional 2D *in vitro* cell culture systems (Chandrasekaran et al. 2016b).

As a promising solution, microfluidic technologies emerged, in this case those also providing 3D microenvironment with microvascular-like perfusion and diffusion between mimicked microvessels and 3D cell culture, which is closer to what cells encounter in real tissues or organs.

Briefly, microfluidics is the field of fluids manipulation in channels with dimensions of tens of micrometers (Whitesides 2006). Microfluidic chips are small platforms comprising channel systems connected to liquid reservoirs by, for example, tubing systems in turn linked to syringes. The size of the channels is in the range of a few micrometers, which greatly facilitates handling of volumes much smaller than a microliter (Bettinger and Borenstein 2010).

In combination with appropriate biological assays and high-sensitivity detection techniques, such systems allow the identification and isolation of individual cells or molecules (Dittrich and

Manz 2006). A great variety of microchip fabrication techniques and materials are available for producing highly sophisticated 2½D and 3D microstructures with integrated modules. Sensors, detectors and optical components can be integrated on chip (Tsao 2016).

With rapid growing on emphasis on translational research and more matured bio-microfabrication technologies, currently in the USA alone there are over three companies using microchips for *in vivo* drug delivery, at least five microchip-based cancer cell/biomarker diagnostic companies and about 10 companies working on microfabricated cell chips for drug screening and discovery (FluidicMEMS 2018).

This book chapter focuses on current and future applications of biomaterials in microfluidics, microfluidics systems in drug discovery and highlights the potential of 'organ-on-a-chip' technology for drug discovery.

8.2 Biomaterials Applied in Microfluidic Systems

Biomaterials are gradually being developed as *in vitro* microenvironments mimicking *in vivo* cell niches, as well as widely used in tissue engineering for regeneration and replication of diverse types of both normal and diseased tissues (Shantanu et al. 2016). The development of 3D culture models with tunable biomaterials has facilitated the investigation of biological phenomena with a higher level of complexity and physiological relevance than the traditional 2D models (Carvalho et al. 2015).

The combination of microfluidic technology and biomaterials now promises to enable the recapitulation of important aspects of the spatial and temporal characteristics of complex native niches, therefore facilitating the identification of new mechanisms of cell regulation and pathogenesis, profoundly impacting drug discovery (Kobel and Lutolf 2011).

Briefly, hydrogels can be derived from natural polymers, including chitosan, alginate, hyaluronic acid (HA), gellan gum (GG), agarose, collagen, gelatin, and fibroin, which have been extensively reported (Radhakrishnan et al. 2017). The use of a variety of hydrogels based on synthetic polymers (e.g. poly(ethylene glycol, PEG), polyvinyl alcohol, poly(N,Ndimethylacrylamide), and methoxy poly(ethylene glycol)-poly(e-caprolactone) have also been widely reported in literature (Lai et al. 2013; Cui et al. 2012; Bichara et al. 2014; de Girolamo et al. 2015; Inagaki et al. 2014).

Cross-linked hydrogel networks can be engineered on a molecular basis to tune for example physical parameters such as mesh size and with it mechanical properties such as stiffness according to the desired characteristics (Zeng et al. 2014). Hydrogel networks, both natural and synthetic, are therefore able to match the physical, chemical, and mechanical properties of native Extracellular Matrix (ECM) molecules, justifying the widespread use of hydrogels as ECM mimics in microfluidic models (Lutolf and Hubbell 2005; Barata et al. 2016). Various types of hydrogels have been used in microfluidic systems (Wan 2012) with collagen (Baker et al. 2013; Toh et al. 2009), poly(ethylene glycol) (PEG) (Dinh et al. 2013; Kobel and Lutolf 2011), agarose (Si et al. 2012), fibrin (Bita et al. 2010; Toh et al. 2009), and Matrigel (Anguiano et al. 2017; Chaw et al. 2007) being some of the most commonly used ones.

Toh et al. (2009) used collagen to develop a microfluidic 3D hepatocyte chip ('3D HepaTox Chip') for *in vitro* drug toxicity testing. The 3D HepaTox Chip is based on multiplexed microfluidic channels where a 3D microenvironment is engineered in each of the eight channels channels to maintain the hepatocytes' synthetic and metabolic functions (Toh et al. 2009). Positively charged methylated collagen and negatively charged HEMA-MMA-MAA terpolymer were chosen to form a localized 3D matrix because they have been shown to support hepatocyte functions.

Dinh et al. (2013) combined PEG with the benefits of microfluidics for precision single-cell handling. The result was a model comprising compartmentalized neuron arraying (CNA) microfluidic circuits that allow high-throughput experimentation (Fig. 8.1)(Dinh et al. 2013).

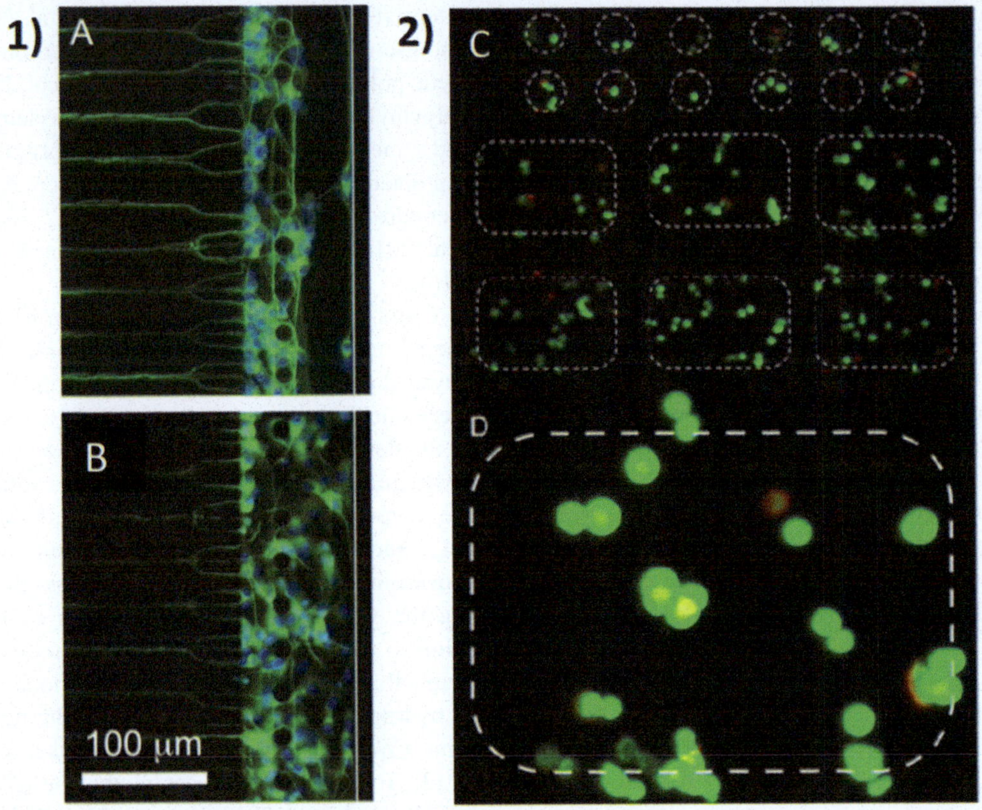

Fig. 8.1 – Biomaterials applied in microfluidic systems. (1) *In situ* neuron patterning. Differentiated human SH-SY5Y neurons cultured for 5 days within the CNA devices on a PL/PLL-g-PEG pattern (**a**) and on a continuous PL coating (**b**). Biomaterial patterning increases neurite guidance into the outgrowth microchannels. (Reprinted with permission from (Dinh et al. 2013). Copyright © 2018 Royal Society of Chemistry). 2 – (**C**) Optical image of hydrogel microstructure arrays containing two phenotypes of cells: HepG2 and A549 cells. HepG2 cells were encapsulated inside cylindrical hydrogel microstructures (78 µm Ø × 50 µm), while A549 cells were encapsulated inside rectangular hydrogel microstructures (330 µm × 217 µm × 50 µm). (**b**) Corresponding fluorescence image for the cell viability expressed by live/dead assay (live cells green, dead cells red). (**d**) Enlarged optical and fluorescence images of one array element. (Reprinted with permission from (Gao et al. 2010). Copyright © 2010 Elsevier B.V)

Similarly, Gao et al. (2010) employed photolithography to geometrically encapsulate human hepatoma HepG2 and human lung epithelial A549 cells in a PEG hydrogel inside a microfluidic device. The cells in the device were exposed to anticancer drugs and apoptosis was studied (Fig. 8.2). Two anticancer drugs (Actinomycin D and methotrexate) exhibited distinct effects on the levels of intracellular glutathione and reactive oxygen species, indicating the selectivity of these drugs on the disturbance of redox balance within cells (Gao et al. 2010).

To address the global challenge of vascularization, Zhang et al. (2013) employed additive manufacturing to build a blood vessel-like microfluidic structure, which was then embedded inside the bulk material. To achieve this, alginate and chitosan hydrogels were used for direct printing of the channels with a wall thickness below 200 µm. The printed microfluidic network was able to sustain media perfusion, both without and with being embedded in bulk hydrogels to support cell viability, representing an improvement towards vascularization models (Zhang et al. 2013).

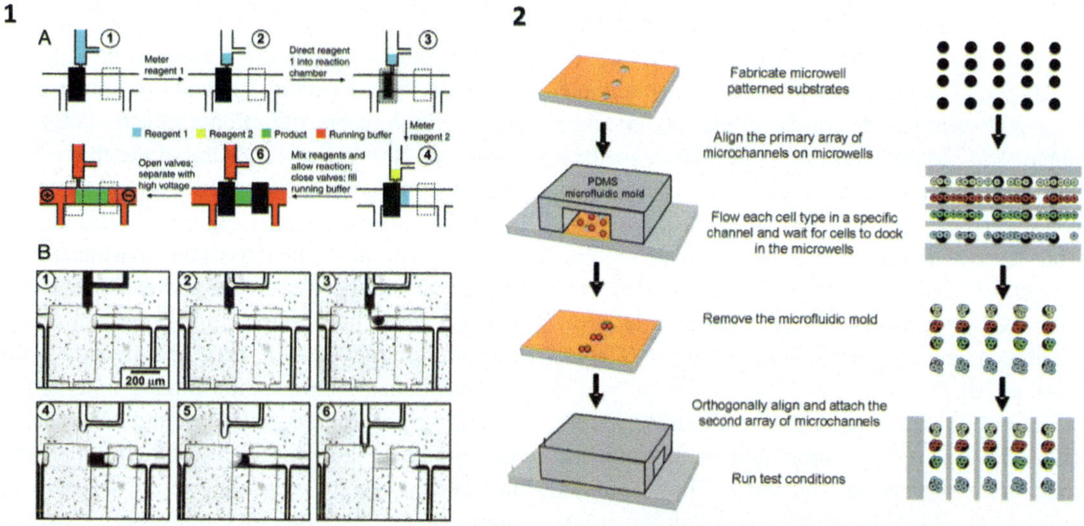

Fig. 8.2 – Methods of microfluidics in drug discovery. (1) On-chip reaction. (**a**) Schematic of the process of on-chip reaction. (**b**) Frames of a video that correspond to the steps of A. A concentrated solution of crystal violet (a pH indicator) in 0.1 N sodium hydroxide solution is loaded into the microchamber with the three-state valve half-open; loading an equal amount of 1 N HCl into the chamber turns the solution from blue to yellow. Reprinted with permission from (Wu et al. 2004). Copyright © 2004 The National Academy of Sciences; (2) Schematic diagram of reversible sealing of microfluidic arrays onto microwell-patterned substrates to fabricate multiphenotype cell arrays. Initially, a PDMS microfluidic mold in the form of an array of microchannels was aligned on an array of microwells. Each cell type was flown through an independent channel. The cells docked onto the corresponding micro-wells, which resulted in a patterned array of cells. To deliver multiple solutions, the PDMS microfluidic mold was removed and replaced with another mold which was placed orthogonally to create multiphenotype cell arrays inside each microchannel. (Reprinted with permission from (Khademhosseini et al. 2005). Copyright © 2018 Copyright Clearance Center, Inc)

Regarding fibrin hydrogels, Carrion et al. (Bita et al. 2010) described a novel 3D microfluidic device as a model system to study the molecular regulation of perivascular stem cell niches. Endothelial cells (ECs) suspended within 3D fibrin hydrogels patterned in the device adjacent to stromal cells (either fibroblasts or human bone marrow-derived mesenchymal stromal cells, hBM-MSCs) executed a morphogenetic process similar to vasculogenesis, maturing into a robust capillary network with well-defined hollow lumens. Both MSCs and fibroblasts formed pericytic associations with the ECs but promoted capillary morphogenesis with distinct kinetics (Bita et al. 2010).

The fields of biomaterials engineering and microfluidics have advanced to impressive levels of sophistication. The pioneering efforts reported above to further integrate biomaterials technology into microfluidic chips tackle pertinent questions in several fields from drug discovery to stem cell research.

ECM derived biomaterials approaches are now being studied for their potential to preserve the tissue-specific biochemical composition as well as the ultrastructure of the ECM (Hussey et al. 2018). The development of biomaterials that promote the formation of functional tissues in clinical applications is therefore of the utmost importance in tissue engineering. So far, many studies have focused on the decellularization of tissues and organs, including small intestinal submucosa, heart valve, blood vessel, skin, nerve, tendon, ligament, bladder, amniotic membrane, heart, liver and lung (Uygun et al. 2010; Choi et al. 2010; Lu et al. 2011). For example, Choi et al. (2010) decellularized an ECM scaffold derived from porcine cartilage and proved it was more efficient in maintain chondrogenesis of rabbit mesenchymal stem cells (rMSCs) *in vitro* and in the nude mouse model *in vivo* than the PGA scaffold used as a control (Choi et al. 2010).

Uygun et al. (2010) generated a transplantable liver graft using perfusion decellularization technique, and further introduced perfusion-seeding and culture techniques for the preparation of recellularized liver matrix for transplantation. The recellularized graft supports liver-specific function including albumin secretion, urea synthesis and cytochrome P450 expression at comparable levels to normal liver *in vitro (Uygun* et al. 2010*)*.

Inorganic compounds such as hydroxyapatite (HA) have also been integrated in microfluidic models for mimicking bone tissues. For example, Jusoh et al. (2015) developed a microfluidic platform that integrates fibrin ECM with the synthetic bone mineral HA nanoparticles to provide *in vivo*-like microenvironments for bone vessel sprouting. The observed that fibrin with 0.2% HA exhibited a higher number of sprouts, and higher length of sprouts (Jusoh et al. 2015).

8.3 Microfluidics in Drug Discovery

Microfluidic technologies represent an improvement to existing technologies in key areas of drug discovery. Microfluidics involves materials and techniques for controlling the movement of minute quantities of fluids, and its ability to miniaturize assays and increase experimental throughput have generated significant interest of the drug discovery and development domain (Kang et al. 2008).

Microfluidics can be applied to drug discovery in a variety of ways. For instance, it can be used in target identification, through platforms for cell culture in which cell's response to a drug in a more *in vivo* like environment is assessed (Barata et al. 2017).

(a) To select specific targets

In drug discovery, the signal transduction pathways and protein-protein interactions within cells must be understood. Microfluidic devices can be used to address this issue. These devices are integrated to manipulate, label, lyse, separate and quantify the protein contents of single cells using single-molecule fluorescence counting (Wu et al. 2004) (Fig. 8.2). One of the many successful examples consists of measuring the number of epitope-tagged human β_2-adrenergic receptors (β_2ARs), an important pharmacologic target in a number of airway and cardiovascular diseases. Another interesting application of microfluidic devices in drug discovery is for ligand binding studies, minimizing interaction times, improving sensitivity and increasing throughput. In this sense, a microfluidic platform has been used to characterize DNA binding energy using four eukaryotic transcription factors to predict their *in vivo* function (Maerkl and Quake 2007).

(b) For hit identification

b1) Compound generation: Another way is through the generation of compounds in microfluidic devices, either at a 2D or 3D device-architectural level. Developments in combinatorial chemistry have greatly enhanced our ability to generate drug candidates (Kang et al. 2008). Merging channel geometries, various reagents can be mixed and biochemical reactions can be induced. Reduction of reaction chamber's size and consequently amount of reagents and time needed for reaction empower this strategy. Mitchell et al. (2001) developed a miniaturized-synthesis/total analysis system that incorporated a chemical microprocessor with time-of-flight mass spectrometry (TOF-MS) which has been used in the generation of compound libraries (Mitchell et al. 2001). This system enabled real-time processing of many reactions due to the integration of continuous-flow synthesis and provision for on-line analysis within a microfabricated structure. In another example, Zhou et al. (2004) developed a microfluidic pico array device for the synthesis and purification of oligonucleotides. This device consisted of fluidic channels and 3698 individual pico-reaction chambers for parallel synthesis of oligonucleotides (Zhou et al. 2004). Microfluidic systems can be also used to understand and assess natural drug candidates. Natural products are undoubtedly one of the major sources of new chemical entities, with an increase in world market growth rate (Shen 2015). Natural molecules are "evolutionarily optimized" as drug-like molecules and remain one of the best sources of drugs leads (Shen 2015). To identify the physiological action of a natural compound out of a complex array of constituent molecules in the source material, microfluidic devices can also be used. Gramowski et al. (2006) applied multielectrode microchips for the screening of herbal medicines. The screening of complex mixtures of neuroactive substances was assessed in the chip, where changes in the spontaneous activity of cultured networks of primary cortical neurons were quantified to evaluate the action of the drugs (Gramowski et al. 2006).

b2) HTS is considered one of the most important methods for identifying "hit" compounds (Chandrasekaran et al. 2016a). Over the last decades, pharmaceutical companies established HTS as a major tool to screen the properties of new chemical entities. A significant number of microfluidic technologies can now be used to enable HTS studies, including multiplexed systems, microwell arrays, plug-based methods and gradient-generating devices (Du et al. 2016).

Khademhosseini et al. (2005) studied the response of multiple cell types (e.g. hepatocytes, fibroblasts, and embryonic stem cells) to different compounds by seeding cells within microwells that were integrated within an array of reversibly sealed microfluidic channels (Fig. 8.2). The ability to position many cell types on a single chip could be useful for studying the effects of a series of compounds on different cell types.

Concentration gradients also have a great role in drug screening and cell-based studies. A gradient-generating microfluidic device can generate a certain range of doses and thereby control different cell behaviors at the same time. By using gradients, cells can be stimulated with controlled temporal and spatial resolution to study the effects of drug concentration on chemotaxis. Occhetta et al. (2015) introduced a branched channel to generate a gradient and culture 3D micromasses of adult hBM-MSCs under continuous and controlled laminar flow perfusion (Occhetta et al. 2015). Moreover, the platform allows for conditioning of the cellular microaggregates with different combinations of growth factors, morphogens or drug molecules, in a high-throughput fashion.

8.4 Organ on Chip and Drug Discovery

Recent scientific developments in areas such as stem cell research, regenerative medicine, biomaterials, tissue engineering and microfluidics have been integrated into new 3D *in vitro* models in order to better mimic human tissues and organs. These "organs-on-chip" devices can provide not only the biological relevance but also the requisite high-throughput applications in drug discovery. In the following, we provide an overview of the developed technologies of organ-on-chip devices and their role on drug testing and discovery.

8.4.1 Lung on a Chip

Huh et al. (2012) developed a biomimetic microdevice that is able to reconstitute organ-level lung functions to create a human disease: pulmonary edema (as well as a model that mimics normal lung function) (Fig. 8.3). In this model, the authors cultured two types of human lung cells (epithelial and endothelial primary lung cells) in stacked parallel microchannels separated by a (porous) membrane. Alike the human lung, the upper alveolar channel was filled with air,

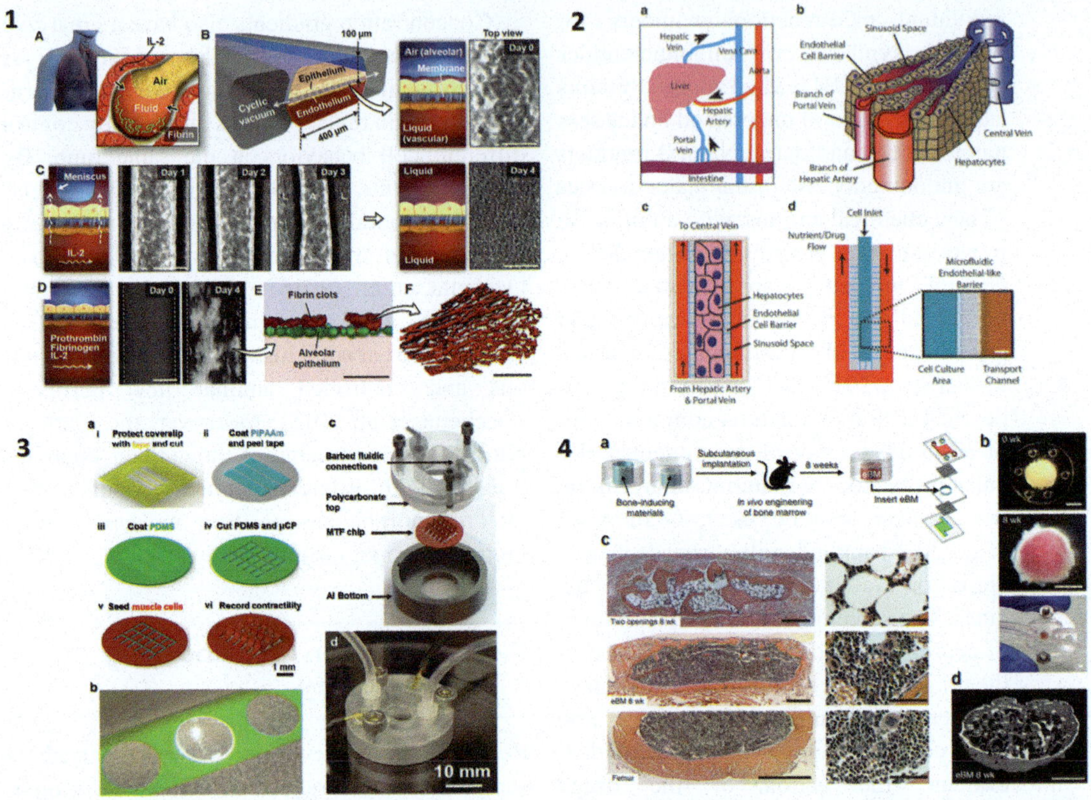

Fig. 8.3 – (1) Wxamples of organs on chips: A microengineered model of human pulmonary edema. (**a**) IL-2 therapy is associated with vascular leakage. (**b**) IL-2–induced pulmonary edema is modeled in a microengineered lung-on-a-chip. The top or "air" portion is the alveolar channel; the bottom or "liquid" portion is the vascular channel. Scale bar represents 200 µm. (**c**) Endothelial exposure to IL-2 (1000 U/ml) causes liquid from the lower, microvascular channel to leak into the upper, alveolar chamber. Scale bars represent 200 µm. (**d**) During IL-2 treatment, prothrombin (100 µg/ml) and fluorescently labeled fibrinogen (2 mg/ml) introduced into the microvascular channel form fluorescent fibrin clots (white) over the course of 4 days. Scale bar represents 200 µm. (**e**) A fluorescence confocal microscopic image shows that the fibrin deposits (red) in (**d**) are found on the upper surface of the alveolar epithelium (green). Scale bar represents 50 µm. (**f**) The clots in (**d**) and (**e**) are highly fibrous networks. Scale bar represents 5 µm. Reprinted with permission from (Huh et al. 2012). (Copyright © 2012, American Association for the Advancement of Science). (2) Microfluidic liver sinusoid. (**a**) The liver receives blood flow from the hepatic artery and the portal vein, which carries xenobiotics such as ingested drugs from the small intestine to be metabolized. (**b**) Schematic of the hepatocyte microenvironment in liver tissue. (**c**) The sinusoid space is bordered by a sheet of highly fenestrated endothelial cells. (**d**) The single microfluidic sinusoid had three fluid terminals: a flow inlet, a flow outlet, and a cell inlet. Scale bar represents 20 µm. Reprinted with permission from (Lee et al. 2007). (Copyright © 2007 Wiley Periodicals, Inc. (3). Higher-throughput heart on a chip and fluidic microdevice). (**a**) Schematic of fabrication process. (**b**) Image of the engraving laser processing of Muscular Thin Films (MTFs) in the 18 mm diameter chip. (**c**) Exploded view of the conception and assembly of the fluidic device which fits the chip. (**d**) Image of an actual device in action. (Reprinted with permission from Agarwal et al. 2013). Copyright © 2018 Royal Society of Chemistry. (4) *In vivo* bone marrow engineering. (**a**) Workflow to generate a bone marrow-on-a-chip system (**b**) PDMS device containing bone-inducing materials in its central cylindrical chamber before implantation (top). (**c**) Low- (left) and high-magnification views (right) of histological hematoxylin-eosin stained sections of the bone marrow formed in the PDMS device with two openings (top) or one lower opening (center) at 8 weeks. (Reprinted with permission from (Torisawa et al. 2014). Copyright © 2014, Springer Nature)

whereas the lower microvascular channel was filled with liquid. Stretch was cyclically applied to the membrane by additional vacuum actuator channels to mimic the breathing movements of the lung. After adding interleukin-2 (IL-2) to the microvascular channel, the fluid started to leak into the air compartment, mimicking pulmonary edema. The authors tested their pulmonary disease model against the pharmacological agent GSK2193874. GSK2193874 blocks ion channels activated by mechanical strain, and inhibited leakage of fluid to the air compartment, suggesting it would be a viable option for patients with pulmonary edema. The next step will be to hook this lung up to other chip-based organs with the goal of being able to rapidly screen many drugs across many organs (Huh et al. 2012).

Jain et al. (2018) developed another lung model on a chip with primary lung alveolar and endothelial cells. This new chip design enables whole human blood to be perfused through the vascular channel without producing thrombus formation, while allowing high-resolution, real-time analysis of interactions between human blood cells and endothelial cells in an *in vivo*-like context. Using this technology in combination with novel analytical tools for quantitation of dynamic platelet-endothelial interactions and clot formation, the authors demonstrated a key role of the epithelium in inflammation-driven vascular thrombosis during lipopolysaccharide endotoxin (LPS)-induced acute lung injury. The study showed that the model can be used to evaluate cytoprotective effects of Protease-activated Receptor-1 (PAR-1), a potential therapeutic with anti-thrombotic and anti-inflammatory activities, *in vitro* (2018).

8.4.2 Liver on Chip

The liver is one of the most relevant organs involved in maintaining physiological homeostasis. Models that could recapitulate functional liver tissue will have an enormous impact in pharmaceutical fields, allowing for toxicity assessment of new drugs (Uygun et al. 2010). To develop such models, hepatocytes with membrane polarity and functional bile canaliculi are essential (Nakao et al. 2011). Lee et al. (2007) created a biologically inspired artificial liver sinusoid with a microfluidic endothelial-like barrier having mass transport properties similar to the liver acinus (Fig. 8.3). This unit consisted of a cord of hepatocytes (50 μm × 30 μm × 500 μm) fed by diffusion of nutrients across the microfluidic endothelial-like barrier from a convective transport vessel. Combining photolithography-based microfluidic technologies and primary cell culture, the team was able to assess liver toxicity of diclofenac, since it has long been implicated in drug-related liver damage (Lee et al. 2007).

Following Lee's work, Nakao et al. (Nakao et al. 2011) presented a microfluidic device comprising an hepatic cord-like structure model with primary rat hepatocytes, which was proven functional: computer simulation results show that the flow velocity in the medium flow channel in the model is 1 mm/s, which is close to the respective blood flow velocity *in vivo*. Moreover, when 5-(and-6)-carboxy-2′,7′-dichloro-fluorescein diacetate (CDFDA) was injected into the device, it was readily absorbed by the cells and metabolized into CDF by esterase. The metabolites were actively excreted into bile canaliculi by the MRP2 protein (Nakao et al. 2011).

8.4.3 Kidney on a Chip

Jang et al. (Jang and Suh 2010) developed a simple multi-layer microfluidic device by integrating a polydimethylsiloxane (PDMS) microfluidic channel (using photolithography technology) and a porous membrane substrate (from polyester; 0.4 μm pore size;, 10 μm thickness) to culture primary rat inner medullary collecting duct (IMCD) cells (Jang and Suh 2010). In order to evaluate molecular transport in renal tubule cells, the team measured water and sodium uptake after hormonal stimulations of the IMCD cells using vasopressin and aldosterone. The cells were subjected to a fluidic shear stress of 1 dyn/cm^2 for the time period of 5 h on the porous membrane, which was found to be sufficient for the IMCD cells to enhance cell polarization and rearrange cytoskeleton and cell junctions (Jang and Suh 2010).

8.4.4 Heart on a Chip

Agarwal et al. (Agarwal et al. 2013) developed anisotropic cardiac microtissues recapitulating the laminar architecture of the heart ventricle, which were engineered on cantilevers. The sub-millimeter-sized thin film cantilevers of soft elastomers were fabricated by a laser-based process (Fig. 8.3) (Agarwal et al. 2013). They used the microdevice seeded with cardiac myocytes from mice to test the positive inotropic effect of isoproterenol on cardiac contractility at dosages ranging from 1 nM to 100 μM. Tissues were exposed to different drug concentrations with the advantage of complete washout between dosages, leading to a fast collection of data from just one chip (Agarwal et al. 2013).

8.4.5 Bone Marrow on a Chip

Torisawa et al. (2014) developed a tissue-engineered bone marrow model that is able to retain hematopoietic stem and progenitor cells in *in vivo*-like conditions for 1 week. In this example, the bone marrow is first generated in mice, and then explanted as a whole and maintained *in vitro* within a microfluidic device fabricated using standard photolithography (Fig. 8.3). Results show this model mimics complex tissue-level responses to Y-radiation toxicity normally observed only *in vivo*, as well as to a therapeutic countermeasure agent (G-CSF) that is known to accelerate recovery from radiation-induced toxicity in cancer patients (Torisawa et al. 2014). Moreover, this research sheds light on the fact that it is crucial to keep the whole hematopoietic niche to obtain an *in vivo* like response, rather than culturing particular cell types (Torisawa et al. 2014).

8.4.6 Intestine on a Chip

When discussing drug development, the small intestine is crucial, given that orally administered drugs are mainly absorbed in the small intestine (Fasinu et al. 2011). There they diffuse across a mucous layer covering an epithelial cell layer lining the intestinal wall. Therefore, it is logical that drugs and chemicals are tested on intestinal cells and tissues to assess pharmacological parameters such as absorption, distribution, metabolism, elimination and toxicity, before any other.

Having this in mind, several groups started developing models destined for absorption and metabolism studies. Realistic models should include certain key components such as cellular (i.e. enterocytes, goblet and vascular endothelial cells), structural (i.e. villi and mucus) and dynamic (i.e. peristalsis) features. Some of these issues were addressed by Kimura et al. (2008). The authors engineered an intestinal model with a membrane and vascular flow simulating the epithelial barrier and the epithelial-endothelial interface. The microfluidic model comprised optical fiber inserts for on-line and real-time pharmacokinetic measurements to detect the amount of fluorescence of rhodamine-123 in culture, working as a tracer substrate to study diffusion (Kimura et al. 2008).

8.4.7 Tumor on a Chip

Cancer remains one of the main causes of death, despite enormous efforts to cure the disease (Stroock and Fischbach 2010; Buchanan and Rylander 2013). It is no longer arguable that we need to have a better understanding of the tumour microenvironment, as well as a more effective means of screening anti-cancer drug leads (Carvalho et al. 2017). The emerging tumour-on-a-chip technology can be of extreme usefulness in these processes.

In an approach to better understand breast cancer metastasis to bone, Jeon et al. (2015) developed a microfluidic model of a human vascularized organ-specific microenvironment, which was used to explore and tune the extravasation process of metastatic tumor cells (Jeon et al. 2015). The experiment comprises an organ-specific model that enables the study of human metastatic breast cancer cell extravasation within a perfusable human microvascularized bone-mimicking microenvironment. The microfluidic model was characterized by anastomoses with the lateral media channels. MDA-MB-231 breast cancer cells were introduced and extravasation events monitored. Once established, the platform was used to understand the anti-metastatic role of adenosine in the human breast cancer metastasis to bone (Jeon et al. 2015).

In its turn, Bischel et al. (2014) integrated a microfluidic co-culture platform with a multi-photon imaging-based technique to determine phenotypic cell behavior and flavin adenine dinucleotide (FAD) fluorescence intensity and fluorescence lifetime simultaneously in the same cell. This platform combines two independent assays normally performed with two different cell populations into a single device, allowing researchers to simultaneously assess both phenotypic cell behavior and enzyme activity (Bischel et al. 2014). The metastasis towards bone was mimicked by differentiating MC3T3-E1 cells (osteoblast precursor cell line) towards osteoblast lineage and seeded on the sidewall of a microchannel. After a collagen I hydrogel coating was applied as extracellular matrix, either Lymph Node Carcinome of the Prostate cells (LNCaP) or (Bone metastatic LNCaP-derivative *C4-2B)* C4-2B cells were seeded into the channel. By engineering this cancer-bone microenvironment, it was demonstrated that the cross-talk between a bone metastatic prostate cancer cell line, C4-2B, and bone stromal cells, MC3T3-E1, increases the invasive behavior of the C4-2Bs, possibly by an increase in reactive oxygen species (ROS)-producing N1-acetyl polyamine oxidase (APAO) activity (Bischel et al. 2014).

Sung et al. (2010) presented a microscale cell culture analog (mCCA), which is basically a microfluidic device based on pharmacokinetics-pharmacodynamics (PK-PD) mathematical models. In this model, multiple cell culture chambers are connected with fluidic channels to mimic multi-organ interactions and test drug toxicity in a pharmacokinetic-based manner (Sung et al. 2010). Three cell lines representing the liver, a tumor and bone marrow were cultured in the three-chamber microCCA to test the toxicity of an anticancer drug, 5-fluorouracil (5-FU). The result was analyzed with the PK-PD model of the device, and compared with the result of the static conditions (cell culture flasks). Each cell type exhibited differential responses to 5-FU, and the responses in the microfluidic environment were different from those in the static environment. The authors showed a relatively new approach based on the combination of mathematical modeling and an *in vitro* experimentation, with improved predictability for testing drug toxicity.

More recently, Wong et al. (2017) exploited a 2.4 × 2.4 cm PDMS-based microfluidic chip to generate droplets and perform drug screens against suspended and adherent cancer cell lines, as well as cells dissociated from primary tumors of human patients. Bortezomib and Vorinostat were chosen as target drugs for leukemia in conjunction with Jurkat cells, whereas Cisplatin and Epirubicin were chosen as target drugs for breast cancer in conjunction with MDA-MB-231 cells. Cell viability was measured by ethidium homodimer-1 staining, enabling the capture of single cell drug response, without compromising population analysis. This method provides a versatile and rapid drug screening method using cells in different states (Wong et al. 2017).

8.5 Conclusions

New micro-engineering-based strategies have significantly improved the drug development process. Microfluidics makes it possible to predict therapeutic activity by enabling high-throughput functional readouts; to conduct high-throughput screens for drug discovery in realistic engineered 3D culture microenvironments; to develop organ-on-a-chip platforms to screen those drugs, which would probably fail during the expensive animal testing and human clinical trials later in the drug discovery process pipeline. Critical challenges remain in making these technological advances practically feasible for real-world pharmaceutical discovery. Thus, the optimal solution for drug discovery may rely on sufficiently complex and realistic technology, while being simple enough to allow for assay readouts, reliability and operational simplicity. In addition, mathematical and computational modeling will guide the design of the next generation of organ-on-chip devices, and nanosensing will enable the real-time and *in situ* monitoring of the cellular metabolic responses.

Regarding biomaterials, *in vitro* platforms are just a scientific attempt at mimicking the natural system. It is therefore important to make the right choice of the function to be mimicked, as this choice will eventually determine whether a system is considered valuable or not. Because of this, there is now a huge demand for new and tunable biomaterials to interface with microfluidic chips for cell biological studies.

However, the reality is that despite rapid development of expertise in designing, fabricating and operating microfluidic systems for nearly 20 years now, no commercially available drugs have been discovered as a unique result of microfluidic technologies, to the best of our knowledge.

Griffith, a world-class scientist in organ-on-a-chip research said that more questions would be answered if Pharma collaborated more with engineers, as Pharma knows what problems it needs to solve, but engineers know how to build chips. Therefore, success in this endeavor must start with alliances between the fields of pharmaceutical industry, engineering, medicine, biology and computational biology. Maybe that is our future answer.

Acknowledgments This chapter is a result of the project FROnTHERA (NORTE-01-0145-FEDER-000023), supported by Norte Portugal Regional Operational Programme (NORTE 2020), under the PORTUGAL 2020 Partnership Agreement, through the European Regional Development Fund (ERDF). Mariana Carvalho acknowledges her PhD scholarship NORTE-08-5369-FSE-000044, funded by Programa Operacional Regional do Norte, Fundo Social Europeu, Norte2020 TERM&SC and EMBO Short-Term Fellowship 7232 J. M. Oliveira thanks FCT for his distinction attributed under the FCT Investigator program (IF/01285/2015).

References

Agarwal A, Goss JA, Cho A, McCain ML, Parker KK (2013) Microfluidic heart on a chip for higher throughput pharmacological studies. Lab Chip 13(18):3599–3608. https://doi.org/10.1039/c3lc50350j

Anguiano M, Castilla C, Maska M, Ederra C, Pelaez R, Morales X, Munoz-Arrieta G, Mujika M, Kozubek M, Munoz-Barrutia A, Rouzaut A, Arana S, Garcia-Aznar JM, Ortiz-de-Solorzano C (2017) Characterization of three-dimensional cancer cell migration in mixed collagen-Matrigel scaffolds using microfluidics and image analysis. PLoS One 12(2):e0171417. https://doi.org/10.1371/journal.pone.0171417

Baker BM, Trappmann B, Stapleton SC, Toro E, Chen CS (2013) Microfluidics embedded within extracellular matrix to define vascular architectures and pattern diffusive gradients. Lab Chip 13(16):3246–3252. https://doi.org/10.1039/c3lc50493j

Barata D, van Blitterswijk C, Habibovic P (2016) High-throughput screening approaches and combinatorial development of biomaterials using microfluidics. Acta Biomater 34:1–20. https://doi.org/10.1016/j.actbio.2015.09.009

Barata D, Spennati G, Correia C, Ribeiro N, Harink B, van Blitterswijk C, Habibovic P, van Rijt S (2017) Development of a shear stress-free microfluidic gradient generator capable of quantitatively analyzing single-cell morphology. Biomed Microdevices 19(4):81. https://doi.org/10.1007/s10544-017-0222-z

Bettinger CJ, Borenstein JT (2010) Biomaterials-based microfluidics for engineered tissue constructs. Soft Matter 6(20):4999–5015. https://doi.org/10.1039/C0SM00247J

Bichara DA, Bodugoz-Sentruk H, Ling D, Malchau E, Bragdon CR, Muratoglu OK (2014) Osteochondral defect repair using a polyvinyl alcohol-polyacrylic acid (PVA-PAAc) hydrogel. Biomed Mater (Bristol, England) 9(4):045012. https://doi.org/10.1088/1748-6041/9/4/045012

Bischel LL, Casavant BP, Young PA, Eliceiri KW, Basu HS, Beebe DJ (2014) A microfluidic coculture and multiphoton FAD analysis assay provides insight into the influence of the bone microenvironment on prostate cancer cells. Integr Biol 6(6):627–635. https://doi.org/10.1039/C3IB40240A

Bita C, Huang CP, Ghajar CM, Kachgal S, Kniazeva E, Jeon NL, Putnam AJ (2010) Recreating the perivascular niche ex vivo using a microfluidic approach. Biotechnol Bioeng 107(6):1020–1028. https://doi.org/10.1002/bit.22891

Buchanan C, Rylander MN (2013) Microfluidic culture models to study the hydrodynamics of tumor progression and therapeutic response. Biotechnol Bioeng 110(8):2063–2072. https://doi.org/10.1002/bit.24944

Carvalho MR, Lima D, Reis RL, Correlo VM, Oliveira JM (2015) Significance of tissue-engineered 3D in-vitro models for cancer research and drug screening. Trends Biotechnol

Carvalho MR, Lima D, Reis RL, Oliveira JM, Correlo VM (2017) Anti-Cancer drug validation: the contribution of tissue engineered models. Stem Cell Rev 13:347–363. https://doi.org/10.1007/s12015-017-9720-x

Chandrasekaran A, Abduljawad M, Moraes C (2016a) Have microfluidics delivered for drug discovery? Expert Opin Drug Discov 11(8):745–748. https://doi.org/10.1080/17460441.2016.1193485

Chandrasekaran A, Abduljawad M, Moraes C (2016b) Have microfluidics delivered for drug discovery? Expert Opin Drug Discovery 11(8):745–748. https://doi.org/10.1080/17460441.2016.1193485

Chaw KC, Manimaran M, Tay FE, Swaminathan S (2007) Matrigel coated polydimethylsiloxane based microfluidic devices for studying metastatic and non-metastatic cancer cell invasion and migration. Biomed Microdevices 9(4):597–602. https://doi.org/10.1007/s10544-007-9071-5

Choi K-H, Choi BH, Park SR, Kim BJ, Min B-H (2010) The chondrogenic differentiation of mesenchymal stem cells on an extracellular matrix scaffold derived from porcine chondrocytes. Biomaterials 31(20):5355–5365. https://doi.org/10.1016/j.biomaterials.2010.03.053

Cui X, Breitenkamp K, Finn MG, Lotz M, D'Lima DD (2012) Direct human cartilage repair using three-dimensional bioprinting technology. Tissue Eng Part A 18(11–12):1304–1312. https://doi.org/10.1089/ten.TEA.2011.0543

de Girolamo L, Niada S, Arrigoni E, Di Giancamillo A, Domeneghini C, Dadsetan M, Yaszemski MJ, Gastaldi D, Vena P, Taffetani M, Zerbi A, Sansone V, Peretti GM, Brini AT (2015) Repair of osteochondral defects in the minipig model by OPF hydrogel loaded with adipose-derived mesenchymal stem cells. Regen Med 10(2):135–151. https://doi.org/10.2217/rme.14.77

Dinh ND, Chiang YY, Hardelauf H, Baumann J, Jackson E, Waide S, Sisnaiske J, Frimat JP, van Thriel C, Janasek D, Peyrin JM, West J (2013) Microfluidic construction of minimalistic neuronal co-cultures. Lab Chip 13(7):1402–1412. https://doi.org/10.1039/c3lc41224e

Dittrich PS, Manz A (2006) Lab-on-a-chip: microfluidics in drug discovery. Nat Rev Drug Discov 5:210–218. https://doi.org/10.1038/nrd1985

Du G, Fang Q, den Toonder JMJ (2016) Microfluidics for cell-based high throughput screening platforms—A review. Anal Chim Acta 903:36–50. https://doi.org/10.1016/j.aca.2015.11.023

Fasinu P, Pillay V, Ndesendo VM, du Toit LC, Choonara YE (2011) Diverse approaches for the enhancement of oral drug bioavailability. Biopharm Drug Dispos 32(4):185–209. https://doi.org/10.1002/bdd.750

FluidicMEMS (2018) List of microfluidics companies. http://fluidicmems.com/list-of-microfluidics-lab-on-a-chip-and-biomems-companies/

Gao D, Liu J, Wei HB, Li HF, Guo GS, Lin JM (2010) A microfluidic approach for anticancer drug analysis based on hydrogel encapsulated tumor cells. Anal Chim Acta 665(1):7–14. https://doi.org/10.1016/j.aca.2010.03.015

Gramowski A, Jugelt K, Stuwe S, Schulze R, McGregor GP, Wartenberg-Demand A, Loock J, Schroder O, Weiss DG (2006) Functional screening of traditional antidepressants with primary cortical neuronal networks grown on multielectrode neurochips. Eur J Neurosci 24(2):455–465. https://doi.org/10.1111/j.1460-9568.2006.04892.x

Hay M, Thomas DW, Craighead JL, Economides C, Rosenthal J (2014) Clinical development success rates for investigational drugs. Nat Biotechnol 32:40. https://doi.org/10.1038/nbt.2786. https://www.nature.com/articles/nbt.2786#supplementary-information

Hughes JP, Rees S, Kalindjian SB, Philpott KL (2011) Principles of early drug discovery. Br J Pharmacol 162(6):1239–1249. https://doi.org/10.1111/j.1476-5381.2010.01127.x

Huh D, Leslie DC, Matthews BD, Fraser JP, Jurek S, Hamilton GA, Thorneloe KS, McAlexander MA, Ingber DE (2012) A human disease model of drug toxicity-induced pulmonary edema in a lung-on-a-chip microdevice. Sci Transl Med 4(159):159ra147. https://doi.org/10.1126/scitranslmed.3004249

Hussey GS, Dziki JL, Badylak SF (2018) Extracellular matrix-based materials for regenerative medicine. Nat Rev Mat 3:159–173. https://doi.org/10.1038/s41578-018-0023-x

Inagaki Y, Kitamura N, Kurokawa T, Tanaka Y, Gong JP, Yasuda K, Tohyama H (2014) Effects of culture on PAMPS/PDMAAm double-network gel on chondrogenic differentiation of mouse C3H10T1/2 cells: in vitro experimental study. BMC Musculoskelet Disord 15:320. https://doi.org/10.1186/1471-2474-15-320

Jain A, Barrile R, van der Meer AD, Mammoto A, Mammoto T, De Ceunynck K, Aisiku O, Otieno MA, Louden CS, Hamilton GA, Flaumenhaft R, Ingber DE (2018) Primary human lung Alveolus-on-a-chip model of intravascular thrombosis for assessment of therapeutics. Clin Pharmacol Ther 103(2):332–340. https://doi.org/10.1002/cpt.742

Jang KJ, Suh KY (2010) A multi-layer microfluidic device for efficient culture and analysis of renal tubular cells. Lab Chip 10(1):36–42. https://doi.org/10.1039/b907515a

Jeon JS, Bersini S, Gilardi M, Dubini G, Charest JL, Moretti M, Kamm RD (2015) Human 3D vascularized organotypic microfluidic assays to study breast cancer cell extravasation. Proc Natl Acad Sci 112(1):214–219. https://doi.org/10.1073/pnas.1417115112

Jusoh N, Oh S, Kim S, Kim J, Jeon NL (2015) Microfluidic vascularized bone tissue model with hydroxyapatite-incorporated extracellular matrix. Lab Chip 15(20):3984–3988. https://doi.org/10.1039/c5lc00698h

Kang L, Chung BG, Langer R, Khademhosseini A (2008) Microfluidics for drug discovery and development: from target selection to product lifecycle management. Drug Discov Today 13(1–2):1–13. https://doi.org/10.1016/j.drudis.2007.10.003

Khademhosseini A, Yeh J, Eng G, Karp J, Kaji H, Borenstein J, Farokhzad OC, Langer R (2005) Cell docking inside microwells within reversibly sealed microfluidic channels for fabricating multiphenotype cell arrays. Lab Chip 5(12):1380–1386. https://doi.org/10.1039/b508096g

Khanna I (2012) Drug discovery in pharmaceutical industry: productivity challenges and trends. Drug Discov Today 17(19):1088–1102. https://doi.org/10.1016/j.drudis.2012.05.007

Kimura H, Yamamoto T, Sakai H, Sakai Y, Fujii T (2008) An integrated microfluidic system for long-term perfusion culture and on-line monitoring of intestinal tissue models. Lab Chip 8(5):741–746. https://doi.org/10.1039/b717091b

Kobel S, Lutolf MP (2011) Biomaterials meet microfluidics: building the next generation of artificial niches. Curr Opin Biotechnol 22(5):690–697. https://doi.org/10.1016/j.copbio.2011.07.001

Lai JH, Kajiyama G, Smith RL, Maloney W, Yang F (2013) Stem cells catalyze cartilage formation by neonatal articular chondrocytes in 3D biomimetic hydrogels. Sci Rep 3:3553. https://doi.org/10.1038/srep03553. http://dharmasastra.live.cf.private.springer.com/articles/srep03553#supplementary-information

Lee PJ, Hung PJ, Lee LP (2007) An artificial liver sinusoid with a microfluidic endothelial-like barrier for primary hepatocyte culture. Biotechnol Bioeng 97(5):1340–1346. https://doi.org/10.1002/bit.21360

Lu H, Hoshiba T, Kawazoe N, Chen G (2011) Autologous extracellular matrix scaffolds for tissue engineering. Biomaterials 32(10):2489–2499. https://doi.org/10.1016/j.biomaterials.2010.12.016

Lutolf MP, Hubbell JA (2005) Synthetic biomaterials as instructive extracellular microenvironments for morphogenesis in tissue engineering. Nat Biotechnol 23(1):47–55. https://doi.org/10.1038/nbt1055

Maerkl SJ, Quake SR (2007) A systems approach to measuring the binding energy landscapes of transcription factors. Science 315(5809):233–237. https://doi.org/10.1126/science.1131007

Mak IW, Evaniew N, Ghert M (2014) Lost in translation: animal models and clinical trials in cancer treatment. Am J Transl Res 6(2):114–118

Mitchell MC, Spikmans V, Mello AJD (2001) Microchip-based synthesis and analysis: control of multicomponent reaction products and intermediates. Analyst 126(1):24–27. https://doi.org/10.1039/B007397K

Mullard A (2011) 2010 FDA drug approvals. Nat Rev Drug Discov 10(2):82–85. https://doi.org/10.1038/nrd3370

Nakao Y, Kimura H, Sakai Y, Fujii T (2011) Bile canaliculi formation by aligning rat primary hepatocytes in a microfluidic device. Biomicrofluidics 5(2):22212. https://doi.org/10.1063/1.3580753

Occhetta P, Centola M, Tonnarelli B, Redaelli A, Martin I, Rasponi M (2015) High-throughput microfluidic platform for 3D cultures of mesenchymal stem cells, towards engineering developmental processes. Sci Rep 5:10288. https://doi.org/10.1038/srep10288. https://www.nature.com/articles/srep10288#supplementary-information

Paul SM, Mytelka DS, Dunwiddie CT, Persinger CC, Munos BH, Lindborg SR, Schacht AL (2010) How to improve R&D productivity: the pharmaceutical industry's grand challenge. Nat Rev Drug Discov 9(3):203–214. https://doi.org/10.1038/nrd3078

Radhakrishnan J, Subramanian A, Krishnan UM, Sethuraman S (2017) Injectable and 3D bioprinted polysaccharide hydrogels: from cartilage to osteochondral tissue engineering. Biomacromolecules 18(1):1–26. https://doi.org/10.1021/acs.biomac.6b01619

Shantanu P, Iman H, CJ M, LE A (2016) Polymeric biomaterials for in vitro Cancer tissue engineering and drug testing applications. Tissue Eng Part B Rev 22(6):470–484. https://doi.org/10.1089/ten.teb.2015.0567

Shen B (2015) A new Golden age of natural products drug discovery. Cell 163(6):1297–1300. https://doi.org/10.1016/j.cell.2015.11.031

Si G, Yang W, Bi S, Luo C, Ouyang Q (2012) A parallel diffusion-based microfluidic device for bacterial chemotaxis analysis. Lab Chip 12(7):1389–1394. https://doi.org/10.1039/C2LC21219F

Stroock AD, Fischbach C (2010) Microfluidic culture models of tumor angiogenesis. Tissue Eng Part A 16(7):2143–2146. https://doi.org/10.1089/ten.TEA.2009.0689

Sung JH, Kam C, Shuler ML (2010) A microfluidic device for a pharmacokinetic-pharmacodynamic (PK-PD) model on a chip. Lab Chip 10(4):446–455. https://doi.org/10.1039/b917763a

Szymański P, Markowicz M, Mikiciuk-Olasik E (2012) Adaptation of high-throughput screening in drug discovery—toxicological screening tests. Int J Mol Sci 13(1):427–452. https://doi.org/10.3390/ijms13010427

Toh YC, Lim TC, Tai D, Xiao G, van Noort D, Yu H (2009) A microfluidic 3D hepatocyte chip for drug toxicity testing. Lab Chip 9(14):2026–2035. https://doi.org/10.1039/b900912d

Torisawa Y-S, Spina CS, Mammoto T, Mammoto A, Weaver JC, Tat T, Collins JJ, Ingber DE (2014) Bone marrow–on–a–chip replicates hematopoietic niche physiology in vitro. Nat Methods 11:663. https://doi.org/10.1038/nmeth.2938. https://www.nature.com/articles/nmeth.2938#supplementary-information

Tsao C-W (2016) Polymer microfluidics: simple, low-cost fabrication process bridging academic lab research to commercialized production. Micromachines 7(12). https://doi.org/10.3390/mi7120225

Uygun BE, Soto-Gutierrez A, Yagi H, Izamis ML, Guzzardi MA, Shulman C, Milwid J, Kobayashi N, Tilles A, Berthiaume F, Hertl M, Nahmias Y, Yarmush ML, Uygun K (2010) Organ reengineering through development of a transplantable recellularized liver graft using decellularized liver matrix. Nat Med 16(7):814–820. https://doi.org/10.1038/nm.2170

Wan J (2012) Microfluidic-based synthesis of hydrogel particles for cell microencapsulation and cell-based drug delivery. Polymers (Basel) 4(2):1084–1108. https://doi.org/10.3390/polym4021084

Whitesides GM (2006) The origins and the future of microfluidics. Nature 442:368–373. https://doi.org/10.1038/nature05058

Wong AH-H, Li H, Jia Y, Mak P-I, Martins RPS, Liu Y, Vong CM, Wong HC, Wong PK, Wang H, Sun H, Deng C-X (2017) Drug screening of cancer cell lines and human primary tumors using droplet microfluidics. Sci Rep 7(1):9109. https://doi.org/10.1038/s41598-017-08831-z

Wu H, Wheeler A, Zare RN (2004) Chemical cytometry on a picoliter-scale integrated microfluidic chip. Proc Natl Acad Sci U S A 101(35):12809–12813. https://doi.org/10.1073/pnas.0405299101

Zeng L, Yao Y, Wang D-A, Chen X (2014) Effect of microcavitary alginate hydrogel with different pore sizes on chondrocyte culture for cartilage tissue engineering. Mater Sci Eng C 34:168–175. https://doi.org/10.1016/j.msec.2013.09.003

Zhang Y, Yu Y, Ozbolat IT (2013) Direct bioprinting of vessel-like tubular microfluidic channels. J Nanotechnol Eng Med 4(2):0210011–0210017. https://doi.org/10.1115/1.4024398

Zheng XT, Yu L, Li P, Dong H, Wang Y, Liu Y, Li CM (2013) On-chip investigation of cell-drug interactions. Adv Drug Deliv Rev 65(11–12):1556–1574. https://doi.org/10.1016/j.addr.2013.02.001

Zhou X, Cai S, Hong A, You Q, Yu P, Sheng N, Srivannavit O, Muranjan S, Rouillard JM, Xia Y, Zhang X, Xiang Q, Ganesh R, Zhu Q, Matejko A, Gulari E, Gao X (2004) Microfluidic PicoArray synthesis of oligodeoxynucleotides and simultaneous assembling of multiple DNA sequences. Nucleic Acids Res 32(18):5409–5417. https://doi.org/10.1093/nar/gkh879

9 Dynamic Culture Systems and 3D Interfaces Models for Cancer Drugs Testing

Diogo C. Fernandes, Raphaël F. Canadas, Rui L. Reis, and Joaquim M. Oliveira

Abstract

The mass use of biological agents for pharmaceutical purposes started with the development and distribution of vaccines, followed by the industrial production of antibiotics. The use of dynamic systems, such as bioreactors, had been already applied in the food industry in fermentation processes and started being used for the development of pharmaceutical agents from this point on. In the last decades, the use of bioreactors and microfluidic systems has been expanded in different fields. The emergence of the tissue engineering led to the development of *in vitro* models cultured in dynamic systems. This is particularly relevant considering the urgent reduction of the total dependence on animal disease models that is undermining the development of novel drugs, using alternatively human-based models to make the drug discovery process more reliable. The failure out coming from animal models has been more prevalent in certain types of cancer, such as glioblastoma multiform and in high-grade metastatic cancers like bone metastasis of breast or prostatic cancer. The difficulty in obtaining novel drugs for these purposes is mostly linked to the barriers around the tumors, which these bioactive molecules have to overcome to become effective. For that reason, the individualized study of each interface is paramount and is only realistic once applying human-based samples (e.g. cells or tissues) in three-dimensions for *in vitro* modeling under dynamic conditions. In this chapter, the most recent approaches to model these interfaces in 3D systems will be explored, highlighting their major contributions to the field. In this section, these systems' impact on increased knowledge in

D. C. Fernandes · R. F. Canadas (✉)
3B's Research Group, I3Bs – Research Institute on Biomaterials, Biodegradables and Biomimetics, University of Minho, Headquarters of the European Institute of Excellence on Tissue Engineering and Regenerative Medicine, AvePark, Parque de Ciência e Tecnologia, Barco, Guimarães, Portugal

ICVS/3B's–PT Government Associate Laboratory, Braga, Guimarães, Portugal
e-mail: raphael.canadas@i3bs.uminho.pt

R. L. Reis · J. M. Oliveira
3B's Research Group, I3Bs – Research Institute on Biomaterials, Biodegradables and Biomimetics, University of Minho, Headquarters of the European Institute of Excellence on Tissue Engineering and Regenerative Medicine, Guimarães, Portugal

ICVS/3B's PT Government Associate Lab, Braga, Guimarães, Portugal

The Discoveries Centre for Regenerative and Precision Medicine, Headquarters at University of Minho, Guimarães, Portugal

J. M. Oliveira, R. L. Reis (eds.), *Biomaterials- and Microfluidics-Based Tissue Engineered 3D Models*, Advances in Experimental Medicine and Biology 1230,
https://doi.org/10.1007/978-3-030-36588-2_9

relevant aspects of cancer aggressiveness as invasive or motile cellular capacity, or even resistance to chemotherapeutic agents will have particular focus. The last section of this chapter will focus on the integration of the tumor interfaces in dynamic systems, particularly its application on high-throughput drug screening. The industrial translation of such platforms will be discussed, as well as the main upcoming challenges and future perspectives.

Keywords
3D interfaces models · *In vitro* tumor models · Drug screening · Bioreactors

Highlights

- The impairment in novel drugs development is mainly due to the failure of results in animal models to translate to humans, particularly in certain cancer types, and the best alternative is the use of *in vitro* models;
- The three-dimensional (3D) interfaces contacting cancer cells work as barriers to therapies effectiveness, becoming essential features in *in vitro* models;
- 3D tumor interfaces can be cellular, molecular or physical and have been modeled in recent years, existing already some examples of industrial application for drug testing purposes.

9.1 Introduction

During the past few decades, a decrease in the number of new pharmaceutical drugs has become evident, with the newly found drugs becoming also increasingly costly (DiMasi et al. 2016). Although recent therapies, such as immunotherapy, have looked promising for different tumors, they have failed in treating certain primary cancer types, such as glioblastoma multiform (Lim et al. 2018) or castration-resistant prostate metastatic cancer (Twardowski et al. 2019).

The lack of innovative drugs is mostly related to the failure during the preclinical stage due to: poor mimicry of human tissue specific barriers such as the blood-brain barrier (BBB) (Lim et al. 2018); and the translation of false-positive results from pre-clinical to clinical stage (Mak et al. 2014). This last one is particularly bound to the fact that 92% of the molecules selected from animal models fail in efficacy when applied to humans (Mak et al. 2014).

Although the most promising alternative to the use of animal models is the use of *in vitro* models (Fig. 9.1) achieving clinically relevant complexity together with reproducibility is the major challenge of this approach. Despite tumor models have increasingly mimicked human physiology over the last decade (Katt et al. 2016), the recreation of general and tissue-specific hallmarks and surrounding barriers is still an unresolved quest in the field.

There are several sources of variability that influence tumor progression and, eventually, the patient's outcome. The surroundings of a tumor site turn into a series of interfaces to which cancer cells contact and interact, undergoing complex feedback loops that have proven relevant in chemo resistance (Castells et al. 2012; Rice et al. 2017) and tumor recurrence (Walens et al. 2019) (Fig. 9.2). The most recent strategies to mimic the several tumor interfaces will be reviewed and analyzed herein. In addition, the use of physiologically relevant dynamic conditions involving bioreactors and microfluidic systems will be discussed and an author's analysis on the future outlook of the field will be presented.

9.2 3D Interface Models Critical Features – Recreation of the Tumor's Interfaces

9.2.1 Structural

Interactions between cells and the surrounding ECM are paramount in cancer development and progression. The molecular assembly of ECM can promote aberrant behavior, even in noncancerous cells, revealing the importance of the

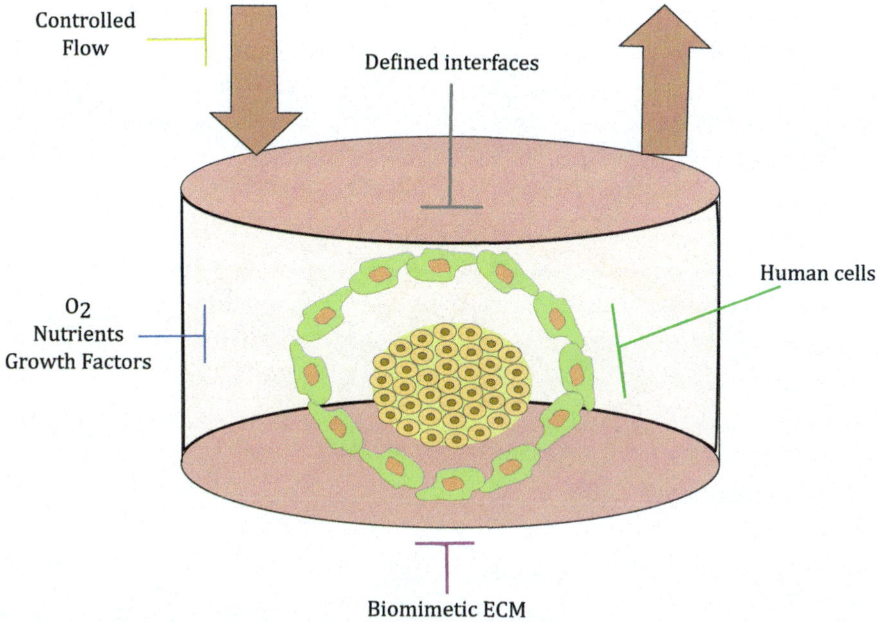

Fig. 9.1 Advantages of *in vitro* models as compared to animal models. Controlled flow rate at physiological shear stress, administration of defined concentrations of soluble factors, biomimetic extracellular matrix (ECM) and the use of human cells are the main reasons why *in vitro* models may outperform animal models as disease models for drug testing

ECM remodeling in the development of malignant phenotype (Paszek et al. 2005). In the following sub-sections, the latest biomimetic strategies to generate and study the feedback loops between cancer cells and structural cues of the surrounding tissue will be addressed.

9.2.1.1 Tumor-Basement Membrane

The basement membrane is an adjacent structure to the epithelium, endothelium and mesothelium that supply architectural support to the tissue. These membranes are steep and discrete interfaces that tumors have to overcome during tissue invasion. The interface between the connective tissue and the cancer and cancer-associated cells is associated with tumor progression and malignancy (Engbring and Kleinman 2003).

Due to its relevance, when designing a 3D tumor *in vitro* model, such interface should be structurally reproduced. Using this strategy, it was reported that this interface can promote, but not induce, malignant phenotype on cancer cells in a model where spheroids were culture contacting an *in vitro* healthy basement membrane. Strikingly, benign tumor cells and regular epithelial cells, when contacting the membrane, had controlled proliferation and the integrity of the membrane was kept. However, when contacting malignant cancer cells, the basement membrane got disrupted and the invasiveness of these cells got exacerbated (Guzman et al. 2017).

Nonetheless, the interplay between a tumor and the surrounding basement membrane is dependent on the cancer grade and sensitive to other cell types surronding the tumor site. This was observed in patient-derived organoids of pancreatic cancer surrounded by basement membrane invasive behavior and consequent membrane disruption induced by the presence of pancreatic stellate cells. Pancreatic adenocarcinoma cells cultured in spheroids encapsulated in Matrigel® developed their own basement membrane maintaining homeostasis. Nonetheless, once pancreatic stellate cells were present, the cancer cells disrupted the basement membrane and invaded the surrounding matrix (Koikawa et al. 2018). This interference on the interplay between cancer cells and the basement mem-

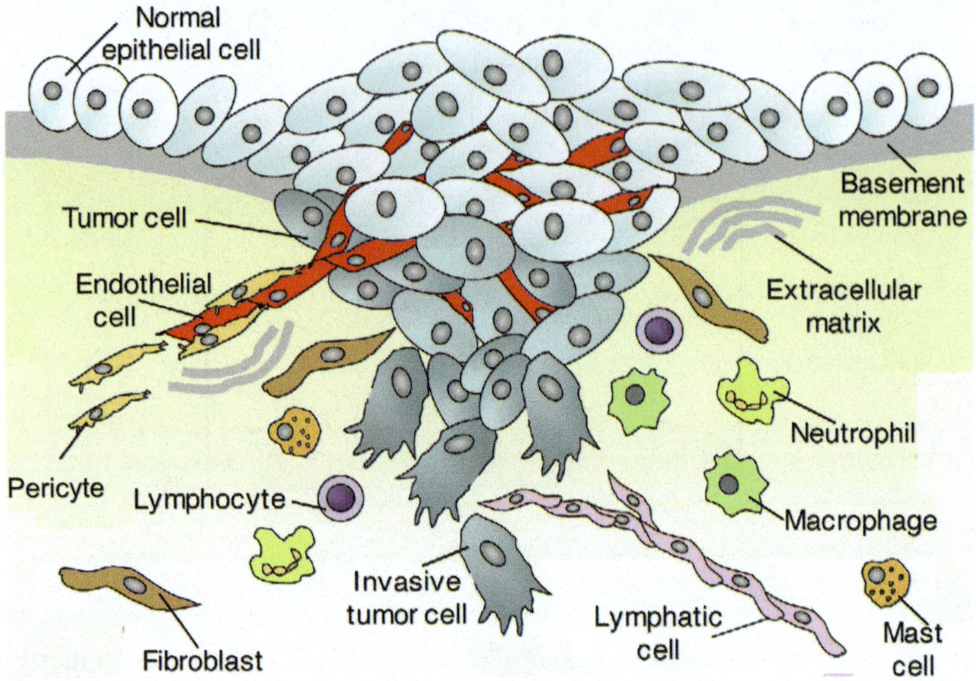

Fig. 9.2 Tumor microenvironment's interfaces. Every cell type, physical and biochemical features surrounding cancer cells establish interplays that affect and condition the tumor's outcome. (Figure adapted from Joyce 2005)

brane differs by other cell types uncovers a level of complexity that urges for increased understanding and following studies.

These results highlighted the importance of the tumor-basement membrane interface in an *in vitro* model. In high-throughput systems, considering it is a discrete 2D homogeneous interface, the addition of this interface would increase biomimetic capacity to the existing models without presenting a major fabrication challenge.

9.2.1.2 Tumor-ECM Interface: Fibrotic *vs.* Healthy

The structural remodeling of the ECM of the tumor's surrounding tissue is a hallmark in several cases, such as pancreatic (Puls et al. 2017) or breast cancers (Vermeulen et al. 2017). Since the most relevant component of the ECM of these tissues is collagen, its reassembly remodels the microenvironment and cell behavior. Thus, most models are based on the fibrotic assemble of this protein.

Recently, models mimicking the changes in tissue fibrosis have been used to uncover its role in the epithelial-to-mesenchymal transition (EMT), namely how the rearrangement of the ECM leads to the development of fibrotic tumor microenvironment (Guzman et al. 2017; Sharma et al. 2017; Puls et al. 2017; Shi et al. 2019; Tanaka et al. 2019). These models, however, are quite diverse in complexity, uncovering different dynamics according to its purpose.

One-dimensional (1D) and two-dimensional (2D) models, simplistic by design, deliver straightforward quantifications that can be useful to account for the contribution of each feature in a more complex model, particularly in 3D.

The 1D breast cancer fibrotic models have been used for example to quantify the correlation between fiber's diameter and cell linear motility. Measuring the average linear velocity of cultured cancer cells onto fibers with diameters ranging from 20 to 0.7 μm, it was found that fibers with smaller diameter, closer to the ones observed in clinical assessments, led to increased migration velocity. Moreover, motility-related nuclear deformation, usually associated with the squeezing process in 3D motion, was found to be related to the forces balance necessary for linear cellular motion (Sharma et al. 2017).

Thus, for the observation of specific phenomena, simpler models enable a clear classification and quantification of particular aspects of the disease, therefore, aiding the assembly of better and more representative 3D models. However, it was already demonstrated that only 3D cancer models can realistically mimic the dynamics of the native tumor tissue, which is critical either for drugs testing or for the optimization of medical devices (Ridky et al. 2010).

To do so, *in vitro* disease models using clinically observed features enable information that could not be achieved *in vivo* without advanced microscopy strategies. For instance, such models allowed the clarification that pancreatic stellate cells, not cancer-associated fibroblasts, are the cells responsible for the fibrotic reassembly of collagen and fibronectin that drives the malignant phenotype in cancer cells, using clinically observed tissue thickness (Tanaka et al. 2019).

However, none of the mentioned models accounted for the effect of stiffness increase related to the increased fibrotic tissue fraction. With that purpose, in a 3D pancreatic cancer model of invasion, three different pancreatic ductal adenocarcinoma cell lines were encapsulated in engineered fibrotic tissue of variable stiffness. Matrigel® was used to counter the increase of stiffness caused by having a thicker fibrotic tissue fraction, maintaining the stiffness constant. It was then found that an increase in fibrotic tissue, with a constant stiffness, led by itself to the overexpression of mesenchymal markers (e.g., vimentin), and a decrease of epithelial markers (Puls et al. 2017). Importantly, these results demonstrate the distinct roles of stiffness and tissue structure in driving tumor malignancy.

Considering the presented models, the importance of the ECM architecture in tissue homeostasis is undeniable. However, it is paramount to consider the variation of biophysical properties with the architectural and chemical changes of the tissue, which can have drastic effects over the tumor's fate, as it will be addressed in the following sub-section.

9.2.2 Mechanical Properties

9.2.2.1 Tumor-Solid Interface

Over the last years it was established that cancer cells undergo EMT upon variations of the matrix stiffness (Liou et al. 2014; Ko et al. 2016; Bauer et al. 2018) and for more than a decade it is known to drive carcinogenesis (Paszek et al. 2005). Nonetheless, only by using novel 3D models, the role of stiffness in chemoresistance was uncovered (Rice et al. 2017).

Polyacrylamide gels with tunable rigidity, but no bioactive role in cell adhesion, induced an increase in mesenchymal markers on pancreatic cancer by three-fold when compressive stiffness increased from 1 kPa to 25 kPa. Furthermore, it was concluded that the increasing stiffness led to an increased resistance to paclitaxel, a chemotherapeutic agent that stabilizes the microtubules cytoskeleton, but not to gemcitabine, a nucleotide analogue that competes with other nucleotides during replication of DNA (Rice et al. 2017). Thus, it was shown that stiffness itself is a sufficient factor to induce resistance to chemotherapeutic agents.

Strikingly, the compressive modulus of the interstitial matrix does not affect uniquely the cancerous cells but also all the tumor microenvironment. Increased stiffness was found to arrest the differentiation of pre-adipocytes cultured in between tumor spheroids (Yue et al. 2018). Therefore, stiffness-related malignancy affects not only cancer cells, modulating their cytoskeleton, but also other cells in the tumor microenvironment and even healthy stromal cells, changing their behavior towards a disease one.

9.2.2.2 Tumor-Liquid Interface

The circulation of fluid from the vascular system through the surrounding tissue and collected in the lymphatic one produces a shear stress effect that is inversely proportional to the distance from the blood vessel. When the tumor promotes angiogenesis or undergoes surrounding tissue invasion, approaching blood vessels, the shear stress increases due to the proximity to the blood flow (Sefidgar and Soltani 2014).

The shear stress felt by cancer cells influences their physiology and is relevant for tumor outcome (Mak 1986; Hillsley and Frangos 1994; Mow et al. 1999; Triantafillu et al. 2017). Once cancer cells undergo endothelial-to-mesenchymal transition (EMT) and start intravasation into surrounding blood vessels, the blood flow and the fluid permeating the tissue exert increased influence over cancer cells. To investigate how important the tumor-blood interface can be on the tissue and blood vessels invasion by cancer cells, *in vitro* models have recently proven that fluid shear stress can promote malignant phenotype in cancer cells. For instance, in ovarian (Hyler et al. 2018) and breast cancer (Bratton et al. 2019) models, low magnitude fluid shear stress drove genomic instability and mesenchymal phenotype, leading to metastasis. In the case of the ovarian cancer model, malignant and non-malignant cells were subjected to fluid shear stress, being cultured while attached and in suspension, it was concluded that suspension culture severely promoted malignant (Hyler et al. 2018).

There are still numerous limitations to the creations of these models. The heterogeneity of the density of capillary vessels within tissues, or the pressure variation with the distance from the heart makes the creation of models for personalized medicine particularly. Nonetheless, these models shall account for the near *in vivo* complexity with the application of advanced fabrication techniques, like bioprinting, that will allow the creation of hierarchical 3D model.

9.2.2.3 Tumor-Air Interface

The air-tissue interface of specific organs is known to be a critical factor in carcinogenesis due to the nefarious effects of certain external factors, like smoking (Van Duuren and Goldschmidt 1976) or pollution (Demetriou and Vineis 2015), or sun radiation (Kripke 1981). Moreover, these interfaces are critical for the physiological behavior of the correspondent tissues, being responsible for the tissue's main function, such as lung, skin or intestines.

In the case of lung, intestines and stomach models, it is necessary to create tissue-air interfaces in conditions of high humidity to recreate the native conditions of the tissue. For that purpose, an *ex vivo* 3D model of colorectal cancer was developed by culturing primary tissue in inner wells, exposed to a humidified atmosphere, while being narrowly submerse by the medium dispensed in the outer well. The liquid moves into the inner well by capillarity and keeping the cells nearly submerged, mimicking the circulation of fluid observed in the native tissue. The model kept the physiology of the biopsied tissue for a minimum of 14 days, either healthy or cancerous, showing to be a robust system for *ex vivo* tissue culture (Usui et al. 2018).

Skin tissue, on the other hand, is the only tissue that contacts directly with air that has not been firstly humidified. Therefore, a non-vascularized melanoma model should include this unique feature of the skin. To mimic that tissue-air interface of skin cancer, melanoma spheroids were inserted in the dermis of the model, consisting of fibroblasts and collagen type I, below a layer of keratinocytes, mimicking the epidermis. This structure was developed into an insert well slightly submerged in medium. Histology results, when compared to explants, showed that the model resembled the native tissue, either in the skin construct, either in the heterogeneity of the tumor, having a proliferative front and necrotic core, which is characteristic of non-vascularized melanomas (Houghton 1987). The generation of this model was performed in multiple simultaneous batches, allowing the production of a high number of air-tissue organotropic replicas within a 17 days period. This way, this model showed a method for a reproducible, mass production of organotypic melanoma models (Müller and Kulms 2018).

This interface, however, is still quite unexplored, particularly in lung cancer where the exogenous stimuli are the main cause of cancer, according to the United Kingdom National Health Institute (NHS) website for causes of lung cancer. Models of skin or the respiratory or gastric tract should include, in following years, this interface and validate the disease with organ function impairment.

9.2.3 Biochemical

The biochemical interactions of interest in a cancer model are diverse. These molecular cues either happen in the form of gradient or as a uniform signal across the tumor mass, solubilized or deposited in the surrounding ECM. Furthermore, they can be composed of small molecules or proteins, thus making the characterization and emulation of the tumor's biochemical interactions particularly challenging. However, over the last few years, a number of promising works have been suggesting innovative designs incorporating these cues.

9.2.3.1 Soluble Factors

The interplay between cells within the tumor microenvironment is often done by the intermediacy of soluble factors. These are the most diverse signals responsible for cell-cell communication and the characterization of the activated pathways by several of these molecules has been extensively reviewed (Kessenbrock et al. 2010; Jung et al. 2015). However, to identify and characterize the role most relevant soluble factors for the development of malignant phenotype to be particularly targeted by new drugs, there is the need to create tumor models where the intercellular relations can be isolated and individually assessed.

Recently, the role of neural Wiskott-Aldrich syndrome protein (N-WASP), an actin cytoskeleton-related protein, in the motile and invasive behavior of pancreatic ductal adenocarcinoma, was studied using a 2D model. Even though the promotion of motile behavior on cancer cells by autocrine stimulation with N-WASP was already known, this model revealed that impairing N-WASP led to unresponsive cancer cells to chemo attractant such as lysophosphatidic acid (Juin et al. 2018). The assessment of these feedback loops that can now be targeted with novel therapies, would not be possible without an *in vitro* simplified construct.

Small molecules and secreted proteins are soluble factors for intercellular communication but they can also be metabolites that indicate an abundance or shortage of nutrients or overpopulation. For example, one of the most well-known and well-established tumor's hallmarks is the oxygen gradient within the tumor mass, that is known to drive a pH gradient known as the Warburg effect (Warburg 1956; Lu et al. 2002). Recently, 2D breast cancer models, using soluble oxygen gradients, allowed a deeper understanding about the mechanisms of hypoxia-related migration (Enokida et al. 2017; Tsuruno et al. 2018). Inducing a gradient of 0–3%, generated cell migration towards the oxygenated region. This migration significantly decreases with the cell density. Thus, oxygen-dependent migration is, actually, a pH-dependent migration, from acidic to more neutral locations (Enokida et al. 2017; Tsuruno et al. 2018).

To further understand the role of hypoxia-related genes on chemo resistance, spheroids of colorectal cancer cell line were categorized correlating size with oxygen levels. Overall, the spheroids that were in hypoxia and developed necrosis presented gene expression levels closer to *in vivo* ones, demonstrating an increasing resistance to 5FU, a chemotherapeutical agent clinically used for colorectal cancer treatment (Däster et al. 2017).

However, some behavioral consequences are still unclear, namely the role of hypoxia and hypoxia-related genes on chemo resistance still requires further investigation, despite the clear relevance of hypoxia on the adaptive capacity of cancer cells to chemotherapy.

9.2.3.2 ECM-Bound Factors

The delivery of biochemical stimuli is often done by fixing molecular factors known to be present surrounding the tumor site to the ECM, therefore producing an enhanced effect with lower concentrations than when present in solution (Hollier et al. 2005). Such factors have shown to be altered in diseased tissues and to drive aggressive behavior in cancer cells (Huleihel et al. 2016).

To replicate this effect two main approaches can be applied: top-down, through decellularization, and bottom-up, by the regular assembly of components *in vitro* (Lee et al. 2011). Despite the most common one is the bottom-up due to the control achieved in every component of the model, the top-down approach has been the one with larger success in recreating this interaction due to its complexity.

Decellularization, as a top-down approach, consists in extracting a part of the tissue, withdraw the cellular part of the excised tissue and use it to create a full *in vitro* model that has the native ECM's biochemical signature.

In vitro models using decellularized tissue evaluate only the effect of ECM-bound motifs, when compared to matrices with the same physical properties. For such properties, a model for colorectal cancer was recently proposed using decellularized matrix allowed to conclude that, when compared with a healthy control, tumor tissue showed a decreased angiogenic potential in chorioallantoic membrane assays and led to the expression of autocrine and paracrine factors characteristic of native tumors by cancer cells after recellularization with a colorectal cancer cell line (Piccoli et al. 2018).

The manifestation of cancer hallmarks by the *in vitro* model validates it as a biomimetic effort. Thus, to verify if a patient-derived decellularized matrix of a glioma triggered native-like behavior on glioblastoma *in vitro,* glioblastoma spheroids were encapsulated in such matrix and their behavior was assessed. The cellular and collective dynamics were quantified using tumor spheroids cultured in decellularized ECM. Taking into account the heterogeneity of the glioblastoma multiform tumor, the shape and expression levels of ECM remodeling proteins, such as matrix metalloproteinases, were assessed and compared to the results obtained in collagen matrices. Using tumor spheroids, glioblastoma's migration patterns were isotropic, meaning that cells did not exhibit a particular or preferable migration direction. Moreover, it was verified that cells in the tumor invasive front were faster, showed an elongated morphology, while the core consisted in round, less motile cells, similarly to what is known from *in vivo* studies (Sottoriva et al. 2013). Tumor cells maintained in the decellularized tissue predominantly presented higher gene expression of ECM remodeling factors when compared to the control group cultured in collagen, exhibiting an invasive profile. These features validate this as a biomimetic modeling approach for aggressive glioblastoma (Koh et al. 2018).

However, this dependence on patient-derived tissues limits the applications of decellularized matrices, since the availability is short and variable from donor to donor. Further research into the composition, particularly the spatial distribution of factors within these matrices is paramount, which can potentially be recapitulated by synthetic/engineered approaches (Yang et al. 2016; Zhu et al. 2018b).

9.2.4 Cellular

The tumor's cell mass is surrounded by stromal, epithelial, endothelial and immune cells that have been proven relevant in carcinogenesis, EMT and metastasis settling (Magdeldin et al. 2017; Chung et al. 2017; Angeloni et al. 2017; Heinrich et al. 2019). However, intercellular crosstalk is complex and the feedback mechanisms are not fully characterized. In animal models is impossible to assure control on the interplay among the cellular niches.

For that reason, *in vitro* models represent a flexible tool to address interfaces and assess the influence of cellular composition over cancer outcome, as it will be addressed in the following sub-sections.

9.2.4.1 Stromal

There have been some models focusing on the study of individual or group interactions among stromal cells and cancer-associated cells over last years. The development of some techniques allowed, in the last few years, the generation of interfaces with controlled features and dimensions, increasing reproducibility and monitoring capacity.

Following this reasoning, metastatic colorectal, ovarian and stomach tumors' microenvironment models were developed by culturing cancer cell lines with primary fibroblasts (Chung et al. 2017).

One of the most relevant interactions to be explored is the paracrine interactions in cancer cells' viability and stromal angiogenic potential. Using *in vitro* models, it was made possible to perceive how the paracrine interactions affect cancer cells' viability and stromal angiogenic potential. By co-culturing of a cancer cell spheroid with a 2D monolayer of fibroblasts, cancer cells cytoplasmic area and number of nuclei were significantly increased when compared to a monoculture of cancer cells (Chung et al. 2017).

Using a microfluidic platform designed for the study of this cancer model paracrine effect on the recruitment of endothelial cells, researchers demonstrated that the *in vitro* cultured ovarian and metastatic colorectal cancers contributed to reduce the angiogenic potential of the fibroblast culture, while the stomach cancer model led to an increase of angiogenesis (Chung et al. 2017). Thus, this model showed that stroma-tumor interaction occurs in a site-dependent manner.

However, though several models were created along the years, the comparison of the results obtained with the *in vitro* construct with *ex-vivo* models cultured in the same conditions is usually absent.

To perform such validation, the migration and invasion patterns of glioblastoma cellular spheroids cultured onto several layers of astrocytes were compared to patient-derived samples. The model exhibited glioblastoma hallmarks also verified in the brain cancer slices, such as individually migrating cells, collective strands extending along blood vessels, and multicellular networks of interconnected glioma cells infiltrating the neuropil. These observations were not present in a control group of spheroids without the surrounding astrocytes. Essentially, this study showed the relevance of the presence of stromal cells around the tumor to mimic the invasion and intravasation is revealed by this study (Gritsenko et al. 2017).

Stromal cells are essential to create representative models for drug testing. However, for that purpose, validation of the results found *in vitro* with *in vivo* or patient-derived samples is crucial to understand the relevance of these models. Moreover, the observation and characterization of these cancer-stromal cells interplay also has become increasingly complete, by using more advanced materials, fabrication and visualization techniques. The following years must include tissue specific hierarchy in comparison to explants to understand how the spatial distribution affects the interplay between the tumor and the surrounding stromal cells.

9.2.4.2 Immune Cells

Recently, the interest on the triple interaction between the immune, stromal and cancer cells has risen, particularly considering the panoply of models lately proposed for the study of these interfaces (Herter et al. 2017; da Costa et al. 2018; Suresh et al. 2018; Pereira et al. 2019). This interest is mainly based on the applications of immunotherapy for cancer treatment (Gopalakrishnan et al. 2018).

Strikingly, several evidences of cooperation among these cells have also been reported both in carcinogenesis and in tumor progression. To understand this cooperation, a 3D bioprinted co-culture model of glioblastoma cells and macrophages showed that this interplay led, firstly, to a polarization of the macrophages towards a cancer-associated macrophage phenotype, which is followed by an increased aggressiveness and invasiveness of the glioblastoma cells. Moreover, when this interplay is disrupted the tumor's growth is diminished and its sensitivity to chemotherapy increased.

Considering this contribution of immune cells for cancer progression, the interplay between stromal and immune cells is also of particular curiosity since it is an interface that has shown related to cancer progression (Heinrich et al. 2019). For instance, to study the influence of cancer-associated fibroblasts and mast cells on carcinogenesis and first cancer progression, a 3D model using patient-derived cancer-associated and healthy fibroblasts, mast cells and healthy prostate epithelial cells was developed sequentially to understand the effects of an established niche over cancer cells. Thus, in an initial stage, fibroblasts, healthy and cancer-associated, were cultured in a 3D-printed scaffold to produce native tissue. Afterwards, healthy prostate epithelial cells were cultured onto the scaffold and morphogenesis was observed and measured. Finally, mast cells were introduced into the model and the morphological changes were assessed. When compared to healthy fibroblasts, cancer-associated cells led to significant morphogenesis towards a cancerous phenotype, which was exacerbated with the presence of mast cells (Pereira et al. 2019).

Furthermore, this synergistic interplay has been observed in some other *in vitro* models (da Costa et al. 2018; Suresh et al. 2018), showing the potential to be a hallmark and a therapeutic target for cancer progression. For example, heterocellular spheroids, containing pancreatic cancer, fibroblast and macrophage cell lines, were proposed as a pancreatic cancer microenvironment model. The authors verified that cancer cells underwent severe morphological changes and significantly increased the expression of stem cell factor when compared to homotypic spheroids (Suresh et al. 2018).

Alternatively, using a bone metastasis prostate cancer model, da Costa et al. 2018 showed that tumor-associated macrophages and cancer-associated fibroblasts were able to modulate the viability, cell proliferation and response to growth factors of patient-derived cancer cells (da Costa et al. 2018). So, the presence of immune cells has been shown to directly influence the progression of varied cancer types.

Despite the evident role of the cells from the immune system in cancer aggressiveness highlighted during the present section, immune cells also have therapeutic potential, which can also be reproduced and tested using *in vitro* models. For the colorectal cancer, to model the immunotherapy fibroblast and cancer cell lines were encapsulated and co-culture were treated with a fibroblast-targeted T-cell biospecific antibody. Primary peripheral mononuclear blood cells were added to the model and their effects over the treatment were assessed by means of measuring viability of cancer cells over time. For that purpose, the mononuclear cells were cultured in suspension and stimulated using cytokines to promote infiltration into the spheroid. Strikingly, the immune cells led to a significant enhancement of the treatment by acting as both permeation and killing agents (Herter et al. 2017).

Overall, different recent studies have been demonstrating that immune cells can drive cancer progression or be molecularly reprogrammed for cancer targeting. However, the manipulation of the immune cells still requires a profound characterization of the effects of the biochemical loops between cancer cells and cancer-associated immune cells. For this purpose, proteomic and secretomic approaches must be assessed in the future when applying *in vitro* models, ideally with real-time monitoring technology.

9.2.4.3 Epithelial

During the metastatic process of extravasation, depending on the site, cancer cells can face a layer of epithelial cells or a bulk of ECM and stromal cells. The importance of the interplay between cancer cells and the epithelial layer and the relevance of the target cells on metastasis formation is still unclear and it is challenging to measure when using animal models. *In vitro* tumor-epithelium models have been helping in the characterization of these interfaces and enabling quantification of their interactions for metastasis formation.

For instance, to study tumor and target tissue interaction, two models have recently revealed the importance of epithelial cells in metastatic cancer cells adhesion (Angeloni et al. 2017; Marques-Torrejon et al. 2018). A simplified *in vitro* model of metastatic breast cancer metastasis in bone was assembled. Adipose stem cells (ASC)-derived osteoblasts onto polyurethane and co-culturing breast cancer in a layer-by-layer procedure onto the osteoblasts. Interestingly, breast cancer cells were not able to adhere and spread onto the polyurethane surface as they did when cultured onto osteoblasts (Angeloni et al. 2017), demonstrating that the tissue's epithelial layer is essential for metastasis formation.

Similar results were achieved after the injection of glioblastoma stem cells into *ex vivo* cultured mouse adult brain slices. Cancer cells particularly engrafted and responded to endothelial stimuli when inserted in a biomimetic neuronal epithelium (Marques-Torrejon et al. 2018).

However, the interplay between metastatic cancer cells and the target tissue's epithelium is more complex than the management of the adhesion of circulating tumor cells, affecting permanently the biochemical profile of the tissue's epithelial cells. To study these changes, the influence of metastatic lung cancer cells over brain, liver and bone was studied using a body-on-a-chip device with astrocytes, hepatocytes or osteocytes cultured in interconnected chambers, respectively (Fig. 9.3). With no endothelial barriers, metastasis can be formed over the three-epithelium layers. However, the newly formed metastasis did not simply adhered and proliferated, having also induced aberrant gene expression in the target tissues epithelium, leading to severe cell damage, similarly to clinically observed metastatic lesions (Xu et al. 2016b).

Overall, the interface in between the tumor and the epithelial layer has demonstrated to be extremely relevant to be reproduced *in vitro,* particularly in the study of the mechanisms underlying metastasis settling. However, the current state-of-the-art still lacks further investigation on quantifying the contribution of cancer cells secreted extracellular vesicles previous to intravasation and cell-cell contact upon extravasation that will require specific and advanced

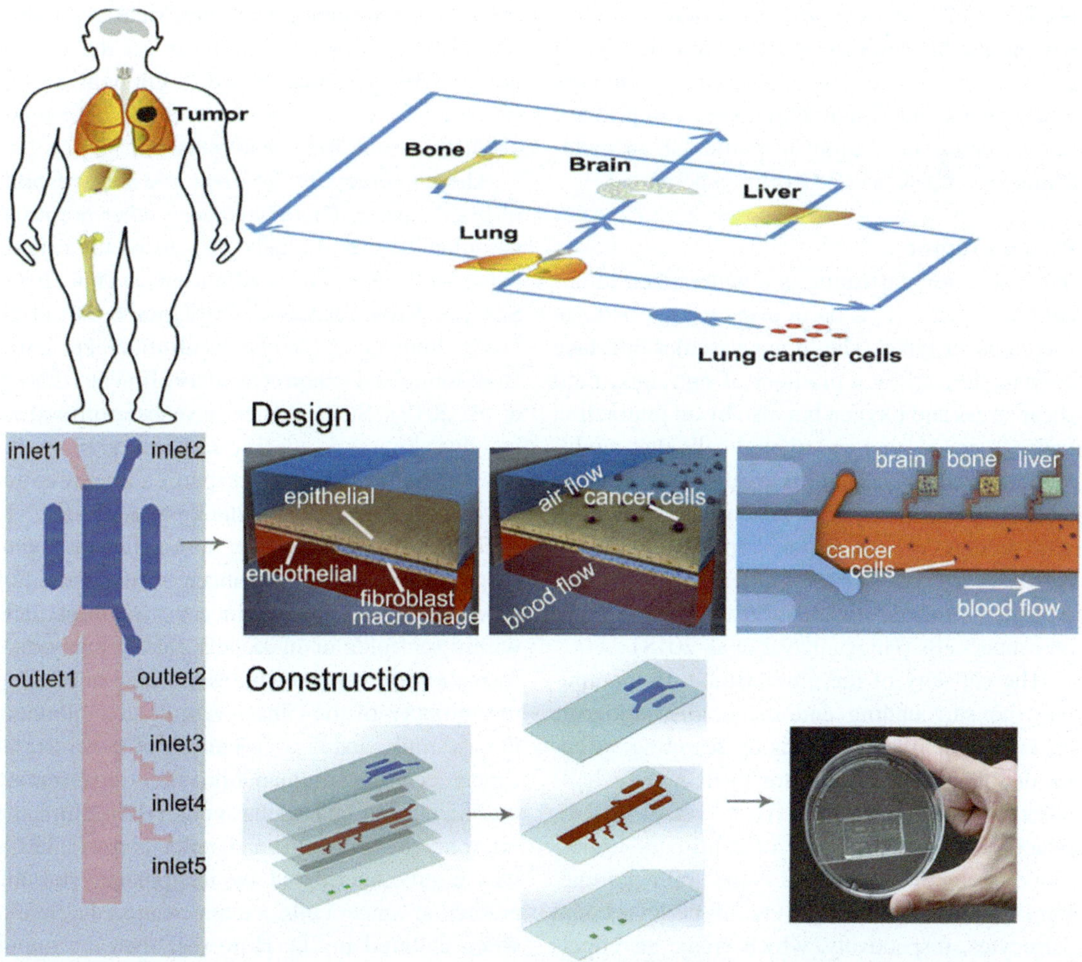

Fig. 9.3 Body-on-a-chip for the affinity assessment of lung metastasis for bone, liver and brain epithelia settling. With interconnected epithelium chambers and metastatic lung cancer cells in circulation, the differential settling of cancer cells is an indication that the preference of metastasis formation sites is related to the interplay of cancer cells and tissue's specific epithelium. (Adapted with the permission of (Xu et al. 2016b). Copyright 2016 American Chemical Society)

analytical methods integrated on epithelial-cancer interface model devices.

9.2.4.4 Vascular

Tumor vascularization is a key parameter in tumor evaluation and it is linked to late diagnosis and consequent poor prognosis. Nonetheless, the vascular system is a central interface in each grade of cancer. Angiogenesis is exacerbated in primary tumor sites, metastatic cancer cells undergo intravasation into the blood stream through these capillaries and undergo extravasation at specific tissue sites, without having a known motif for it. To study each of the particularities of this interface, *in vitro* vascular models with an established cancer grade were developed and contributed over last years to understand the mechanisms underlying angiogenesis in primary tumors and extavasation, in particular, in metastatic cancer as it will be explored herein.

Primary Tumor

Increased vascularization is a generalized tumor hallmark, but it is a particular characteristic in some cancer types. The tumor-vascular interface is characterized by a gradient of nutrients, fluid shear stress and oxygen but also by an interaction between cancer and endothelial cells that modulate tumor's progression. Although it is difficult to distinguish the contribution of each effect, *in vitro* models revealed that a simple cellular interplay reduced the expression of endothelial markers by endothelial cells and increased proliferation on cancer cells (Heggannavar et al. 2018).

The capacity of the glioblastoma to manipulate the surrounding endothelia, disrupting its characteristic tight junctions by secreting vascular endothelial growth factor (VEGF) is a hallmark of this tumor type (Herter et al. 2017), which is a particular case of a primary tumor-blood interaction. The BBB is the brain endothelium covered by monolayers of pericytes and astrocytes, respectively, which create an impermeable membrane to most compounds based on the tight junctions that glioblastoma cells are known to disrupt.

Despite this tumor-blood interaction, the cancer-destabilized vessels are still not permeable enough for an efficient drug delivery. For that reason, *in vitro* drug tests for glioblastoma have to take into account the impermeability of the surrounding vascularization of the tumor, requiring a BBB model based on human cells and interfacing the *in vitro* tumor model. This way, the models can recreate the biggest obstacle to the drug delivery.

Using a tumor-blood interface, glioblastoma targeting drug delivery systems were effective in overcoming BBB with a value of transendothelial electrical resistance of 250 Ω/cm^2 *in vitro* (Heggannavar et al. 2018). Despite being promising, these results were obtained using a BBB presenting a transendothelial electrical resistance one order of magnitude inferior to the one of native BBB measured in mice (Butt et al. 1990), revealing the urge to create tumor and BBB models more similar to the native tissues.

These interactions, however, are not particular of brain cancer. There have been other proposed vascular cancer models of different origins (Magdeldin et al. 2017; Klimkiewicz et al. 2017; Sánchez-Rodríguez et al. 2017; Racordon et al. 2017; Jung et al. 2017; Roudsari et al. 2018; Boussommier-Calleja et al. 2019; Jiménez-Torres et al. 2019). Such studies have been providing insights about type-specific and general biomarkers and hallmarks that are particularly relevant for diagnosis and therapeutic development.

The 3D models of tumor vasculature is crucial to properly mimic the cancer environment for new drug development, since the 3D constructs within scaffolds or hydrogels enable the spatial formation of vascular-like segments, improving the mimicry of the cancer-vasculature interface. For example, using a co-culture of a colorectal cancer cell line, human fibroblasts and human umbilical vein endothelial cells (HUVECs), the interplay between stromal cells and HUVECs was highly overdriven by the presence of the colorectal cancer cells, a classic cancer hallmark, when cultured in 2D. However, once the tumor started growing in 3D, cancer cells decreased their expression of CK20, a primary colorectal cancer molecular biomarker, while increasing their invasive phenotype, disrupting the newly formed vessels (Magdeldin et al. 2017).

As a different 3D approach, spheroids are particularly useful 3D models to create biochemical gradients. Thus, when combining spheroids with endothelial cells, it enables the study of angiogenesis in a tumor mass, allowing the study of the loops that regulate such interaction, with special interest for drug discovery. To further do so, lung (Roudsari et al. 2018), liver (Jung et al. 2017) and skin (Klimkiewicz et al. 2017) cancers were modulated using heterotypic spheroids of angiogenesis to compare primary cancer types of different grades.

From the lung cancer model it was possible to verify that lung cancer spheroids with more metastatic phenotype recruited twice the endothelial cells that benign tumor spheroids did, showing that with a grade increase, the pro-angiogenic potential also increases (Roudsari et al. 2018).

One of the most important roles of angiogenesis is the control of oxygen levels within a tissue that is particularly relevant in cancer, since it is correlated to tumor aggressiveness. Since hypoxia is key aspect on cancer aggressiveness, a hypoxic tumor model containing a surrounding endothelium was designed to understand how the balance between hypoxia and normoxia is controlled by the tumor.

Specifically, a melanoma model was developed to study the effects of hypoxia over cancer cells recruitment of endothelial cells. It was found that the combined effect of the co-culture and hypoxia conditions led to the recruitment and posterior stabilization of the population of endothelial and epithelial cancer stem cells, showing a tissue-like behavior similar to *in vivo* observations (Klimkiewicz et al. 2017). Nonetheless, this led to the hypothesis of a cancer-mediated oxygen balance by endothelial cells recruitment.

To assess the angiogenesis-mediated oxygen equilibrium in tumors, liver cancer spheroids were co-cultured with endothelial cells at 2% of the total number of cells. HUVECs not only reduced the necrotic core of the spheroid but also induced EMT transition-related genes and angiogenesis-related genes in cancer cells, leading to a gene expression profile closer to the one verified in patient samples (Jung et al. 2017). These models showed that the interplay between cancer and endothelial cells regulates the behavior of both cell types when in co-culture. However, the extent of these alterations and how permanent they are is still unclear.

To better understand the consequences of cancer-endothelium interface over the behavior of endothelial cells, patient-derived renal tumor tissue and healthy endothelia were cultured in a microfluidic device. The tumor-derived endothelial cells generated an increased number of sprouts from the cultured vessels when compared to healthy tissue derived ones, exhibiting an increased angiogenic potential. This pro-tumorigenic behavior of recruited endothelial cells is something that can play a role in cancer reappearance, and should be targeted in next generation therapies (Jiménez-Torres et al. 2019).

Despite the fact that tumor cells alter the physiology of tumor surrounding endothelial cells, the aberrant response of immune cells may also be responsible for a significant part of tumor's physiological angiogenesis (Shi et al. 2019). In a validated vascular model for monocyte differentiation and maturation, where melanoma and breast cancer cells were introduced, it was verified that while monocytes stopped cell progression and, particularly, extravasation, monocyte-derived macrophages did not produce any significant effect over cancer cell viability (Boussommier-Calleja et al. 2019). This model showed immune cell type dependent modulation of the tumor-blood interface, highlighting the urge of deeper understanding of the immune system role on tumor angiogenesis processes.

The creation of vascular models to study primary cancer behavior is important to understand the balance of oxygen and nutrients supply within the tumor microenvironment. This balance between what is necessary to allow cancer cell viability and what is required to induce malignant phenotype may then be targeted in future therapies and significantly increase treatments efficiency.

Metastasis

Metastasis are of tumors formed in secondary sites, within a different tissue from the primary site, after dissemination of metastatic cancer cells through the blood stream and extravasation at a specific site and tissue. The cellular consequences that the extravasation process induces and the selectivity behind it are still unclear and particular challenging to observe *in vivo* due to the spatio-temporal unpredictability of the process.

The existence of a capillary network is a key feature to observe biochemical alterations promoted by the extravasation process. However, the creation of these capillary networks are rather challenging and, consequently, uncommon in *in vitro* models. This issue was overcome in a breast cancer metastatic in bone model was developed where the bone ECM and vascularization were mimicked. To do so, bone mesenchymal stem cells (BMSCs) were co-cultured with endothelial cells at an optimized flow rate and oxygen concentration to achieve a perivascular bone model. After so, cancer cells were introduced to the model and extravasation occurred, as predicted, validating it as a metastasis model and grew in low-proliferative state when compared to static culture. However, once exposed to sunitinib, a chemotherapeutic agent, the settled metastasis cultured in dynamic conditions, post-extravasation, showed resistance to the drug, exhibiting a decreased impairment of its growth rate. This way, this platform not only showed to be a valid model for breast cancer metastatic in bone as it showed to reveal mechanisms behind chemo resistance of these tumors.

Some metastasis models have been developed over recent years. An interesting work stood out by the increased potential for allow studying of the mechanisms of cancer-tissue selective extravasation.

Using a blood-brain barrier model built in a microfluidic device, malignant cancer cell lines originated from lung, breast, skin and liver cancer were circulated in contact with the vascular barrier. Once it contacted the model, the invasion speed was measured. Lung cancer was the fastest to invade, followed by breast cancer and melanoma, respectively, while liver cancer was not able to perform extravasation across the BBB. Thus, the *in vitro* model led to significant differences among cancer cell types in BBB extravasation, which are relevant for the development of novel therapeutics (Xu et al. 2016a).

Metastatic models like the one described in this sub-section will be used in following years for further understanding the molecular mechanisms behind tissue specific recognition, even though these developments are dependent on advanced imaging techniques coupled with transparent devices.

9.2.4.5 Microbiological

Several tissues present a symbiotic relationship with microbes, making them a crucial part not only of tissues function, but also of systemic homeostasis (Gopalakrishnan et al. 2018; Ma et al. 2018; Mrázek et al. 2019), and consequently a relevant variable to be considered for *in vitro* models. For instance, the intestinal mucosa and the skin are tissues where the microbiome significantly contributes for healthy physiology (Chng et al. 2016; Daniel et al. 2017).

Deregulations in the profile of these microbiomes have been implied in several diseases and their presence in healthy tissue *in vitro* models has been increasing over last years. This way, several probiotics have been proposed for example to fight intestinal infections, substituting, at least in early stages, the use of antibiotics (Hao et al. 2015). Since some microbes, namely bacteria, were enriched in tumor sites, the hypothesis of having a probiotic disrupting this cooperative situation has been proposed for possible future cancer therapy.

To investigate this possibility, *Lactobacillus rhamnosus*, a common probiote, was co-cultured with *Fusobacterium nucleatum*, bacteria shown to be particularly present in colorectal cancer, and together with spheroids of a colorectal cancer cell line.

Interestingly, *Lactobacillus rhamnosus* considerably impaired the attachment of *Fusobacterium nucleatum* when cultured simultaneously and in the same cell density, mainly due to its binding capacity. In protective and competitive models, the probiotic was able to reduce in 80% the binding capacity of *Fusobacterium nucleatum*, where *Lactobacillus rhamnosus* was cultured previous and simultaneously to *Fusobacterium nucleatum*, respectively. Nonetheless, in displacement assays, mimicking an eventual treatment of previously infected tumor tissue with *Fusobacterium nucleatum*, the removal only achieved 20% (Mokhtar et al. 2019).

Still, tumor models incorporating microbiomes are yet underdeveloped (Villenave et al. 2017; Jalili-Firoozinezhad et al. 2018; Réquilé et al. 2018; Zhu et al. 2018a; Shin et al. 2019). In following years, the incorporation of the microbiome in 3D models will be done by means of establishing the most representative bacteria in native microbiome and their influence over cancer cells viability and proliferation, identifying what can be the largest contributors for microbe-induced cancer development.

9.3 3D Models Integration in Dynamic Culture Systems

The presence of fluid flow influences the behavior of tissue culture *in vitro*. Nonetheless, it is only relevant to model if it is used to recapitulate *in vivo* dynamics. There are interfaces, namely the tumor-blood, where the fluid shear stress can be highly important for the progression of the disease.

The importance of the establishment of a dynamic culture of 3D tumor interfaces will be addressed in the first part of this section, with a special focus on the relevance of its application in anti-cancer drugs standardized screenings.

The second part of this section will be directed mostly at volume scales and how scale-up strategies have been designed in the development of engineered tumor interfaces.

Standard bioreactors are usually classified according to its geometry, stirring, and feeding mode. In classical bioreactors classification, there were two main types of systems – tank and tubular (Cooney 1983). Regarding the stirring, the knowledge gathered about the effect of shear stress over the phenotypic homeostasis of sensitive cell types, namely mammalian cells and, particularly, the development of *in vitro* models using biomimetic strategies have almost retired the idea of classical stirring using propellers (Zhao et al. 2016).

Recently, dynamic systems have been miniaturized and adapted to tissue engineering applications. Particularly in the field of *in vitro* models, microfluidic devices, a miniaturized bioreactor concept, has become increasingly popular (Zhao et al. 2016). These devices allow a controlled culture, real-time visualization and the multiplex measurements with small sample volume.

Nonetheless, microfluidic platforms are produced using different architectures, dependent on the target mimicked tissue and the balance between complexity and processability. Besides the discussion about this dichotomy, this section will also address some innovative bioreactors designed for 3D tumor interfaces that are examples of the balance between the achievement of a representative model and the reproducibility necessary for the clinical translation.

9.3.1 Dynamic vs. Static

The theoretical effects of physiological fluid shear stress over tumor progression have been previously discussed, particularly the role of 3D models in the characterization of this interface. However, under a biomimetic perspective, the practical use of dynamic conditions is particularly relevant because the human body is a construct of organs perfused by blood indirectly through vessels. This perfusion occurs by a leakage from the vessels, with the fluid being collected by the lymphatic system.

The differential of pressure between these two systems, vascular and lymphatic, drives the fluid over cells with a flux that decreases proportionally to the distance from the blood vessel. The flux that flows through the tissue varies quite significantly according to the activity of the organism. All the mentioned features can be potentially mimicked using diverse dynamic systems, namely by applying advanced microfabrication techniques.

Advanced designs allow the establishment of constructs with well-defined interfaces under real-time visualization of the interactions along the interface. Cao et al. (2019) developed a perfusion bioreactor that mimicked the near tumor blood and the lymphatic systems, siding the tumor mass encapsulated in photo-crosslinked methacrylated gelatin. The vessels were built using 3D printing of photo-crosslinkable poly(ethylene glycol) gels, instantly gelified using ultraviolet light. This design increased the fluid flux through the tumor mass creating easily tunable dynamics, enabling high-throughput fabrication and near *in vivo* flow dynamics relevant for drug testing over different tumor models (Cao et al. 2019).

However, the commitment with the simplification of processes for reproducibility is a major challenge for widespread dissemination. Li et al. (2017) developed a platform for ovarian cancer behavior assessment when contacting the peritoneal membrane, as it often happens in this cancer type. However, this device differs from others due to the generation of physiological interstitial fluid shear stress. Strikingly, the design and fabrication process are particularly reproducible, since the visualization is optimized by the use of the minimalistic design arresting cells just by cell adhesion to fibronectin or to the mesenthelial cells using a layer-by-layer approach. Thus, this device shows enormous potential for future standardized high-throughput drug screening (Li et al. 2017).

The translation of these devices to the pharmaceutical market is hugely dependent on the capacity of up scaled standardized production, as well as on the robustness of disease models production. For that purpose, these systems must ally advanced microfabrication techniques with real-time assessment capacity. Carvalho et al. 2019 developed a platform for standardized colorectal targeting drug testing through a tumor-blood interface. Using an advanced design, this platform allowed the creation of a centered tumor mass encapsulated in Matrigel® surrounded by a circular HUVEC blood vessel by a set of channels and baffles. As a proof-of-concept of the use of this platform for drug screening, the effect of a drug-loaded advanced delivery system, through the tumor-blood interface, over the tumor viability was assessed and quantified using image processing techniques (Carvalho et al. 2019).

This device allowed the verification of a cell viability decrease as the distance from the mimicked blood vessel increases, as was predictable from a mass-transfer perspective. The gene expression levels of tumor progression-related biomarkers like metalloproteinase and proliferation markers also verified the effects of the drug, showing the robustness of this model. Although further details about time of fabrication and processing would be required to classify this as a high-throughput system, the robustness of the device makes it a candidate for industrial use.

The use of these channel-based microfluidic devices require a total change in the protocols of the pharmaceutical industry. Most industrial protocols are still optimized for well plates-based drug screening.

Also, the translation to the widespread use by the pharmaceutical industry requires more than the mere achievement of the design parameters. Within the parameters mentioned as the key ones for platform design, uniformity is the main goal to achieve. As a relevant example for dynamic systems, flow pumping can cause differential pressure along the device, triggering fluctuations and non-conformity.

To face this situation, Chen et al. (2019) created a high-throughput pumpless system for drug testing on a tumor-stromal interface that allows a uniform shear stress over the entire tissue, using a swinging well-plate design. Using arrays of 4 wells per condition, the device is a simple system, which avoids the use of external pumps, therefore more translational and user-friendly. Regarding the tissue model, it consists of a 2-layer mode, one of cancer other of stromal cells. Though there are interactions in 3D, the system do not recapitulate 3D heterogeneities or gradients and cells do not form an architecture that characterizes *in vivo* tumors (Chen et al. 2019).

There is still, however, a large market need for tumor interfaces, with simplified protocols, automation, and up scalable features.

In the following sub-section, the most recently industrially translated devices that were used for 3D interfacial tumor models will be reviewed. This analysis will allow better understanding of the features that characterize the industrial systems and what makes them different from most of the academic outcomes.

9.3.2 Industrial Transition

Several devices containing 3D tumor models are often considered promising in a field, but not suitable for industrial drug screening. The main reason behind it relates to the equilibrium among mimicry, reproducibility, productivity, and industrial validation, which are necessary for industrial production and general use on the pharmaceutical industry.

Disregarding the fluid dynamics, models as those presented along this chapter often differ from each other in one or several parameters (e.g. cell origin or type; biomimetic stimuli, such as growth factors or metabolic cues; mechanical properties of the supporting material or shape of the 3D model, design of the bioreactor platform when existing). However, while the increasing diversity of approaches is paramount for the evolution of the research field, it also represents a delay on standardization for industrial uptake.

Devices for tumor-targeting drug high-throughput screenings shall have designs that consider productivity, increasing conformity and standardization simultaneously to biological validation. To exhibit the robustness for widespread drug testing using tumor models, the system must proof to work regardless of the cancer type or kind of interface.

With that intent, Kiyatec© validated the perfusable system 3DKUBE® for the culture of patient-derived xenografts for immuno-oncology, glioblastoma (Ashley et al. 2018), breast (Guo et al.), and ovarian cancers (Desrochers et al. 2015; Shuford et al. 2019). Moreover, in each model, cancer cells were isolated and cultured with exogenous stromal, endothelial and immune cells, usually present at the tumor site, and the models were validated by molecular characterization and chemo resistance to known compounds.

However, the recreated interfaces are not actually discrete nor the dimensions were controlled since the system was developed for a cultured tumor mass. It has only a chamber with no specific design interface models.

The development of interface models often requires the existence of two inlets or a valves system. For the creation of gradients, that can drastically alter cancer cell dynamics as was verified in the biochemical interfaces sub-section, the use of multiple channels is the most common and robust strategy. For that reason, Mimetas© developed the OrganoPlate®, a system for the purpose of having perfusable tumor interface models (Fig. 9.4). This bioreactor is a pumpless device, with contacting channels that can provide 364 simultaneous replicas. The number and position of channels can be tuned according to desired model and it has been applied in healthy (van Duinen et al. 2019) and in tumor tissue (Lanz et al. 2017; Ogmundsdottir et al. 2018; Queiroz et al. 2018) modeling, validating its usefulness in standardized drug testing.

However, none of these systems allow the establishment of a more complex microenvironment that can be insufficient for the replacement of drug testing in animal models. For that purpose, there is the need to create dynamic models with the mimicking of the tissue's hierarchical structure that vary enormously among tissues. Therefore, such system must have a specific design interface constructs that allow the use of optimized material methods, such as the use scaffolds or hydrogels, and the real-time assessment of all the components of the interfacial model. A dual-chamber 6 well-plate sized bioreactor (Canadas et al. 2017) showed to be a robust strategy for interfacial models. This innovative system can allow the creation of gradients and interfaces, either solid or soluble in dynamic conditions (Canadas et al. 2018).

Nonetheless, the currently increasing industrially translated devices constitute robust upgrades to 2D and static screenings, opening-up ways for even more realistic systems. There is still, however, the market need to combine efforts to build models of tumor interfaces with defined features, simplified protocols allowing automation and upscale of the ones currently existent in the academia.

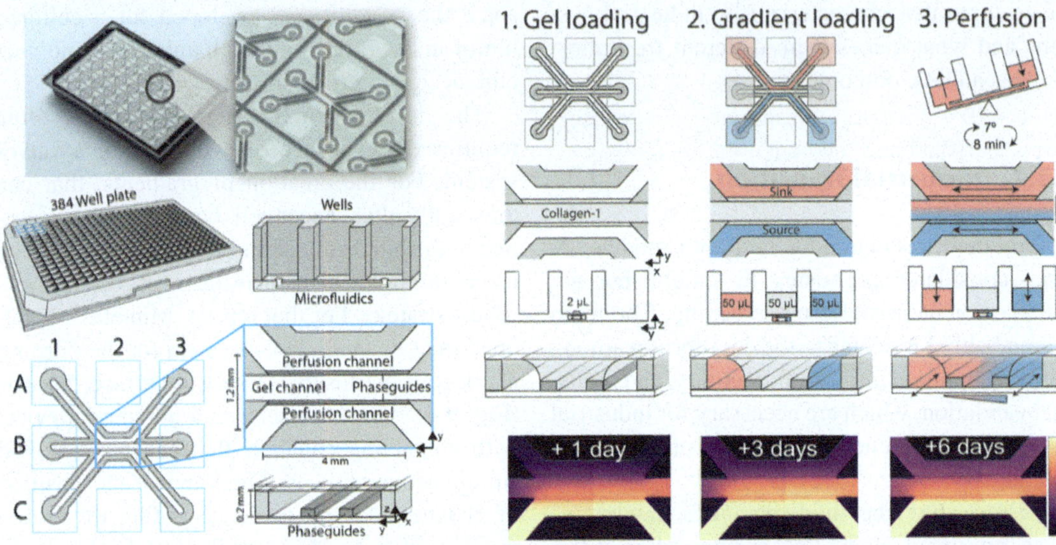

Fig. 9.4 Example of OrganoPlate® for interface formation. (**a**) Design and placement of 3 lane OrganoPlate®, where (**b**) two different gradients can be formed within a central channel while perfusing two different medias in the outside lanes. (A and B were adapted from van Duinen et al. 2019)

9.3.3 Concluding Remarks and Future Outlook

The combination of 3D cues of the tumor's interfaces in a dynamic platform contributes for the recapitulation of shear stress, which is extremely relevant for drug effectiveness. Finally, the translation of the academic platforms to the industry was discussed. For that purpose, the reviewed models clearly show the progress margin and the milestones to be achieved in the following years. Importantly, bioreactors and microfluidic systems are now tending to avoid pumping systems due the industry's demand for user-friendly, simpler setups.

Specifically from the tissue modeling perspective, the most recent interface tumor models are, in general, already developed as 3D constructs. The increased complexity when compared to 2D contributes to a better representation of native tissue. To do so, advanced microfabrication techniques are applied, which makes the developed models more biomimetic when compared to 2D and homogeneous 3D, since it represents a larger number of interactions among cancer cells and surrounding ECM. On the other hand, these processes are often increasingly complex, diversified and expensive, which creates important barriers for industrial application.

Strikingly, when specifically mimicking metastatic cancer models, either based on organism models, like body-on-a-chip, or just focusing on one vascular model of barrier to monitor intravasation/extravasation, the current approaches are often too simplistic. Frequently, the state-of-the-art still neglects models able to replicate systemic effects, a requirement to replace animal experimentation in the future. The models that address systematic consequences are often built on the basis of proof-of-concept approaches, without using features with physiological and pathological values. The validation with commercially successful and non-successful drug candidates is also uncommon, being also a main requirement to prove the model's robustness.

Despite that, the recent advances represent an increased effort to replace or minimize animal models and improve the predictability of false/positive results before the clinical trials. Furthermore, the regulatory pressure to use *in vitro* systems started with the cosmetics industry, and will expand across the pharma industry in the following years. So, the continuous evolution of material sciences, microfabrication techniques and measurement tools, altogether with the introduction of artificial intelligence-based data processing, allow the anticipation that the *in vitro* models will outperform the animal ones for drug testing, personalized medicine, and diagnostics, including systemic effects, in the upcoming decades.

Acknowledgments The authors would like to acknowledge the financial support provided by the Portuguese Foundation for Science and Technology (FCT) through the projects B-FABULUS (PTDC/BBB-ECT/2690/2014) and 3BioMeD (FCT/4773/4/5/2017/S). This work has been funded under the FCT doctoral programme for Advanced Therapies for Health (PATH) and supported by the Frontiers of technology for theranostics of cancer, metabolic and neurodegenerative diseases (FROnTHERA) Structured Project NORTE-01-0145-FEDER-000023, supported by Norte of Portugal Regional Operational Programme (NORTE 2020), under the PORTUGAL 2020 Partnership Agreement, through the European Social Fund. Raphaël F. Canadas thanks to FCT for the Postdoc fellowship in the scope of the Project PTDC/EMD-EMD/29139/2017 funded by FEDER program. The FCT distinction attributed to J. Miguel Oliveira (IF/01285/2015) under the Investigator FCT program is also greatly acknowledged.

References

Angeloni V, Contessi N, De Marco C et al (2017) Polyurethane foam scaffold as in vitro model for breast cancer bone metastasis. Acta Biomater 63:306–316. https://doi.org/10.1016/j.actbio.2017.09.017

Ashley MS, Lauren O, Lillia H et al (2018) Profiling patient-specific glioblastoma drug response in vitro using complex 3D microtumors. Cancer Res 78:1121

Bauer J, Staudacher JJ, Mancinelli G et al (2018) Increased stiffness of the tumor microenvironment in colon cancer leads to an increase in Activin and metastatic potential. Cancer Res 78:177 LP – 177

Boussommier-Calleja A, Atiyas Y, Haase K et al (2019) The effects of monocytes on tumor cell extravasation in a 3D vascularized microfluidic model. Biomaterials 198:180–193. https://doi.org/10.1016/j.biomaterials.2018.03.005

Bratton M, Byrne CE, Rahman SM et al (2019) Using an in vitro fluidics approach to model the evolution of metastatic breast cancer reveals shear stress as a possible driver of genomic instability and somatic mutation. FASEB J 33:lb175–lb175. https://doi.org/10.1096/fasebj.2019.33.1_supplement.lb175

Butt AM, Jones HC, Abbott NJ (1990) Electrical resistance across the blood-brain barrier in anaesthetized rats: a developmental study. J Physiol 429:47–62. https://doi.org/10.1113/jphysiol.1990.sp018243

Canadas RF, de Oliveira JMA, Marques AMP, dos Reis RLG (2017) Multi-chambers bioreactor, methods and uses. PCT/IB2015/057210

Canadas RF, Ren T, Tocchio A et al (2018) Tunable anisotropic networks for 3-D oriented neural tissue models. Biomaterials 181:402–414. https://doi.org/10.1016/j.biomaterials.2018.07.055

Cao X, Ashfaq R, Cheng F et al (2019) A tumor-on-a-chip system with bioprinted blood and lymphatic vessel pair. Adv Funct Mater 29:1807173. https://doi.org/10.1002/adfm.201807173

Carvalho MR, Barata D, Teixeira LM et al (2019) Colorectal tumor-on-a-chip system: A 3D tool for precision onco-nanomedicine. Sci Adv 5:eaaw1317. https://doi.org/10.1126/sciadv.aaw1317

Castells M, Thibault B, Delord J-P, Couderc B (2012) Implication of tumor microenvironment in Chemoresistance: tumor-associated stromal cells protect tumor cells from cell death. Int J Mol Sci 13:9545–9571. https://doi.org/10.3390/ijms13089545

Chen Z, He S, Zilberberg J, Lee W (2019) Pumpless platform for high-throughput dynamic multicellular culture and chemosensitivity evaluation. Lab Chip 19:254–261. https://doi.org/10.1039/C8LC00872H

Chng KR, Tay ASL, Li C et al (2016) Whole metagenome profiling reveals skin microbiome-dependent susceptibility to atopic dermatitis flare. Nat Microbiol 1:16106. https://doi.org/10.1038/nmicrobiol.2016.106

Chung M, Ahn J, Son K et al (2017) Biomimetic model of tumor microenvironment on microfluidic platform. Adv Healthc Mater 6:1700196. https://doi.org/10.1002/adhm.201700196

Cooney CL (1983) Bioreactors: design and operation. Science (80-) 219:728–733. https://doi.org/10.1126/science.219.4585.728

da Costa FHB, Navone NM, Farach-Carson MC, Carson DD (2018) Abstract 3799: Heparan sulfate modifiers and signaling between tumor-associated macrophages and cancer-associated fibroblasts in a biomimetic hydrogel model of bone-metastatic prostate cancer. In: Immunology. American Association for Cancer Research. Chicago, pp 3799–3799

Daniel SG, Ball CL, Besselsen DG et al (2017) Functional changes in the gut microbiome contribute to transforming growth factor β-deficient colon cancer. mSystems 2. https://doi.org/10.1128/mSystems.00065-17

Däster S, Amatruda N, Calabrese D et al (2017) Induction of hypoxia and necrosis in multicellular tumor spheroids is associated with resistance to chemotherapy treatment. Oncotarget 8. https://doi.org/10.18632/oncotarget.13857

Demetriou CA, Vineis P (2015) Carcinogenicity of ambient air pollution: use of biomarkers, lessons learnt and future directions. J Thorac Dis 7:67–95. https://doi.org/10.3978/j.issn.2072-1439.2014.12.31

Desrochers T, Shuford S, Mattingly C et al (2015) Abstract LB-282: ex vivo 3d drug response profiling of primary human ovarian cancer differentiates treatment-naive and relapsed patients and molecular subtypes. In: Molecular and cellular biology. American Association for Cancer Research. Philadelphia, p LB-282-LB-282

DiMasi JA, Grabowski HG, Hansen RW (2016) Innovation in the pharmaceutical industry: new estimates of R&D costs. J Health Econ 47:20–33. https://doi.org/10.1016/j.jhealeco.2016.01.012

Engbring JA, Kleinman HK (2003) The basement membrane matrix in malignancy. J Pathol 200:465–470. https://doi.org/10.1002/path.1396

Enokida Y, Tsuruno Y, Okubo K, Yamaoka Y, Takahashi E (2017) Directional migration of MDA-MB-231 cells under O2/pH gradients. In: Halpern H, LaManna J, Harrison D, Epel B (eds) Oxygen transport to tissue XXXIX. Advances in experimental medicine and biology, vol 977. Springer, Cham

Gopalakrishnan V, Spencer CN, Nezi L et al (2018) Gut microbiome modulates response to anti–PD-1 immunotherapy in melanoma patients. Science (80-) 359:97–103. https://doi.org/10.1126/science.aan4236

Gritsenko P, Leenders W, Friedl P (2017) Recapitulating in vivo-like plasticity of glioma cell invasion along blood vessels and in astrocyte-rich stroma. Histochem Cell Biol 148:395–406. https://doi.org/10.1007/s00418-017-1604-2

Guo Q, Millard M, Lotstein A, et al Complex, patient-derived, multi-cell type, 3d models of breast cancer for personalized prediction of therapeutic response. http://kiyatec.com/wp-content/uploads/2019/03/kiyatec_poster_Breast-Model-v2.pdf

Guzman A, Sánchez Alemany V, Nguyen Y et al (2017) A novel 3D in vitro metastasis model elucidates differential invasive strategies during and after breaching basement membrane. Biomaterials 115:19–29. https://doi.org/10.1016/j.biomaterials.2016.11.014

Hao Q, Dong BR, Wu T (2015) Probiotics for preventing acute upper respiratory tract infections. Cochrane Database Syst Rev. https://doi.org/10.1002/14651858.CD006895.pub3

Heggannavar GB, Hiremath CG, Achari DD et al (2018) Development of doxorubicin-loaded magnetic silica–Pluronic F-127 Nanocarriers conjugated with transferrin for treating glioblastoma across the blood–brain barrier using an in vitro model. ACS Omega 3:8017–8026. https://doi.org/10.1021/acsomega.8b00152

Heinrich MA, Bansal R, Lammers T et al (2019) 3D-bioprinted mini-brain: a glioblastoma model to study cellular interactions and therapeutics. Adv Mater 31:1806590. https://doi.org/10.1002/adma.201806590

Herter S, Morra L, Schlenker R et al (2017) A novel three-dimensional heterotypic spheroid model for the assessment of the activity of cancer immunotherapy agents. Cancer Immunol Immunother 66:129–140. https://doi.org/10.1007/s00262-016-1927-1

Hillsley MV, Frangos JA (1994) Review: bone tissue engineering: the role of interstitial fluid flow. Biotechnol Bioeng 43:573–581. https://doi.org/10.1002/bit.260430706

Hollier B, Harkin D, Leavesley D, Upton Z (2005) Responses of keratinocytes to substrate-bound vitronectin:growth factor complexes. Exp Cell Res 305:221–232. https://doi.org/10.1016/j.yexcr.2005.01.004

Houghton AN (1987) Phenotypic heterogeneity of melanoma. Relation to the differentiation program of melanoma cells. J Exp Med 165:812–829. https://doi.org/10.1084/jem.165.3.812

Huleihel L, Hussey GS, Naranjo JD et al (2016) Matrix-bound nanovesicles within ECM bioscaffolds. Sci Adv 2:e1600502. https://doi.org/10.1126/sciadv.1600502

Hyler AR, Baudoin NC, Brown MS et al (2018) Fluid shear stress impacts ovarian cancer cell viability, subcellular organization, and promotes genomic instability. PLoS One 13:e0194170. https://doi.org/10.1371/journal.pone.0194170

Jalili-Firoozinezhad S, Prantil-Baun R, Jiang A et al (2018) Modeling radiation injury-induced cell death and countermeasure drug responses in a human Gut-on-a-Chip. Cell Death Dis 9:223. https://doi.org/10.1038/s41419-018-0304-8

Jiménez-Torres JA, Virumbrales-Muñoz M, Sung KE et al (2019) Patient-specific organotypic blood vessels as an in vitro model for anti-angiogenic drug response testing in renal cell carcinoma. EBioMedicine 42:408–419. https://doi.org/10.1016/j.ebiom.2019.03.026

Joyce JA (2005) Therapeutic targeting of the tumor microenvironment. Cancer Cell 7:513–520. https://doi.org/10.1016/j.ccr.2005.05.024

Juin A, Spence HJ, Martin KJ et al (2018) Chemotaxis in pacreatic ductal adenocarcinoma metastasis: an unexpected role of NWASP in maintaining self-generated gradients and LPA receptor recycling. SSRN Electron J. https://doi.org/10.2139/ssrn.3275292

Jung H-Y, Fattet L, Yang J (2015) Molecular pathways: linking tumor microenvironment to epithelial-mesenchymal transition in metastasis. Clin Cancer Res 21:962–968. https://doi.org/10.1158/1078-0432.CCR-13-3173

Jung H-R, Kang HM, Ryu J-W et al (2017) Cell spheroids with enhanced aggressiveness to mimic human liver Cancer in vitro and in vivo. Sci Rep 7:10499. https://doi.org/10.1038/s41598-017-10828-7

Katt ME, Placone AL, Wong AD et al (2016) In vitro tumor models: advantages, disadvantages, variables, and selecting the right platform. Front Bioeng Biotechnol 4. https://doi.org/10.3389/fbioe.2016.00012

Kessenbrock K, Plaks V, Werb Z (2010) Matrix metalloproteinases: regulators of the tumor microenvironment. Cell 141:52–67. https://doi.org/10.1016/j.cell.2010.03.015

Klimkiewicz K, Weglarczyk K, Collet G et al (2017) A 3D model of tumour angiogenic microenvironment to monitor hypoxia effects on cell interactions and cancer stem cell selection. Cancer Lett 396:10–20. https://doi.org/10.1016/j.canlet.2017.03.006

Ko P, Kim D, You E et al (2016) Extracellular matrix rigidity-dependent Sphingosine-1-phosphate secretion regulates metastatic cancer cell invasion and adhesion. Sci Rep 6:21564. https://doi.org/10.1038/srep21564

Koh I, Cha J, Park J et al (2018) The mode and dynamics of glioblastoma cell invasion into a decellularized tissue-derived dimensional tumor model. Sci Rep 8:1–12. https://doi.org/10.1038/s41598-018-22681-3

Koikawa K, Ohuchida K, Ando Y et al (2018) Basement membrane destruction by pancreatic stellate cells leads to local invasion in pancreatic ductal adenocarcinoma. Cancer Lett 425:65–77. https://doi.org/10.1016/j.canlet.2018.03.031

Kripke ML (1981) Immunologic mechanisms in UV radiation carcinogenesis. Adv Cancer Res 34:69–106. Elsevier, Frederick, Maryland

Lanz HL, Saleh A, Kramer B et al (2017) Therapy response testing of breast cancer in a 3D high-throughput perfused microfluidic platform. BMC Cancer 17:709. https://doi.org/10.1186/s12885-017-3709-3

Lee K, Silva EA, Mooney DJ (2011) Growth factor delivery-based tissue engineering: general approaches and a review of recent developments. J R Soc Interface 8:153–170. https://doi.org/10.1098/rsif.2010.0223

Li S-S, Ip CKM, Tang MYH et al (2017) Modeling ovarian Cancer multicellular spheroid behavior in a dynamic 3D peritoneal microdevice. J Vis Exp. https://doi.org/10.3791/55337

Lim M, Xia Y, Bettegowda C, Weller M (2018) Current state of immunotherapy for glioblastoma. Nat Rev Clin Oncol 15:422–442. https://doi.org/10.1038/s41571-018-0003-5

Liou Y-R, Torng W, Kao Y-C et al (2014) Substrate stiffness regulates filopodial activities in lung cancer cells. PLoS One 9:e89767. https://doi.org/10.1371/journal.pone.0089767

Lu H, Forbes RA, Verma A (2002) Hypoxia-inducible factor 1 activation by aerobic glycolysis implicates the Warburg effect in carcinogenesis. J Biol Chem 277:23111–23115. https://doi.org/10.1074/jbc.M202487200

Ma C, Han M, Heinrich B, et al (2018) Abstract A02: Gut microbiome controls growth of liver tumors. In: Cancer and the microbiome. American Association for Cancer Research. Boston, pp A02–A02

Magdeldin T, López-Dávila V, Pape J et al (2017) Engineering a vascularised 3D in vitro model of cancer progression. Sci Rep 7:44045. https://doi.org/10.1038/srep44045

Mak AF (1986) The apparent viscoelastic behavior of articular cartilage—the contributions from the intrinsic matrix viscoelasticity and interstitial fluid flows. J Biomech Eng 108:123–130. https://doi.org/10.1115/1.3138591

Mak IW, Evaniew N, Ghert M (2014) Lost in translation: animal models and clinical trials in cancer treatment. Am J Transl Res 6:114–118

Marques-Torrejon MA, Gangoso E, Pollard SM (2018) Modelling glioblastoma tumour-host cell interactions using adult brain organotypic slice co-culture. Dis Model Mech 11:dmm031435. https://doi.org/10.1242/dmm.031435

Mokhtar N, Wong K, Raja A, Ali R (2019) IDDF2019-ABS-0227 manipulation of Gut microbiota in vitro model of colorectal cancer: strong adherence ability of Lactobacillus Rhamnosus. Gut 68:A23 LP-A23. https://doi.org/10.1136/gutjnl-2019-IDDFAbstracts.43

Mow VC, Wang CC, Hung CT (1999) The extracellular matrix, interstitial fluid and ions as a mechanical signal transducer in articular cartilage. Osteoarthr Cartil 7:41–58. https://doi.org/10.1053/joca.1998.0161

Mrázek J, Mekadim C, Kučerová P et al (2019) Melanoma-related changes in skin microbiome. Folia Microbiol (Praha) 64:435–442. https://doi.org/10.1007/s12223-018-00670-3

Müller I, Kulms D (2018) A 3D Organotypic melanoma spheroid skin model. J Vis Exp. https://doi.org/10.3791/57500

Ogmundsdottir H, Kowal N, Cherek P, Olafsdottir E (2018) PO-420 mechanisms of action of anti-proliferative lichen compound protolichesterinic acid. In: PO-416 Cytotoxicity of Ru (II) and Ru (III) salen complexes against breast and colorectal cancer cell lines. BMJ Publishing Group Ltd, pp A187.3–A1A189

Paszek MJ, Zahir N, Johnson KR et al (2005) Tensional homeostasis and the malignant phenotype. Cancer Cell 8:241–254. https://doi.org/10.1016/j.ccr.2005.08.010

Pereira BA, Lister NL, Hashimoto K et al (2019) Tissue engineered human prostate microtissues reveal key role of mast cell-derived tryptase in potentiating cancer-associated fibroblast (CAF)-induced morphometric transition in vitro. Biomaterials 197:72–85. https://doi.org/10.1016/j.biomaterials.2018.12.030

Piccoli M, D'Angelo E, Crotti S et al (2018) Decellularized colorectal cancer matrix as bioactive microenvironment for in vitro 3D cancer research. J Cell Physiol 233:5937–5948. https://doi.org/10.1002/jcp.26403

Puls TJ, Tan X, Whittington CF, Voytik-Harbin SL (2017) 3D collagen fibrillar microstructure guides pancreatic cancer cell phenotype and serves as a critical design parameter for phenotypic models of EMT. PLoS One 12:e0188870. https://doi.org/10.1371/journal.pone.0188870

Queiroz K, Habani Y, Lanz H, et al (2018) PO-449 3d glioma-on-a-chip models for personalised medicine in organoplates®. In: Poster presentation: translational research. BMJ Publishing Group Ltd, p A408.1–A4A408

Racordon D, Valdivia A, Mingo G et al (2017) Structural and functional identification of vasculogenic mimicry in vitro. Sci Rep 7:6985. https://doi.org/10.1038/s41598-017-07622-w

Réquilé M, Gonzàlez Alvarez DO, Delanaud S et al (2018) Use of a combination of in vitro models to investigate the impact of chlorpyrifos and inulin on the intestinal microbiota and the permeability of the intestinal mucosa. Environ Sci Pollut Res 25:22529–22540. https://doi.org/10.1007/s11356-018-2332-4

Rice AJ, Cortes E, Lachowski D et al (2017) Matrix stiffness induces epithelial–mesenchymal transition and promotes chemoresistance in pancreatic cancer cells. Oncogene 6:e352–e352. https://doi.org/10.1038/oncsis.2017.54

Ridky TW, Chow JM, Wong DJ, Khavari PA (2010) Invasive three-dimensional organotypic neoplasia from multiple normal human epithelia. Nat Med 16:1450–1455. https://doi.org/10.1038/nm.2265

Roudsari LC, Jeffs SE, West JL (2018) Lung adenocarcinoma cell responses in a 3D in vitro tumor angiogenesis model correlate with metastatic capacity. ACS Biomater Sci Eng 4:368–377. https://doi.org/10.1021/acsbiomaterials.7b00011

Sánchez-Rodríguez C, Cruces KP, Riestra Ayora J et al (2017) BCG immune activation reduces growth and angiogenesis in an in vitro model of head and neck squamous cell carcinoma. Vaccine 35:6395–6403. https://doi.org/10.1016/j.vaccine.2017.10.008

Sefidgar M, Soltani M (2014) Interstitial flow in cancerous tissue: effect of considering remodeled capillary network. J Tissue Sci Eng 01. https://doi.org/10.4172/2157-7552.S4-003

Sharma VP, Williams J, Leung E, et al (2017) Abstract 1883: Fibrillar 1D tumor microenvironment is the key driver associating high-speed tumor cell motility with nuclear shape in breast cancer. In: Tumor biology. American Association for Cancer Research, pp 1883–1883

Shi Z, Fultz RS, Engevik MA et al (2019) Distinct roles of histamine H1- and H2-receptor signaling pathways in inflammation-associated colonic tumorigenesis. Am J Physiol Liver Physiol 316:G205–G216. https://doi.org/10.1152/ajpgi.00212.2018

Shin W, Hinojosa CD, Ingber DE, Kim HJ (2019) Human intestinal morphogenesis controlled by transepithelial morphogen gradient and flow-dependent physical cues in a microengineered Gut-on-a-Chip. iScience 15:391–406. https://doi.org/10.1016/j.isci.2019.04.037

Shuford S, Wilhelm C, Rayner M et al (2019) Prospective validation of an ex vivo, patient-derived 3D spheroid model for response predictions in newly diagnosed ovarian Cancer. Sci Rep 9:11153. https://doi.org/10.1038/s41598-019-47578-7

Sottoriva A, Spiteri I, Piccirillo SGM et al (2013) Intratumor heterogeneity in human glioblastoma reflects cancer evolutionary dynamics. Proc Natl Acad Sci 110:4009–4014. https://doi.org/10.1073/pnas.1219747110

Suresh M, Mattheolabakis G, Singh A, Amiji M (2018) In vitro model of inflammatory, hypoxia, and cancer stem cell signaling in pancreatic cancer using heterocellular 3-dimensional spheroids. bioRxiv 454397. https://doi.org/10.1101/454397

Tanaka HY, Kitahara K, Sasaki N et al (2019) Pancreatic stellate cells derived from human pancreatic cancer demonstrate aberrant SPARC-dependent ECM remodeling in 3D engineered fibrotic tissue of clinically relevant thickness. Biomaterials 192:355–367. https://doi.org/10.1016/j.biomaterials.2018.11.023

Triantafillu UL, Park S, Klaassen NL et al (2017) Fluid shear stress induces cancer stem cell-like phenotype in MCF7 breast cancer cell line without inducing epithelial to mesenchymal transition. Int J Oncol 50:993–1001. https://doi.org/10.3892/ijo.2017.3865

Tsuruno Y, Okubo K, Fujiwara T et al (2018) An in vitro model for determining tumor cell migration under metabolic gradients. Adv Exp Med Biol 1072:201–205

Twardowski P, Wong JYC, Pal SK et al (2019) Randomized phase II trial of sipuleucel-T immunotherapy preceded by sensitizing radiation therapy and sipuleucel-T alone in patients with metastatic castrate resistant prostate cancer. Cancer Treat Res Commun 19:100116. https://doi.org/10.1016/j.ctarc.2018.100116

Usui T, Sakurai M, Umata K et al (2018) Preparation of human primary colon tissue-derived organoid using air liquid interface culture. In: Current protocols in toxicology. Wiley, Hoboken, pp 22.6.1–22.6.7

van Duinen V, Zhu D, Ramakers C et al (2019) Perfused 3D angiogenic sprouting in a high-throughput in vitro platform. Angiogenesis 22:157–165. https://doi.org/10.1007/s10456-018-9647-0

Van Duuren BL, Goldschmidt BM (1976) Cocarcinogenic and tumor-promoting agents in tobacco Carcinogenesis. J Natl Cancer Inst 56:1237–1242. https://doi.org/10.1093/jnci/56.6.1237

Vermeulen MA, Slaets L, Cardoso F et al (2017) Pathological characterisation of male breast cancer: results of the EORTC 10085/TBCRC/BIG/NABCG International Male Breast Cancer Program. Eur J Cancer 82:219–227. https://doi.org/10.1016/j.ejca.2017.01.034

Villenave R, Wales SQ, Hamkins-Indik T et al (2017) Human gut-on-A-Chip supports polarized infection of Coxsackie B1 virus in vitro. PLoS One 12:e0169412. https://doi.org/10.1371/journal.pone.0169412

Walens A, DiMarco AV, Lupo R et al (2019) CCL5 promotes breast cancer recurrence through macrophage recruitment in residual tumors. elife 8. https://doi.org/10.7554/eLife.43653

Warburg O (1956) On the origin of cancer cells. Science 123:309–314

Xu H, Li Z, Yu Y et al (2016a) A dynamic in vivo-like organotypic blood-brain barrier model to probe metastatic brain tumors. Sci Rep 6:36670. https://doi.org/10.1038/srep36670

Xu Z, Li E, Guo Z et al (2016b) Design and construction of a multi-organ microfluidic chip mimicking the in vivo microenvironment of lung cancer metastasis. ACS Appl Mater Interfaces 8:25840–25847. https://doi.org/10.1021/acsami.6b08746

Yang M, Nelson R, Ros A (2016) Toward analysis of proteins in single cells: a quantitative approach employing isobaric tags with MALDI mass spectrometry realized with a microfluidic platform. Anal Chem 88:6672–6679. https://doi.org/10.1021/acs.analchem.5b03419

Yue X, Nguyen TD, Zellmer V et al (2018) Stromal cell-laden 3D hydrogel microwell arrays as tumor microenvironment model for studying stiffness dependent stromal cell-cancer interactions. Biomaterials 170:37–48. https://doi.org/10.1016/j.biomaterials.2018.04.001

Zhao J, Griffin M, Cai J et al (2016) Bioreactors for tissue engineering: an update. Biochem Eng J 109:268–281. https://doi.org/10.1016/j.bej.2016.01.018

Zhu TH, Zhu TR, Tran KA et al (2018a) Epithelial barrier dysfunctions in atopic dermatitis: a skin-gut-lung model linking microbiome alteration and immune dysregulation. Br J Dermatol 179:570–581. https://doi.org/10.1111/bjd.16734

Zhu Y, Clair G, Chrisler WB et al (2018b) Proteomic analysis of single mammalian cells enabled by microfluidic Nanodroplet sample preparation and ultrasensitive NanoLC-MS. Angew Chem Int Ed 57:12370–12374. https://doi.org/10.1002/anie.201802843

Nanoparticles and Microfluidic Devices in Cancer Research

10

F. Raquel Maia, Rui L. Reis, and Joaquim M. Oliveira

Abstract

Cancer is considered the disease of the century, which can be easily understood considering its increasing incidence worldwide. Over the last years, nanotechnology has been presenting promising theranostic approaches to tackle cancer, as the development of nanoparticle-based therapies. But, regardless of the promising outcomes within *in vitro* settings, its translation into the clinics has been delayed. One of the main reasons is the lack of an appropriate *in vitro* model, capable to mimic the true environment of the human body, to test the designed nanoparticles. In fact, most of *in vitro* models used for the validation of nanoparticle-based therapies do not address adequately the complex barriers that naturally occur in a tumor scenario, as such as blood vessels, the interstitial fluid pressure or the interactions with surrounding cells that can hamper the proper delivery of the nanoparticles into the desired site. In this reasoning, to get a step closer to the *in vivo* reality, it has been proposed of the use of microfluidic devices. In fact, microfluidic devices can be designed on-demand to exhibit complex structures that mimic tissue/organ-level physiological architectures. Even so, despite microfluidic-based *in vitro* models do not compare with the reality and complexity of the human body, the most complex systems created up to now have been showing similar results to *in vivo* animal models. Microfluidic devices have been proven to be a valuable tool to accomplish more realistic tumour's environment. The recent advances in this field, and in particular, the ones enabling the rapid test of new therapies, and show great promise to be translated to the clinics will be overviewed herein.

Keywords

Nanoparticles · Microfluidic *in vitro* models · Cancer therapies · Drug delivery

F. R. Maia (✉) · R. L. Reis · J. M. Oliveira
3B's Research Group, I3Bs – Research Institute on Biomaterials, Biodegradables and Biomimetics, University of Minho, Headquarters of the European Institute of Excellence on Tissue Engineering and Regenerative Medicine, Guimarães, Portugal

ICVS/3B's PT Government Associate Lab, Braga, Guimarães, Portugal

The Discoveries Centre for Regenerative and Precision Medicine, Headquarters at University of Minho, Guimarães, Portugal
e-mail: raquel.maia@i3bs.uminho.pt

J. M. Oliveira, R. L. Reis (eds.), *Biomaterials- and Microfluidics-Based Tissue Engineered 3D Models*, Advances in Experimental Medicine and Biology 1230,
https://doi.org/10.1007/978-3-030-36588-2_10

Highlights

- New nanoparticle-based therapies have been failing before reaching the clinic due to the lack of a proper *in vitro* model capable of mimic human body reality and efficiently validate those therapies.
- The development of more and more complex microfluidic systems, capable to emulate tumours' environment have been showing results similar with *in vivo* animal models.
- Microfluidic-based 2D and 3D *in vitro* models may help boost decrease the cost of development and testing of nanomedicines.

10.1 Introduction

Cancer is already designated the disease of the century, which can be easily understood considering its increasing incidence worldwide (Bray et al. 2018). In fact, cancer cells have the capacity to proliferate continuously resisting death, induce angiogenesis, suffer constant mutations and metastasize forming a second tumor (Fouad and Aanei 2017). Altogether, these features have been hampering the efficacy of current theranostic strategies and cancer has become a global burden.

Over the last years, nanotechnology has been presenting promising theranostic approaches to tackle cancer, as the development of nanoparticles for the detection of circulating cancer cells (Song et al. 2019) and for drug delivery (Khanna et al. 2019). Nanoparticles present very interestingly characteristics. One of the most appealing is the possibility to be designed with different sizes (in the range of 1–100 nm) and shapes, allowing to take advantage of the enhanced permeation and retention (EPR) effect for cancer treatment approaches (Yuancheng et al. 2018). In this sense, nanoparticles can cross blood vessels, transporting therapeutic agents until the disease site and diminishing systemic side effects. These features associated with the possibility to modify the nanoparticles' surface with molecules of interest, allowing a targeted therapy, easily makes these nanostructures one of the most promising tools for the development of new and more efficient cancer pharmacotherapies. Additionally, nanoparticles present some therapeutic features that were not possible to obtain with the present therapeutic approaches. In fact, nanoparticles can transport different therapeutic agents simultaneously, even the ones that present low water solubility, and can be designed to control the therapeutic agents' release over time. But, regardless of all these advantages, the translation of nanoparticles into the clinics have been delayed. One of the main reasons is the lack of an appropriate *in vitro* model, capable to mimic the true environment of the human body, to test the designed nanoparticles. In one hand, most of the models widely used nowadays (*e.g.*, two dimensional standard cultures or cell spheroid cultures in static conditions) do not address adequately the complex barriers that naturally occur in a tumor scenario, as such as blood vessels, the interstitial fluid pressure or the interactions with surrounding cells that can hamper the proper deliver of the nanoparticles into the desired site (Ozcelikkale et al. 2017). On the other hand, since these same models do not replicate the true *in vivo* environment, the obtained results are not accurate enough nor reliable. This has urged the scientists to develop new *in vitro* models to more accurately test the nanoparticles-based therapies, as photodynamic therapy (Chudy et al. 2018; Wang et al. 2018).

In this reasoning, to bridge the gap between the typically two-dimensional standard cultures and the *in vivo* reality, it has been proposed of the use of microfluidic devices. Microfluidic devices can be designed on-demand to exhibit microchannels and/or microchambers that mimic tissue/organ-level physiological architectures (Zhang et al. 2018; Xu et al. 2014). Moreover, these devices require only a few microliters of volume, resulting in lower amounts of reagents and consequently lower associated costs, which makes them very attractive models (Yun et al. 2013). In the special case of nanoparticle-based therapeutics, it allows evaluating the nanoparticle targeted accumulation, cytotoxicity, transport and uptake, resulting in more relevant data.

Herein, we aim to discuss in depth the evaluation of new nanoparticle-based therapies for cancer research using microfluidic devices, highlighting the most relevant and recent examples. Two main models are overviewed, *i.e.* microfluidic-based two-dimensional (2D) *in vitro* models, where cells are cultured in a monolayer inside the device, and microfluidic-based three-dimensional (3D) *in vitro* models, where cells are cultured as tumour spheroids or within matrices.

10.2 Microfluidic-Based 2D *in vitro* Models

Nanoparticle-based therapies have been widely studied to tackle cancer. In fact, the idea of enabling the delivery of high amounts of therapeutic agents directly into a cancer environment, without damaging any healthy tissue/cells has been very attractive to scientists (Han et al. 2016). Nevertheless, many drug delivery therapies failed during clinical trials and only a few ends up approved by Food and Drug Administration (FDA) and translated into the clinics. One of the main reasons is the lack of a suitable model to accurately predict nanoparticle-based therapies' performance once within the human body. In fact, despite the incredible value of the static 2D standard cultures during the first steps of therapies' study, it cannot emulate the *in vivo* environment and the barriers normally present. For example, the blood vessels' wall, the blood flow within the vascular environment that induces fluid shear stress, and the interstitial flow within cancer's environment. Altogether, these barriers have been preventing the accumulation of the necessary amounts of therapeutic agents in situ. Within this reasoning, microfluidic devices have been proposed to develop a more physiological relevant model and this way overcome the limitations observed within the 2D standard cultures. In the special case of drug delivery systems, body fluids are expected to interfere with the interaction of drug-delivery nanoparticles and cells, and ultimately with the proper release of drugs. For so, many scientists have been assessing the influence of fluid flow into the cellular internalization of designed nanoparticles (Tokarova et al. 2013; Sarfati et al. 2011). For example, Sarfati et al. (2011) used microfluidic devices to culture lung cancer cells in monolayer and test the cellular uptake of nanoparticles modified with specific peptides to target these cells, under a constant flow (Fig. 10.1). As control, the authors used nanoparticles modified with scramble peptides. Within this setting, the authors observed that the flow within the microchannels prevented the nanoparticle precipitation, in opposite to what usually happened in 2D standard cultures. Moreover, they detected that in the absence of flow, cells internalized both nanoparticles, nanoparticles with target peptide and scramble peptide, as depicted in Fig. 10.1 E. But when they deliver the nanoparticles under a flow rate, cells internalized higher amounts of the nanoparticles modified with the target peptide (Figs. 10.1 C, D and E).

Besides fluid flow influence, blood vessels' wall can as well influence the effectiveness of new therapies, namely the endothelial cells can hinder its success. In this sense, microfluidic devices have been used to study nanoparticles' interactions with endothelial cells. In fact, therapeutic nanoparticles are supposed to show no effect on healthy endothelial cells and only interact with cancer cells (Kang et al. 2016; Khor et al. 2018). In the work of Kang et al. (2016), they cultured endothelial cells inside a microchannel and observed that the shear stress promoted by the fluid flow influenced the alignment of actin stress fibers inside endothelial cells, which hampered the internalization of nanoparticles. Interestingly, in the work of Khor et al. (2018) they also took advantage of the environment promoted by the microchannel that mimics the blood vessels, to optimize nanoparticles design. In this sense, they were able to tune the surface of nanoparticles to control its interaction with endothelial cells.

In a different example, Bagley et al. (2015) also used the similarity of microchannels with blood vessels to test the use of plasmonic nanoparticles as an improved drug delivery strategy. In fact, these nanoparticle's can convert near-infrared light into localized heat, perturbing

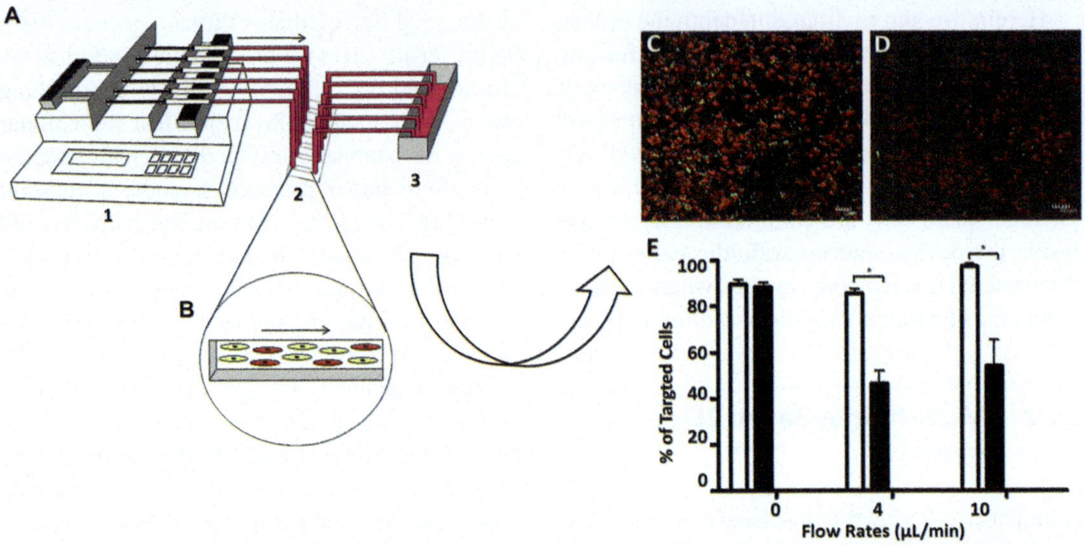

Fig. 10.1 Lung cancer cells uptake of targeted nanoparticles within a microfluidic device. (**a**) Scheme of a microfluidic device, composed of (1) 6 syringes, which are connected by tubes to (2) microchannel and finally to the (3) reservoir; (**b**) Monolayer of cells cultured inside the microchannel; (**c**) Cancer cells with 1,1′-Dioctadecyl-3,3,3′,3′-Tetramethylindocarbocyanine Perchlorate (DiL) stain (red) and internalized nanoparticles modified with target peptide (green); D) Cancer cells with DiL stain (red) and internalized nanoparticles modified with scramble peptides (green); and E) Internalization of nanoparticles modified with target peptide (white column) or scramble peptide (black column) at different flow rates. (* represents $p < 0.001$). (Reprinted with permission from (Sarfati et al. 2011) Copyright © 2010 Elsevier Ltd. All rights reserved)

the vasculature within the cancer environment, and allowing the successful penetration of the therapeutic agents. In this study, the authors produced endothelialized channels inside the microfluidic device and subjected cells to several heating cycles to unravel the contribution of the endothelium to the vascular thermotolerance in a cancer setting (Bagley et al. 2015). At the first cycle of high temperatures, the endothelial monolayer showed some disruption, allowing an increase of permeability as expected. But, upon a second cycle of treatment, the disruption decreased and less permeability was observed. With this study, it was clear that endothelial cells were able to adapt to the high temperatures, which can hamper the success of plasmonic nanoparticles for therapies delivery.

Furthermore, microfluidic devices can be used to visualize nanoparticles' internalization and its effect on cellular behavior, as cell death, making it a promising tool for the validation and optimization of new nanoparticle-based therapies as demonstrated with several works (Carvalho et al. 2017; Zhai et al. 2017). For example, in the work of Carvalho et al. (2017) the authors developed fluorescent-labeled nanoparticles to easily monitor different cancer cells' fate. For this study, the authors compared microfluidic devices with 2D standard settings. When tested the efficiency of internalization, they observed that different cells presented different internalization rates. Nevertheless, the internalization rate was higher in microfluidic devices than in 2D standard settings. Moreover, they observed that cells presented higher sensitivity to nanoparticles' internalization in microfluidic devices, resulting in higher amounts of dead cells. While in 2D conditions less dead cells were observed. Since no therapeutic agents were added to these nanoparticles, the authors hypothesized that the fluorescent label could be cytotoxic at high concentrations, once cell death was higher in microfluidic devices where more nanoparticles were internalized. In fact, the number of nanoparticles' internalized is one important parameter since it can influence new strategies efficiency. It was with this in mind that Zhai et al. (2017) associated surface-enhanced Raman spectroscopy (SERS) to microfluidic devices to validate newly designed nanoparticles. In this sense, they were able to observe the exact amount of nanoparticles that cells internalized and consequently associate with its therapeutic effect on cancer cells.

Ultimately, microfluidic devices present another very appealing feature that is the possibility to perform simultaneously the evaluation of multiple therapeutic nanoparticles as done by Mytxelena-Iribarren et al. (2019). They developed a microfluidic device with multiple microchannels, where they culture osteosarcoma cells. In each microchannel, they submitted cells to nanoparticles loaded with different therapeutic agents, expediting the selection of more efficient therapies. By substituting the cell line tested by patients' cells it is easy to realize that it would be a powerful tool to develop personalized therapies.

Nevertheless, despite the importance of these microfluidic-based 2D *in vitro* models for the validation of new therapies, it lacks the three dimensionality found in vivo. That limitation has urged the development of new and improved three-dimensional (3D) models on a microfluidic device.

10.3 Microfluidic-Based 3D *in vitro* Models

Currently, much attention has been paid to 3D, especially in the development of new *in vitro* models for the validation of new cancer therapies. In the work of Carvalho et al. (2019b) it was clear the importance of three dimensional (3D) features for the assessment of new therapies to fight cancer. They studied the effect of new targeted nanoparticles in 2D and 3D culture settings. Besides the decrease of nanoparticles' internalization in the 3D cultures, as compared to 2D standard cultures, they also observed that the therapeutic effect decreased by 50% in 3D cultures after 72 h in culture.

Additionally, it is widely accepted that by generating a more specific cancer microenvironment through the mimicry of the physiological conditions, it is possible to obtain more reliable data concerning the validation of new nanoparticle-based therapies. In this reasoning, microfluidic devices have been playing important roles in the mimicry of the *in vivo* conditions. Besides the features mentioned in the previous point, microfluidic devices allow the 3D culture of cancer and healthy cells, in the form of spheroids or within a matrix (Yang et al. 2015; Zuchowska et al. 2017; Kalinowska et al. 2019). For example, Yang et al. (2015) co-cultured breast cancer cells with adipose-derived stem cells within a microfluidic for the assessment of the efficiency of photodynamic therapy (PDT) associated with therapeutic agents. PDT is a very attractive therapy since it presents fewer side effects than the typical therapies, radiotherapy and chemotherapy. In fact, it uses photosensitizing agents that upon the exposition to visible light radiation produces reactive oxygen species, which kills only the adjacent cells. To test this therapy, the authors developed a 3D model within a microfluidic device. For that, cells were allowed to proliferate and produce their extracellular matrix, creating a natural 3D environment inside the microfluidic device. Then, gold nanoparticles were perfused along the microchannels and its effects associated with PDT treatment were observed. The results showed that cancer cells cultured in the developed 3D environment within the microfluidic device were more resistant than 2D standard cultures. While in the 3D environment the treatment was efficient in only 50% of cells, in 2D environment, the treatment was 100% efficient. In a different study, the efficacy of PDT was assessed on lung cancer cells' spheroids cultured within a microfluidic device, as depicted in Fig. 10.2 (Zuchowska et al. 2017). Upon 5 days of treatment, cancer cells showed higher resistance to PDT than the observed in the previous example. This difference can be due to the fact that in the first example cells were cultured in a 3D environment less dense than the environment presented by the spheroids. In fact, in the case of spheroids it is possible that the photosensitizing agents did not reach to all cells, as well as the oxygen, essential for the success of this therapy.

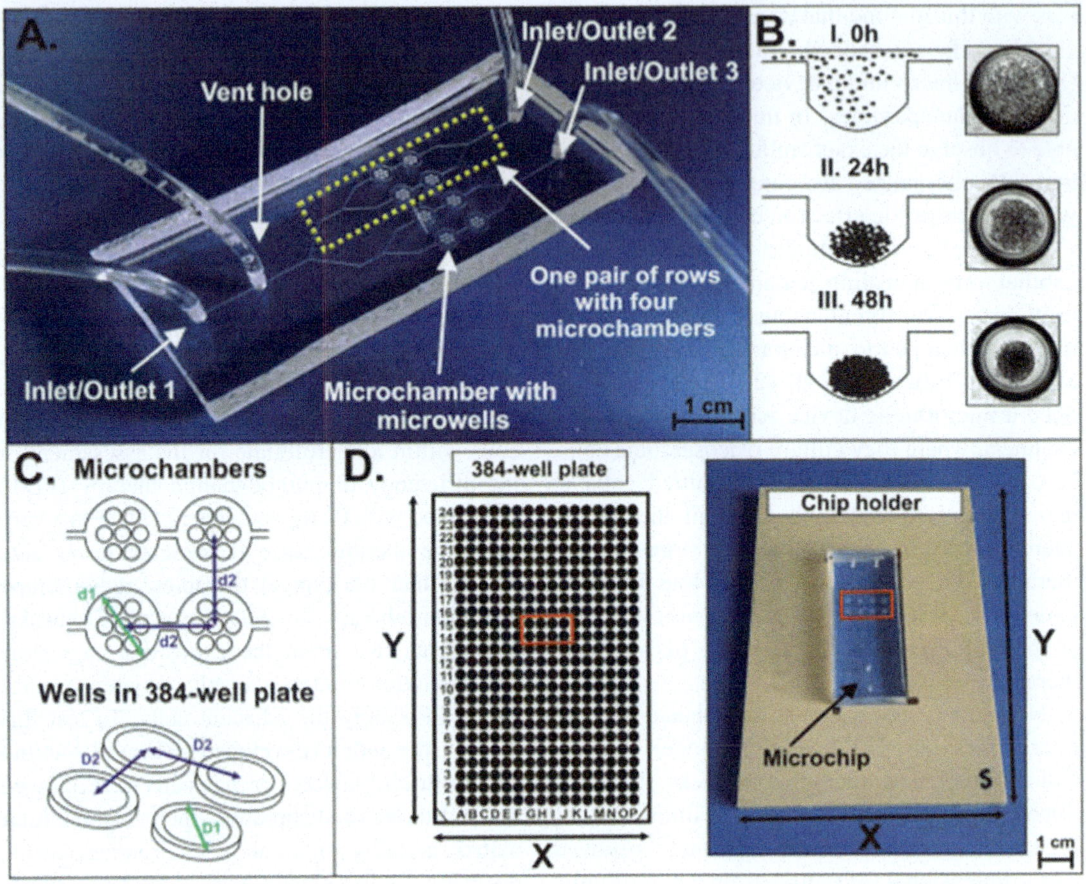

Fig. 10.2 Microfluidic device for the assessment of PDT efficacy on lung spheroids. (**a**) Picture of the microfluidic device. (**b**) Schematic representation of spheroid formation inside the microfluidic device along the time. (**c**) Picture of microwells dimensions, which are the same as a microwell of a 384-well plate. (**d**) Microfluidic device incorporated in holder with the same dimensions as a 384-well plate. (Adapted with permission from (Zuchowska et al. 2017) © 2017 Elsevier B.V. All rights reserved)

Noteworthy, this assay was performed on a microfluidic device incorporated in a holder with the same dimensions as a 384-well plate, which allowed to easily monitor the culture using a spectrofluorimetric microplate reader (Fig. 10.2c and d). Altogether, the developed model seems to be a promising tool for new nanoparticle-based therapies assessment and optimization.

More recently, spheroids cultured inside a microfluidic device were used to test other therapies as photothermal therapy (PTT), an alternative to PDT (Kalinowska et al. 2019). In fact, PTT allows for a deeper tissue penetration in opposite to PDT, due to the use of near-infrared radiation, which makes spheroids a good model to test. Additionally, the microfluidic device allows for a more realistic setting, since it allows to mimic the distribution of gold nanoparticles along the blood vessels until the cancer tissue. To test this therapy, the authors cultured cancer spheroids or healthy spheroids inside the device and gold nanoparticles were perfused, mimicking the *in vivo* reality. Upon irradiation, the authors successfully observed cells death and reduction of the spheroid, demonstrated that PTT may be more powerful for tumour treatment than PDT.

Despite the relevance of previously described models, it is widely accepted that for the success of new nanoparticle-based therapies, it is essential to fully understand cancer environment and mimic all the building blocks. In this sense, the system should not only mimic the tumour or the blood vessels, but should allow their co-culture, envisioning the screening of new drugs in a more realistic setting (DeWitt and Rylander 2018; Carvalho et al. 2019a). Dewitt and Rylander prepared a protocol where they demonstrated how they prepared an optically clear model to enable the visualization of nanoparticles transport along an endothelialized channel and along with the tumour tissue (DeWitt and Rylander 2018). Additionally, this model also allowed for the post-culture analysis of the uptake of nanoparticles by tumour and endothelial cells, which is a plus for the validation of new nanoparticles-based therapeutic strategies. Carvalho et al. (2019a) as well developed a model that allowed the visualization of the effect of new nanoparticles-based treatments and the posterior cell recovery for gene expression evaluation, envisioning its use for personalized treatment development. With the intent to better mimic the tumour environment, Chen et al. (2018) developed a breast tumour model in a microfluidic device where they incorporated the mimicry of blood vessels wall, the matrix that surrounds the tumour and the tumour itself. The device was incorporated within a well-plate for the real-time monitoring of nanoparticles-based therapies efficiency. In this sense, the authors were able to observe with high spatiotemporal resolution the transport of the drug delivery system across the microvascular wall, then the matrix, and finally its infiltration within the spheroid. Interesting, with this system, the authors observed different infiltration rates into spheroids composed of different cancer cells' types. In a different example, Jarvis et al. (2017) used, as well, a microchannel colonized with endothelial cells, but instead of studying just tumour tissues, they also studied healthy tissues in a 3D arrangement, to better mimic the tumour environment and more reliably test new nanoparticle-based therapeutic systems. Additionally, they introduced the nanoparticles under physiologically relevant shear stress. With this setting, they were able to test different nanoparticles for the delivery of super hydrophobic drugs and realized that despite the superficial infiltration of nanoparticles within the tumour tissue, drugs were able to penetrate further, as demonstrated by the increase on the cytotoxicity along the throughout the tumour.

Despite the complexity obtained by introducing new building blocks of the tumour environment, as such as blood vessels, tumour and healthy tissues, altogether within a microfluidic device, it is also possible to control the features of such building blocks, as done by Prabhakarpandian et al. and later on by Vu et al. (Prabhakarpandian et al. 2015; Vu et al. 2019). In the first example they cultured an endothelial cell lined leaky capillary vessel with a 3D spheroid tumour (Prabhakarpandian et al. 2015). This co-culture emulated the typical cervical cancer environment normally composed by leaky vessels within a solid tumour. This setting showed similar results as compared with *in vivo* data, validating the capability of using this device to predict the real efficiency of new drugs *in vivo*. In the second example they developed a model composed of a tumour tissue within a leaky microvascular network (Vu et al. 2019). For that they activated a regulator of the endothelial junction's formation, which promoted the reorganization of the cytoskeleton and disrupted the junctions resulting in a leak endothelium. In this sense they were able to study the delivery of new nanoparticle-based therapies with different sizes and surface chemistries throughout the blood vessels and its accumulation within the tumour tissue in a more realistic manner. In fact, they conclude that it is not only the size of the particles that influences the success of the delivery process but also the surface chemistry. Depending on the surface chemistry, the nanoparticles interacted with the endothelial cells hindering its interaction with the tumour cells. Interestingly, Tang et al. (2017) showed that by close emulate the *in vivo* tumour scenario, i.e. culture a vascular network under shear stress and in contact with highly metastatic tumours is enough to simulate the enhanced permeability and retention (EPR) effect. For that, they created the vascular network by using an image database of *in vivo* networks and comprised a tumour tissue (Fig. 10.3). Additionally, they included shear stress in the

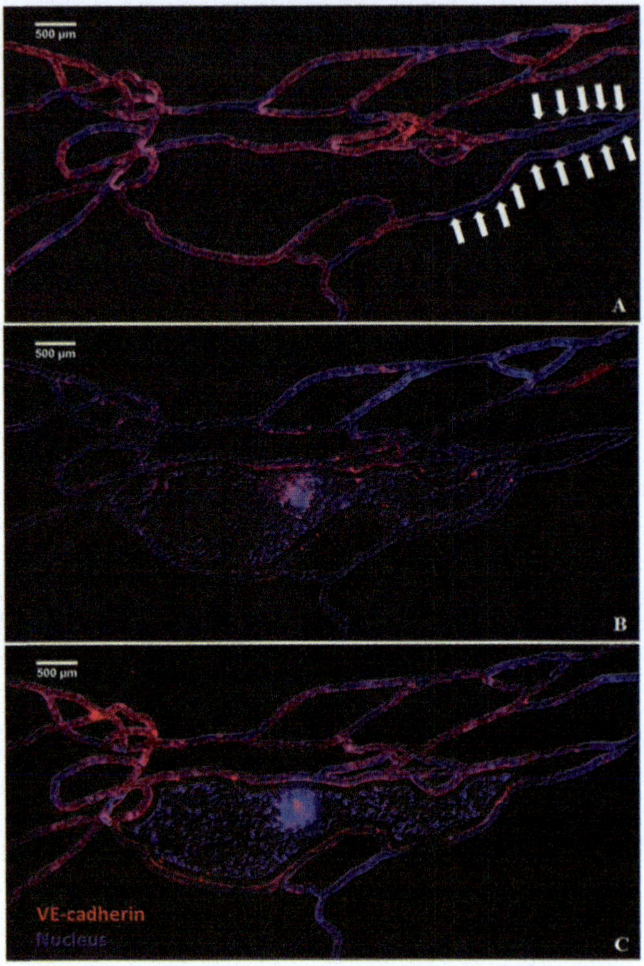

Fig. 10.3 Vascular network comprising a tumour tissue within a microfluidic device. (**a**) Low levels of VE-cadherin (white arrows) were observed under low shear stress; (**b**) VE-cadherin is inhibited in the presence of highly metastatic tumour cells; and (**c**) No differences in the VE-cadherin levels were observed in the presence of less metastatic cells. (Adapted with permission from (Tang et al. 2017) under a Creative Commons Attribution 4.0 International License)

model, which showed to be essential for the formation of endothelial cellular junctions, shown by the low levels of vascular endothelial cadherin (VE-cadherin, the molecule responsible for the adhesion of endothelial cells) under low shear stress (Fig. 10.3 A). With this realistic model, they observed that the extravasation of nanoparticles along the endothelial wall to tumour tissue increased in the presence of metastatic tumour cells. Additionally, the weakening of the endothelial cell barrier function showed similar permeability values as the ones found *in vivo* for similar tumour animal models. Noteworthy, when studying less metastatic tumour cells, no permeability was observed.

Finally, microfluidic devices can as well be designed with different features that are difficult to naturally include in a model, as such as different cut-off pore size, interstitial fluid pressures and tissue microstructures as done by Kwak et al. (2014). For that study, the authors developed a model composed of microchannels that mimic not only, the blood vessels, where the nanoparticles were introduced at physiological relevant velocity and pressure, but also, the lymphatic vessels. In fact, despite the importance of blood vessels for the delivery of therapies, lymphatic vessels are also key factors for the success of those therapies since lymphatic vessels should present reduced drainage to limit the clearance of such delivery therapies from the tumour. In this reasoning, the authors designed a model where blood and lymphatic vessels were surrounding a tumour tissue composed of cancer cells within a collagen matrix. In this sense, they were able to mimic the pore size of tumour blood vessels, the high interstitial fluid pressure and the dense extracellular matrix of the tumour and the high cell density of tumour tissues. At the end they were able to obtain a model where it was possible to change all the previously mentioned parameters but maintain the interactions between the vessels and the tumour, revealing the effect of the physiological conditions of the therapeutic nanoparticles' transport.

10.4 Conclusions and Future Perspectives

The study of new therapies to tackle tumour have been the main focus of researchers worldwide. One of the most studied approaches is the use of nanoparticle-base therapies. Nevertheless, most of the developed therapies fail along the process of validation. For so, only a few nanoparticle-based therapies reach clinics. Part of the reason lays behind the lack of models capable to predict truthfully the effect of those therapies once inside the human body.

Currently, microfluidic devices have emerged as privileged devices since they can be designed on demand. Microfluidic devices can include appealing features as mimicking of blood vessels, by the endothelialisation of the microchannels and addition of shear stress, or by the possibility to mimic the tumour itself, by culture of tumour cells within the microchambers. Nevertheless, many assays, performed within this setting, are still two dimensional which has been shown to underestimate the effect of nanoparticles. To overcome this limitation, several efforts have been made to increase the complexity and include three-dimensionality within the model. Many times by the inclusion of spheroids tumour tissues or by the culture of tumour cells within a matrix, which can be in contact with endothelial cells.

Despite the superiority demonstrated as compared with 2D settings, it was still missing many of the building blocks that compose the tumour microenvironment, which influences the efficiency of new therapies, as such as vascular networks or permeable blood vessels. These same building blocks are not only important for the validation of new therapies, but also playing important role in the optimization of the novel nanoparticle-based therapies. In fact, the delivery of nanoparticles-based therapies is a process that comprises several steps. First nanoparticles are injected intravenously, then they should cross through the blood vessels endothelium and penetrate into the tumour tissue, finally, they should be internalized by the tumour cells without damaging the healthy cells in the way. But all these steps comprise several problems, as such as rapid clearance by the lymphatic vessels, reduced penetration, and unexpected organ uptake and accumulation. Each one of those steps has different requisites and need nanoparticle-based therapies with features that enable to fulfil those requisites for optimal transport. Additionally, tumour tissues differ from patient to patient and within the same tissue it is possible to find a heterogeneous population of tumour cells, which as well impairs the successful effect of new therapies. Even so, the most complex systems created by now within the microfluidic devices have been showing similar results to *in vivo* animal models, which is a step forward the better *in vitro* model for nanoparticle-based therapies validation. However, they do not yet compare with the reality of the human body because of the most complex system created by now do not replicate the interactions of tumour tissues and distant organs. In fact, it is difficult to emulate the signals pathways between tumours and distant organs, and even harder to emulate the signals between tumours and the endocrine and immune system. For so, it looks like that the connection of tumour and distant organs within one device or several devices may be the next step in the development of more physiologically relevant models *in vitro*. Overall, microfluidic devices seem to be the setting for the production of more and more realistic tumour's environment, enabling the rapid test of new therapies, increasing the probability to be translated into the clinical scenario.

Acknowledgments The authors thank the funds obtained through the FROnTHERA (NORTE-01-0145-FEDER-0000232) project supported by Norte Portugal Regional Operational Program (NORTE 2020), under the PORTUGAL 2020 Partnership Agreement, through the European Regional Development Fund (ERDF). FRM acknowledges Portuguese Foundation for Science and Technology (FCT) for her contract under the Transitional Rule DL 57/2016 (CTTI-57/18-I3BS(5)), JMO thanks FCT for the distinction attributed under the Investigator FCT program (IF/01285/2015).

References

Bagley AF, Scherz-Shouval R, Galie PA, Zhang AQ, Wyckoff J, Whitesell L, Chen CS, Lindquist S, Bhatia SN (2015) Endothelial Thermotolerance impairs nanoparticle transport in tumors. Cancer Res 75(16):3255–3267. https://doi.org/10.1158/0008-5472.can-15-0325

Bray F, Ferlay J, Soerjomataram I, Siegel RL, Torre LA, Jemal A (2018) Global cancer statistics 2018: GLOBOCAN estimates of incidence and mortality worldwide for 36 cancers in 185 countries. CA Cancer J Clin 68(6):394–424. https://doi.org/10.3322/caac.21492

Carvalho MR, Maia FR, Silva-Correia J, Costa BM, Reis RL, Oliveira JM (2017) A semiautomated microfluidic platform for real-time investigation of nanoparticles' cellular uptake and cancer cells' tracking. Nanomedicine (Lond) 12(6):581–596. https://doi.org/10.2217/nnm-2016-0344

Carvalho MR, Barata D, Teixeira LM, Giselbrecht S, Reis RL, Oliveira JM, Truckenmüller R, Habibovic P (2019a) Colorectal tumor-on-a-chip system: A 3D tool for precision onco-nanomedicine. Sci Adv 5(5):eaaw1317. https://doi.org/10.1126/sciadv.aaw1317

Carvalho MR, Carvalho CR, Maia FR, Caballero D, Kundu SC, Reis RL, Oliveira JM (2019b) Peptide-modified dendrimer nanoparticles for targeted therapy of colorectal Cancer. Adv Therapeutics 0(0):1900132. https://doi.org/10.1002/adtp.201900132

Chen Y, Gao D, Wang Y, Lin S, Jiang Y (2018) A novel 3D breast-cancer-on-chip platform for therapeutic evaluation of drug delivery systems. Anal Chim Acta 1036:97–106. https://doi.org/10.1016/j.aca.2018.06.038

Chudy M, Tokarska K, Jastrzębska E, Bułka M, Drozdek S, Lamch Ł, Wilk KA, Brzózka Z (2018) Lab-on-a-chip systems for photodynamic therapy investigations. Biosens Bioelectron 101:37–51. https://doi.org/10.1016/j.bios.2017.10.013

DeWitt MR, Rylander MN (2018) Tunable collagen microfluidic platform to study nanoparticle transport in the tumor microenvironment. In: Sirianni RW, Behkam B (eds) Targeted drug delivery: methods and protocols. Springer, New York, pp 159–178. https://doi.org/10.1007/978-1-4939-8661-3_12

Fouad YA, Aanei C (2017) Revisiting the hallmarks of cancer. Am J Cancer Res 7(5):1016–1036

Han B, Qu C, Park K, Konieczny SF, Korc M (2016) Recapitulation of complex transport and action of drugs at the tumor microenvironment using tumor-microenvironment-on-chip. Cancer Lett 380(1):319–329. https://doi.org/10.1016/j.canlet.2015.12.003

Jarvis M, Arnold M, Ott J, Pant K, Prabhakarpandian B, Mitragotri S (2017) Microfluidic co-culture devices to assess penetration of nanoparticles into cancer cell mass. Bioeng Trans Med 2(3):268–277. https://doi.org/10.1002/btm2.10079

Kalinowska D, Grabowska-Jadach I, Liwinska M, Drozd M, Pietrzak M, Dybko A, Brzozka Z (2019) Studies on effectiveness of PTT on 3D tumor model under microfluidic conditions using aptamer-modified nanoshells. Biosens Bioelectron 126:214–221. https://doi.org/10.1016/j.bios.2018.10.069

Kang T, Park C, Lee B-J (2016) Investigation of biomimetic shear stress on cellular uptake and mechanism of polystyrene nanoparticles in various cancer cell lines. Arch Pharm Res 39(12):1663–1670. https://doi.org/10.1007/s12272-016-0847-0

Khanna V, Kalscheuer S, Kirtane A, Zhang W, Panyam J (2019) Perlecan-targeted nanoparticles for drug delivery to triple-negative breast cancer. Fut Drug Discovery 1(1):FDD8. https://doi.org/10.4155/fdd-2019-0005

Khor SY, Vu MN, Pilkington EH, Johnston APR, Whittaker MR, Quinn JF, Truong NP, Davis TP (2018) Elucidating the influences of size, surface chemistry, and dynamic flow on cellular association of nanoparticles made by polymerization-induced self-assembly. Small (Weinheim an der Bergstrasse, Germany) 14(34):1801702. https://doi.org/10.1002/smll.201801702

Kwak B, Ozcelikkale A, Shin CS, Park K, Han B (2014) Simulation of complex transport of nanoparticles around a tumor using tumor-microenvironment-on-chip. J Control Release 194:157–167. https://doi.org/10.1016/j.jconrel.2014.08.027

Mitxelena-Iribarren O, Zabalo J, Arana S, Mujika M (2019) Improved microfluidic platform for simultaneous multiple drug screening towards personalized treatment. Biosens Bioelectron 123:237–243. https://doi.org/10.1016/j.bios.2018.09.001

Ozcelikkale A, Moon HR, Linnes M, Han B (2017) In vitro microfluidic models of tumor microenvironment to screen transport of drugs and nanoparticles. Wiley Interdiscip Rev Nanomed Nanobiotechnol 9(5):e1460. https://doi.org/10.1002/wnan.1460

Prabhakarpandian B, Shen MC, Nichols JB, Garson CJ, Mills IR, Matar MM, Fewell JG, Pant K (2015) Synthetic tumor networks for screening drug delivery systems. J Control Release 201:49–55. https://doi.org/10.1016/j.jconrel.2015.01.018

Sarfati G, Dvir T, Elkabets M, Apte RN, Cohen S (2011) Targeting of polymeric nanoparticles to lung metastases by surface-attachment of YIGSR peptide from laminin. Biomaterials 32(1):152–161. https://doi.org/10.1016/j.biomaterials.2010.09.014

Song Y, Shi Y, Huang M, Wang W, Wang Y, Cheng J, Lei Z, Zhu Z, Yang C (2019) Bioinspired engineering of a multivalent aptamer-functionalized Nanointerface to enhance the capture and release of circulating tumor cells. Angew Chem Int Ed 58(8):2236–2240. https://doi.org/10.1002/anie.201809337

Tang Y, Soroush F, Sheffield JB, Wang B, Prabhakarpandian B, Kiani MF (2017) A biomimetic microfluidic tumor microenvironment platform mimicking the EPR effect for rapid screening of drug delivery systems. Sci Rep 7(1):9359. https://doi.org/10.1038/s41598-017-09815-9

Tokarova V, Pittermannova A, Kral V, Rezacova P, Stepanek F (2013) Feasibility and constraints of particle targeting using the antigen-antibody interaction. Nanoscale 5(23):11490–11498. https://doi.org/10.1039/c3nr04340a

Vu MN, Rajasekhar P, Poole DP, Khor SY, Truong NP, Nowell CJ, Quinn JF, Whittaker M, Veldhuis NA, Davis TP (2019) Rapid assessment of nanoparticle extravasation in a microfluidic tumor model. ACS Appl Nano Mat 2(4):1844–1856. https://doi.org/10.1021/acsanm.8b02056

Wang Y, Cuzzucoli F, Escobar A, Lu S, Liang L, Wang S (2018) Tumor-on-a-chip platforms for assessing nanoparticle-based cancer therapy. Nanotechnology 29(33):332001. https://doi.org/10.1088/1361-6528/aac7a4

Xu X, Farach-Carson MC, Jia X (2014) Three-dimensional in vitro tumor models for cancer research and drug evaluation. Biotechnol Adv 32(7):1256–1268. https://doi.org/10.1016/j.biotechadv.2014.07.009

Yang Y, Yang X, Zou J, Jia C, Hu Y, Du H, Wang H (2015) Evaluation of photodynamic therapy efficiency using an in vitro three-dimensional microfluidic breast cancer tissue model. Lab Chip 15(3):735–744. https://doi.org/10.1039/c4lc01065e

Yuancheng L, Kai Z, Xiao L, Yu Shrike Z (2018) Blood-vessel-on-a-Chip platforms for evaluating nanoparticle drug delivery systems. Curr Drug Metab 19(2):100–109. https://doi.org/10.2174/1389200218666170925114636

Yun H, Kim K, Lee WG (2013) Cell manipulation in microfluidics. Biofabrication 5(2):022001. https://doi.org/10.1088/1758-5082/5/2/022001

Zhai Z, Zhang F, Chen X, Zhong J, Liu G, Tian Y, Huang Q (2017) Uptake of silver nanoparticles by DHA-treated cancer cells examined by surface-enhanced Raman spectroscopy in a microfluidic chip. Lab Chip 17(7):1306–1313. https://doi.org/10.1039/C7LC00053G

Zhang H, Zhu Y, Shen Y (2018) Microfluidics for cancer nanomedicine: from fabrication to evaluation. Small (Weinheim an der Bergstrasse, Germany) 14(28):1800360. https://doi.org/10.1002/smll.201800360

Zuchowska A, Jastrzebska E, Chudy M, Dybko A, Brzozka Z (2017) 3D lung spheroid cultures for evaluation of photodynamic therapy (PDT) procedures in microfluidic lab-on-a-Chip system. Anal Chim Acta 990:110–120. https://doi.org/10.1016/j.aca.2017.07.009

Index

J. M. Oliveira, R. L. Reis (eds.), *Biomaterials- and Microfluidics-Based Tissue Engineered 3D Models*, Advances in Experimental Medicine and Biology 1230,
https://doi.org/10.1007/978-3-030-36588-2

The manufacturer's authorised representative in the EU is Springer Nature Customer Service Centre GmbH, Europaplatz 3, 69115 Heidelberg, Germany. If you have any concerns regarding our products, please contact ProductSafety@springernature.com

Printed and bound by CPI Group (UK) Ltd, Croydon, CR0 4YY
28/09/2026
02224085-0020